CW00536881

# Pedagogy and the Practice of Science

**Inside Technology**

edited by Wiebe E. Bijker, W. Bernard Carlson, and Trevor Pinch

A list of the series appears on page 415.

# Pedagogy and the Practice of Science

## Historical and Contemporary Perspectives

edited by David Kaiser

The MIT Press
Cambridge, Massachusetts
London, England

MIT Press books may be purchased at special quantity discounts for business or sales promotional use. For information, please email special_sales@mitpress.mit.edu or write to Special Sales Department, The MIT Press, 5 Cambridge Center, Cambridge, MA 02142.

Set in Stone sans and Stone serif by The MIT Press. Printed and bound in the United States of America.

Library of Congress Cataloging-in-Publication Data

Pedagogy and the practice of science : historical and contemporary perspectives / edited by David Kaiser.
p. cm. — (Inside technology)
Includes bibliographical references and index.
ISBN 0-262-11288-4
1. Science—Study and teaching (Higher)—History—19th century. 2. Science—Study and teaching (Higher)—History—20th century. 3. Science—Philosophy—History—19th century. 4. Science—Philosophy—History—20th century. I. Kaiser, David. II. Series.
Q181.P344 2005
507.1'1—dc22 2004061364

10 9 8 7 6 5 4 3 2 1

# Contents

## Acknowledgments

The essays in this volume originated in a pair of workshops held at the Massachusetts Institute of Technology during January and September 2002. In addition to the contributors here, it is a great pleasure to thank the other workshop participants, who helped to make these meetings so stimulating: Myles Jackson, Ursula Klein, Robert Kohler, Michael Lynch, Mary Jo Nye, Christopher Ritter, Ana Simões, and Reed Stevens. I would also like to thank Shane Hamilton for research assistance, Stacey Nichols for editorial assistance, and Diane St. Laurent, Lois Folstein, and especially Kris Kipp for administrative assistance. The workshops were generously supported by grants from the Spencer Foundation (grant 200200064), from the National Science Foundation (grant SES-0118165), and from the Provost's Fund for Humanities, Arts, and Social Sciences at MIT.

Pedagogy and the Practice of Science

## Introduction
## Moving Pedagogy from the Periphery to the Center

**David Kaiser**

Scientists are not born, they are made. The ways in which this happens bear the marks of time and place: becoming a scientist in Wilhelmine Germany or in Victorian Britain was not the same as becoming a scientist in Cold War America. Despite the centrality of pedagogical concerns to the modern scientific professions, scholars in science studies have not dedicated much systematic attention to the topic. Questions of pedagogy and training—the crafting of scientific practices and of the practitioners who put them to work—have tended to fall between the cracks separating traditional institutional and disciplinary studies on the one hand, and intellectual or conceptual studies on the other. This volume puts in the center what has usually fallen to the periphery by highlighting how, where, and why questions of scientists' training should fit into our studies of the history, sociology, and anthropology of science. What do we stand to gain by using education and pedagogy as our window onto how sciences have changed and how scientific roles have evolved? Why have such questions not drawn systematic attention from the science studies community in the recent past?

Pedagogy has long been a major concern of the modern scientific professions. Hundreds of millions of dollars were spent on science pedagogy and its reforms throughout the nineteenth and twentieth centuries, both in the United States and in Europe. Whether focused on the introduction of laboratory techniques for teaching in the late nineteenth century, on the highly touted multimedia productions of the Physical Science Study Committee (PSSC), the Biological Sciences Curriculum Study (BSCS), and "Project Physics" during the Cold War, or on more recent efforts to incorporate computer simulations and more direct question-and-answer feedback into large introductory lectures, pedagogy has hardly been taken for granted by scientists themselves.[1]

Historians and sociologists of education have likewise long emphasized that schooling and education are anything but passive or neutral activities. What counts as "appropriate" or "acceptable" pedagogy in a particular setting is always conditioned by decisions about what skills students should acquire, and why, and by related concerns about labor supplies and the flow of human capital into and beyond instructional settings.[2] Education scholars have also highlighted various "hidden curricula" at work: alongside their formal subjects of instruction, educational institutions serve as the crucible for reproducing cultural, political, and moral values. With varying degrees of subtlety and effectiveness, schools prepare students to become good citizens and forge appropriate identities and roles in society.[3] Of course, what counts as "good citizenship," just like what counts as "appropriate skills," always reflects active decisions (and often fraught controversy and bitter negotiations) in given contexts, and show telling variation across time and space. Pedagogy is where the intellectual rubber meets the politico-cultural road.

The relative absence of such questions from the recent literature of science studies is therefore surprising. For one thing, questions of pedagogy, training, and education were heralded as central by well-known science studies scholars several decades ago. The most obvious source, and one to whom several contributors in the volume return, is Thomas Kuhn. Kuhn famously highlighted the importance of scientists' training. In fact, when he first turned to the term "paradigm"—nowadays so much maligned—it was to talk about science students' work on exemplary problems or exercises, by means of which they could practice the skills and tools they would need in order to engage in research later in their careers. Only later, just before writing *The Structure of Scientific Revolutions* (1962), did Kuhn expand the notion of "paradigm," freighting the term with its much more familiar associations of worldviews and reigning conceptual structures.[4] Kuhn is among the best-known historians and philosophers of science to have emphasized the need for proto-scientists to practice the skills of their trade, but he was not alone. Michael Polanyi—another scientist turned philosopher of science, also writing in the 1950s and the 1960s—similarly emphasized time and again the centrality of "tacit knowledge" and the acquisition of craft-like or artisanal skills to scientific work. One must develop a "feel" for a research problem or a scientific instrument, Polanyi argued—a "hands-on" knowledge that no amount of formal, written instructions could ever replace.[5]

Kuhn's and Polanyi's analyses left many questions still open. In particular, they did not explore in any historical or anthropological detail how such exemplars emerged, how students in various generations actually learned to solve exemplars and build upon them in their own research, or whether students at different training centers learned about and leaned upon exemplars in distinct ways. Yet if Kuhn and Polanyi hardly exhausted all the questions we might ask about scientists' training, they certainly tried to put the issues on the table. Pedagogy, education, training are hardly absent from the science studies literature of yesteryear.

Yet, while more recent trends in science studies have surely built upon themes related to scientific practices and skills, they have tended to leave the educational component out.[6] The current fascination with local practices, material culture, replication, and metrology—a fascination that has opened up a tremendous amount of interesting and important work within our field since the 1980s—has tended to focus on the practices of already-established scientists. Case studies have been overwhelmingly concerned with peer-to-peer interactions among working scientists, rather than examining how up-and-coming scientists and engineers have grappled with what Andrew Pickering calls "the mangle of practice."[7] Many of the most prominent studies within this genre, whether treating the replication of modern-day lasers or of seventeenth-century air pumps, the Victorian manufacture of standardized electrical units of measure or the "big science" armamentarium of twentieth-century particle physics, reveal example after example of the difficulties that already-established scientists and instrument makers have had in getting new instruments to work properly.[8] These studies, and many others like them, have certainly opened our eyes anew to the difficulties of acquiring new skills and deploying locally honed practices. Such studies ought to lead *directly* to questions of pedagogy and training: if skills and practices are difficult to master and yet central to the doing of science, then how do scientists in training master them in the first place? Practices, after all, have to be *practiced*—and we are thereby necessarily thrust into questions of scientists' training and learning.

Another exciting theme within recent science studies concerns the crafting of scientific roles or personae and the negotiation of "moral economies" regulating accepted norms and values within and among scientific communities.[9] Once again, we are faced immediately with questions of scientists' education and socialization: at least in the modern period, during which the age of the "gentlemanly amateur" of science has become but a distant memory, no one has just "done" science. Young recruits must be *trained* to become working scientists. Part of this training has always involved learning what it is to *be* a scientist. What is the proper role or self-image? What are the accepted norms, values, and behaviors? These have surely changed over time and across space. If these roles and moral economies are not fixed or merely "natural," then some active work must be done in a given setting to establish and reinforce these roles. One of the most important processes in which such roles and systems of acceptable behavior get worked out and reinforced is in training new generations to "grow into" them. The problem of generational reproduction—how one generation of scientists and engineers brings up its successors—is always shot through with decisions about norms, values, roles, and personae.

Yet we must be careful to avoid a truism. If we stop by noting only that (at least since the late nineteenth century) all scientists and engineers have by necessity gone through some form of training, then we have explained nothing. Think of the hollowness of the terms "urbanization" and "modernization," or of the historians' ruse that "the middle class is always rising." Such formulations have become so broad as to carry little explanatory power. Moreover, it will do us little good to slip into a facile "educational determinism," seeing all scientific or technical results as nothing but the blind recapitulation of previously learned methods. Such a monocausal story line is doomed to failure from the start; tinkering and improvisation are always important parts of scientific practice. Rather than simply stating that all scientists and engineers must somehow have been trained, therefore, this volume pursues substantive links between types of training—as they have been negotiated and enacted in specific times and places—and the resulting scientific practices. Are there particular educational institutions or mechanisms with which we can associate specific styles or approaches within the sciences? In other words, does the *form* of training matter to the *content* of science—and if so, how, in what ways, and to what extent? In pursuing these questions, this volume treats pedagogy not in the narrow sense of classroom teaching techniques—though these are certainly relevant—but more broadly: what are the institutions of training by means of which young recruits have become working scientists and engineers?

## The Practice-Practitioner Dyad

The essays in this volume interrogate these links between scientists' pedagogical formation, research strategies, and scientific identities, criss-crossing between episodes in Europe, Asia, and the United States during the nineteenth and twentieth centuries. Drawing on interrelated examples from the physical sciences, engineering, and technology provides a baseline for comparison among the far-ranging cases, highlighting common themes and methodological concerns. The volume is organized into four parts: "Teaching Practices, Transferring Skills," "Pedagogical Cultures in Collision," "The Action of Textbooks," and "Generational Reproduction." Parts I and III focus on the question of how skills and practices can be transferred to scientists in training (the "practices" half of the practice-practitioner dyad); parts II and IV look more closely at how roles, norms, and values can become inculcated and new generations of practitioners produced.

### Teaching Practices, Transferring Skills

The essays in part I address the interplay between techniques of instruction and changing research strategies. Michael Gordin charts how the nineteenth-century chemist Friedrich Beilstein constantly adapted his teaching techniques and research goals to his changing institutional settings. Out of Beilstein's peripatetic pedagogy came a new *Handbuch* to help instantiate his vision of how young chemists should enter the field, organize their research, and communicate their findings. David Kaiser looks at the various pedagogical processes by which new pencil-and-paper research tools—Feynman diagrams, introduced by the American theorist Richard Feynman—spread throughout postwar physics. New ways to calculate were accelerated by new institutions for training the postwar generation; new skills became "second nature" against a backdrop of older practices that had already become routine. Hugh Gusterson examines a kind of arteriosclerosis in the ways that new skills have been inculcated in one particular branch of modern science: nuclear weapons science. With the rise of bureaucracy and the end of nuclear testing, questions of pedagogy again loom large as the secret weapons laboratories struggle to transfer the old guard's skills and experience to newer recruits.

## Pedagogical Cultures in Collision

Twenty-five years ago, historians and sociologists of science fastened onto scientific controversies as a potent means of bringing to the surface what usually remained tacit among scientists: their taken-for-granted epistemic and social norms. So, too, can scholars in science studies learn about prevailing pedagogical presuppositions by studying episodes of tension: controversies among scientists and engineers over what should count as "appropriate" or "effective" training highlight the pedagogical patterns on each side. Graeme Gooday examines how competing groups staked their claims to expertise over the new (and potentially lucrative) domain of electrotechnical machinery in the late nineteenth century by invoking practitioners' training: would the "university men" drilled in Cambridge University's Mathematical Tripos or the "practical men" schooled in personal apprenticeships retain ultimate authority? Kenji Ito asks how the new science of quantum mechanics spread beyond its originators: did the ever-fabled "Copenhagen *Geist*" float freely from Niels Bohr's institute to the laboratories and universities of interwar Japan? Ito scrutinizes what Japanese physicists such as Nishina Yoshio brought back from their extended stays in Europe, and how they adapted these pedagogical approaches for their home environment. Cyrus Mody studies how members of industrial laboratories and West Coast universities began to work with a new type of instrument during the 1980s and the 1990s. The two groups betrayed distinct expectations over what would count as scanning probe microscopy: from what would count as a probe microscope to who would be deemed a competent microscopist. Mody shows that the two groups' differing pedagogical visions underlay their different responses to and uses of the new instruments.

### The Action of Textbooks

When scholars in science studies have written about pedagogy at all, it has often been to examine (and then disparage) scientific textbooks. Textbooks, so the conventional wisdom has agreed, represent the last stage of scientific creativity, the final desiccation of scientific discovery into routine, banal, taken-for-granted knowledge. Challenging this narrow view, Antonio García-Belmar, José-Ramón Bertomeu-Sánchez, and Bernadette Bensaude-Vincent highlight the historicity of textbooks as a genre: our notion of the scientific textbook emerged in the decades after the French Revolution, taking form in the midst of major overhauls in governmental oversight of curricula and wide-ranging changes in the book-publishing industry. Far from being the end stage of scientific creativity, these authors argue, nineteenth-century chemistry textbooks offered their authors and publishers an influential venue for expressing creativity, both in how they chose to organize their material and how they approached the prevailing scientific and philosophical debates of their day. Karl Hall picks up on this theme, focusing on the world-famous textbooks by the Soviet physicists Lev Landau and Evgenii Lifshitz. Forged as a product of contrasting expectations and requirements within postwar Soviet culture, in which the task of writing textbooks received newfound scrutiny from the state, the Landau-Lifshitz books challenged prevailing generic conventions while framing a distinct social identity for young theoretical physicists. Careful editing and selective translation then helped to spread the books to students all over the world. Buhm Soon Park likewise highlights the choices that scientists make as they write their textbooks, crafting their books as armaments in ongoing theoretical debates. The quantum chemists Charles Coulson and Michael Dewar adopted strongly contrasting pedagogical styles within their postwar textbooks, yet each used his books to help turn the tide of practicing chemists toward his methodological approach. Textbooks and teaching styles proved crucial in setting the pace for theory change in modern chemistry.

### Generational Reproduction

Scientists and engineers do more than pass down skills and knowledge to younger generations; they also strive to inculcate norms, roles, and personae. Kathryn Olesko studies the process of canon formation in physics. Practical manuals such as Friedrich Kohlrausch's embody cognitive and practical preferences, epistemological guidelines, and social norms—elements that later generations both adopt and adapt in the course of their training. Sharon Traweek reveals how decisions about what laboratories to establish and what equipment to build are often, at root, decisions about what kinds of students to train. Various sites within contemporary Japan serve as incubators for different kinds of new physicists. Today, as throughout the nineteenth and twentieth centuries, decisions about how to generate new knowledge are always interwoven with choices about how to generate new knowledge makers.

In the concluding chapter, Andrew Warwick and David Kaiser subject the well-known work of Thomas Kuhn and Michel Foucault to critical scrutiny, drawing out a series of conceptual resources with which further studies of scientific pedagogy might fruitfully be undertaken. In all these ways, the essays in this volume consider the constitutive roles played by pedagogy in making modern science and engineering happen.

## Notes

1. On the introduction of laboratory-based teaching, see Owen Hannaway, "The German model of chemical education in America: Ira Remsen at Johns Hopkins (1876–1913)," *Ambix* 23 (1976), November: 145–164; Larry Owens, "Pure and sound government: Laboratories, playing fields, and gymnasia in the nineteenth-century search for order," *Isis* 76 (1985), June: 182–194; Robert Kohler, "The PhD machine: Building on the collegiate base," *Isis* 81 (1990): 638–662; Klaus Hentschel, *Mapping the Spectrum: Techniques of Visual Representation in Research and Teaching* (Oxford University Press, 2002), chapter 9. On Cold War science curricular reforms, see David Donahue, "Serving students, science, or society? The secondary school physics curriculum in the United States, 1930–65," *History of Education Quarterly* 33 (1993), autumn: 321–352; John Rudolph, *Scientists in the Classroom: The Cold War Reconstruction of American Science Education* (Palgrave, 2002); David Kaiser, "Cold war requisitions, scientific manpower, and the production of American physicists after World War II," *Historical Studies in the Physical and Biological Sciences* 33 (2002), fall: 131–159; John Krige, "NATO and the strengthening of Western science in the post-Sputnik era," *Minerva* 38 (2000): 81–108. For reviews of some recent reforms in undergraduate science teaching, see Eric Mazur, "Peer instruction," in *The Changing Role of Physics Departments in Modern Universities*, ed. Edward Redish and John Rigden (American Institute of Physics, 1997), volume 2, 981–988; and Catherine Crouch and Eric Mazur, "Peer instruction: Ten years of experience and results," *American Journal of Physics* 69 (2001): 970–977.

2. See, e.g., Joel Spring, *The Sorting Machine Revisited: National Educational Policy since 1945*, second edition (Longman, 1989); Herbert Kliebard, *Schooled to Work: Vocationalism and the American Curriculum, 1876–1946* (Teachers College Press, 1999).

3. Pierre Bourdieu and Jean-Claude Passeron, *Reproduction in Education, Society, and Culture*, tr. Richard Nice (Sage, 1977); Michael Apple, ed., *Cultural and Economic Reproduction in Education: Essays on Class, Ideology, and the State* (Routledge, 1982); Richard Arum and Irenee Beattie, eds., *The Structure of Schooling: Readings in the Sociology of Education* (Mayfield, 2000). For interesting examples beyond the United States, see E. Thomas Ewing, "How Soviet teachers taught: Classroom practices and Stalinist pedagogy, 1931 to 1939," *East/West Education* 15 (1994), fall: 117–152; Brian Puaca, Drafting Democracy: Education Reform in American-Occupied Germany, 1945–49 (MA thesis, University of North Carolina, 2001).

4. Thomas Kuhn, *The Essential Tension: Selected Studies in Scientific Tradition and Change* (University of Chicago Press, 1977), xix–xx; idem, *The Structure of Scientific Revolutions*, third edition (University of Chicago Press, 1996 [1962]).

5. Michael Polanyi, *Personal Knowledge* (Routledge and Kegan Paul, 1958); idem, *The Tacit Dimension* (Anchor, 1967).

6. Telling exceptions include Sharon Traweek, *Beamtimes and Lifetimes: The World of High-Energy Physicists* (Harvard University Press, 1988); Kathryn Olesko, *Physics as a Calling: Discipline and Practice in the Königsberg Seminar for Physics* (Cornell University Press, 1991); Gerald Geison and Frederic L. Holmes, eds., *Research Schools*, published as *Osiris* 8 (1993); Robert Kohler, *Lords of the Fly: Drosophila Genetics and the Experimental Life* (University of Chicago Press, 1994); Rudolph, *Scientists in the Classroom*; and Andrew Warwick, *Masters of Theory: Cambridge and the Rise of Mathematical Physics* (University of Chicago Press, 2003).

7. Andrew Pickering, *The Mangle of Practice: Time, Agency, and Science* (University of Chicago Press, 1995).

8. Harry Collins, "The TEA set: Tacit knowledge and scientific networks," *Social Studies of Science* 4 (1974): 165–186; Steven Shapin and Simon Schaffer, *Leviathan and the Air-Pump: Boyle, Hobbes, and the Experimental Life* (Princeton University Press, 1985); Simon Schaffer, "Late Victorian metrology and its instrumentarium: A manufactory of ohms," in *Invisible Connections: Instruments, Institutions, and Science*, ed. Robert Bud and Susan Cozzens (SPIE Optical Engineering Press, 1992), 23–56; Peter Galison, *How Experiments End* (University of Chicago Press, 1987); idem, *Image and Logic: A Material Culture of Microphysics* (University of Chicago Press, 1997).

9. Traweek, *Beamtimes and Lifetimes*; Steven Shapin, "'A scholar and a gentleman': The problematic identity of the scientific practitioner in early modern England," *History of Science* 29 (1991): 279–327; Mario Biagioli, *Galileo, Courtier: The Practice of Science in the Culture of Absolutism* (University of Chicago Press, 1993); Kohler, *Lords of the Fly*; Lorraine Daston, "The moral economy of science," *Osiris* 10 (1995): 3–24; Hugh Gusterson, *Nuclear Rites: A Weapons Laboratory at the End of the Cold War* (University of California Press, 1996); and Lorraine Daston and Otto Sibum, eds., *Scientific Personae and their Histories*, published as *Science in Context* 16 (2003), June: 1–269.

# I Teaching Practices, Transferring Skills

# 1 Beilstein Unbound: The Pedagogical Unraveling of a Man and His *Handbuch*

Michael D. Gordin

Friedrich Konrad Beilstein is the most famous scientist you have never heard of. Unless, that is, you are a chemist, in which case he is the most famous chemist you know absolutely nothing about. Beilstein is a name constantly on the lips of essentially every practicing organic chemist in the world—and has been for over a century—but the amount of information widely known about his life pales in comparison to other less prominent figures such as Bunsen, Davy, and Pauling. Upon his death in 1906, Beilstein elicited the highest praise:

> The name *Beilstein* awakes in the currently living generation of chemists feelings of the sincerest gratitude. There is scarcely any work that so often lies open on the chemist's table or in his laboratory as the Handbuch that carries Beilstein's name and that is the result of his life's work. It has become an indispensable guide to all those who are active in the area of organic chemistry—"a guidebook for the land of organic chemistry"—and one can scarcely even think now about a "Beilsteinless" era.[1]

And there is the source of our ignorance: Beilstein is not known as a person, but rather as a book, a reference work for the properties of all known organic compounds. Beilstein is unknown because historians have not asked "Who is Beilstein?" They have scarcely even asked "What is Beilstein?"

Let us start with *what* Beilstein is. The "Beilstein" (more precisely the *Handbuch der Organischen Chemie*) was first published in 1881–1883 in two volumes comprising 2,200 pages and 15,000 organic compounds. In 1885 the second edition began to appear (completed in 1889 in three volumes), and the 1899–1906 third edition comprised four volumes. These first editions were compiled and edited by Beilstein alone (except the third, with which he had supplementary assistance). By 1981, on the hundredth anniversary of the *Handbuch*, the staff of the Beilstein Institute in Frankfurt am Main stood at 160. As it was simply but emphatically put in a commemorative volume that year: "*The Beilstein Handbook is the most extensive collection of physical data in printed form in the world!*"[2] The Beilstein *Handbuch* represents the first pure reference handbook of chemistry, and it became the site of that science's first large-scale communal literary project.[3]

Yet I come not to praise Beilstein, but to exhume him. I propose that by breaking apart the monolith of the *Handbuch* and looking at the pedagogical roots of its formation, we can come to a better understanding of the multiple ways in which pedagogy was implicated in the development of even the most mundane tools of organic chemistry. The *Handbuch* is an object of pedagogy, having spawned numerous guides inducting students into its proper use, but I will focus on it as a *subject* of pedagogy.[4]

*Who* was Beilstein?[5] Even naming him becomes complicated, since for Germans he was Friedrich Konrad Beilstein and for Russians he was Fedor Fedorovich Beil'shtein. (I adopt the former for the sake of convenience.) Later in his life, he would describe himself thus:

> Beilstein (Friedrich, Conrad) born in St. Petersburg on 17 (5) February 1838. Completed the German high school in St. Petersburg & then moved to the University of Heidelberg. Studied besides that in Munich (with Jolly and Liebig), in Göttingen (with Wöhler) & in Paris (with Wurtz). Became in 1859 an Assistant in Prof. Löwig's University Laboratory in Breslau. Went in 1860 (early in the year) in the same capacity to the Laborat. of Prof. Wöhler in Göttingen. Received doctorate in Göttingen in 1858 (Dissertation: on murexide). Habilitated in Göttingen in chemistry in 1860 and became an extraordinary professor there in 1865. Moved in autumn to Petersburg as professor of chemistry at the Technological Institute. Became in 1867 a teacher of chemistry at the Military-Engineering Academy there and in 1868 a chemist of the Department of Trade and Manufactures in the Imperial Russian Ministry of Finances.
>
> First scientific work: On the Diffusion of Liquids, conducted in 1855/6 in the physical cabinet of Prof. Jolly in Munich.
>
> Books: Introduction to Qualitative chemical analysis, 5 editions, Chemical Great-Industry at the World Exposition in Vienna 1873,
>
> Handbook of organ. Chemistry. In 2 editions as of 1885.[6]

Even a cursory glance at Beilstein's account shows how he perceived his life to be saturated with pedagogy. What mattered was who trained him, where he trained others, and what texts he published to that end.[7] I propose we take Beilstein at his word.

I argue that the *Handbuch* emerged out of a long and diverse set of interactions between Beilstein and pedagogies in various contexts. Specifically, I will situate Beilstein as a teacher at Göttingen, as editor of the *Zeitschrift für Chemie*, as professor at the Technological Institute in St. Petersburg, and as editor of the *Handbuch*, and show how in each case the notion of what it meant to train students into practicing chemists was always tied to a particular receptive environment. As his environment in St. Petersburg turned hostile in 1880, Beilstein began to direct his pedagogy more abstractly. Instead of focusing on actual students, Russian or German, he concentrated on building a Germanophonic international community of standardized thinkers.

## Student Teacher: Beilstein in Germany

Beilstein's paternal line hailed from Darmstadt, and his mother's line, the Rutsches, originated in Baden, both of solidly *Mittelstand* stock. His great uncle, Konrad Rutsch, was a Protestant from the small village of Dühren, which he left at age 16, settling two years later in St. Petersburg in 1810. There he opened a grocery store at the corner of Malaia Morskaia and Nevskii Prospekt, roughly two blocks from the Winter Palace.[8] In 1838, the year Beilstein was born, Konrad died, turning his business over to his niece Katharina Margarete Rutsch and her husband Karl Friedrich Beilstein. Our Beilstein was the first of their seven children (five boys, two girls), most of whom retained their status as German citizens. He was educated at the Protestant St. Petrischule in St. Petersburg, where he received extremely high marks and upon graduation at age 15 embarked for further study to Germany. An uncle paid for the trip.

In Germany, Beilstein traveled to most of the various educational centers for the modern chemist.[9] He studied with Robert Wilhelm Bunsen in Heidelberg for two years, then Justus von Liebig in Munich, and became close friends on another stay in Heidelberg with young *Privatdozent* August Kekulé. Beilstein took courses and learned laboratory skills from each of these figures before obtaining his doctoral degree at Göttingen under Friedrich Wöhler in February 1858, a few days shy of his twentieth birthday.[10] Thus armed with academic credentials, he traveled to Paris and worked in Adolphe Wurtz's laboratory, adopting the latter's program of investigating aldehydes. Paris resolved Beilstein on an academic career in chemistry, and he assumed a laboratory assistant position in Breslau in autumn 1859, under the fearsome charge of Carl Jacob Löwig. Breslau, and particularly Löwig's centralized and dismissive attitude towards his students, were not to Beilstein's taste.[11] In 1860 Wöhler offered Beilstein a position in Göttingen as a laboratory assistant with better pay (and more congenial working conditions).

Göttingen marked a pedagogically important moment for Beilstein, since this was when he was first expected to train future chemists. His main duties were to conduct practical laboratory instruction, an educational innovation stressed decades earlier by Liebig in his previous incarnation as the doyen of Giessen. After Liebig moved to Munich in the aftermath of the 1848–49 unrest in the German states, his model of chemical education spread essentially everywhere.[12] This type of laboratory-specific teaching put heavy demands on the pedagogue's time. As Beilstein reported to Russian chemist Aleksandr M. Butlerov: "If you consider, however, that I only have Saturday on which to work; on the other days instruction in the laboratory absorbs my entire activity, then you would make allowances for me and find it excusable if my patience is completely gone."[13] Nevertheless, despite his complaints, his six years in Göttingen can be considered the most scientifically productive and socially happiest of his life. Of particular note was his work on isomerism. In a series of seminal studies, Beilstein demonstrated that various organic compounds previously considered isomers were in fact the *same* compound, facts he used to expand the credibility of August Kekulé's structure theory. As Beilstein wrote to the latter: "My critics reproach me for having accomplished little of real significance along this line, but it was necessary to show beforehand that there is only *one* benzoic acid, that benzyl chloride and chlorotoluene are different, and so forth, before these could bring to your theory that range and that significance which it had from the beginning."[14] This established Beilstein's reputation as a gifted organic chemist.

Beilstein was not, of course, alone in the teaching labs, and his colleagues, von Uslar and Rudolph Fittig, frequently irritated him. Von Uslar was deplorable as a pharmaceutical chemist with no theoretical interests. Fittig's sins were graver: "[Fittig] is a pedant through and through and treads so closely in the footsteps of his great teacher Limpricht that I consider him to be the most insufferable man I have ever met. . . . The present lecture assistant has fallen ill and Fittig has temporarily taken over his job so I have this monster in my place the whole day."[15] So much for first impressions: Fittig would soon become Beilstein's inseparable colleague (at one point they shared an apartment) and a crucial factor—together with Hans Hübner, another Göttingen *Privatdozent*—in shaping an emerging pedagogical style that stressed collegial cooperation and complementary division of labor among teaching duties in a well-equipped laboratory.[16] Beilstein's magnetic personality made him the clear center of the group. As Fittig commented in his diary:

> I respect his knowledge of organic chemistry, and I cannot fail to recognize that he, especially in the matter of such important laboratory studies, has stimulated me many times. . . . He also stands before the laboratory students in a proper but good relation, he tells them jokes and anecdotes . . . and allows the difference between teacher and student to vanish completely. The students like this, as well as his joviality; his sharp humor makes him their darling. I am in this respect entirely different, and I will happily admit that I am abrupt, that I am too much a teacher, too schoolmasterly. . . .[17]

This teaching style—using humor and individual attention to motivate students—like so many other aspects of pedagogy, was site-specific. When Beilstein left Göttingen, this casual demeanor became harder to maintain.

As Fittig and Beilstein became more closely acquainted, they discovered other complementary aspects of their characters besides pedagogy. It has often been asserted with respect to the *Handbuch* that Beilstein was anti-theoretical, and thus was predisposed towards empirical compilation. Otto Krätz, for example, in his excellent study of Beilstein's correspondence with Heidelberg (later Munich) chemist Emil Erlenmeyer, contrasts Beilstein as the "man of Praxis" to Erlenmeyer's "man of Theory."[18] Perhaps this is true, but next to Erlenmeyer almost any chemist would seem to be a "man of Praxis." A less skewed contrast is between Fittig and Beilstein, where Fittig is classed as the talented experimentalist, and Beilstein is his theorist complement with his exceptional command of the chemical literature.[19] Beilstein in fact repeatedly stressed the importance of "indispensable" theories that were "supported by noteworthy discoveries," such as Kekulé and Butlerov's structure theory, and was quick to criticize Marcellin Berthelot's textbook as overemphasizing facts to the detriment of theory. Given that Beilstein's reputation was built on the success of structure theory, this conclusion should not strike us as odd, despite Beilstein's occasional later disclaimers of all theoretical speculation.[20] Beilstein was able to stake out this middle ground in his research *and* his teaching because Fittig provided enough experimental guidance to balance the students' training.

Beilstein's insistence on the mutual dependence of theory and experiment made him a pedagogue in high demand. In April 1865, St. Petersburg University attempted to hire Beilstein. Wöhler and others lobbied Hannover not to let such a prize teacher go. As the grand old man of Göttingen chemistry wrote to the Ministry in 1865, Beilstein "is the most talented and knowledgeable [of the Assistants], he possesses above all of them the most multifaceted scientific education, and his name known as among the most advantageous through his achievements in the most difficult part of chemistry, organic chemistry, on which he has been giving lectures each semester for 3 years already."[21] In the end, Göttingen counter-offered to make Beilstein an extraordinary professor at the ripe young age of 27, and threw in a sizable salary to boot—which Beilstein accepted after failing to coax similar blandishments from A. A. Voskresenskii in Petersburg.[22] Beilstein couched his final acceptance to Hannover in pedagogical terms: "My entire goal and striving is directed so as to conduct my science, which I place above everything, completely and independently. No place has offered more and better opportunity than here in Göttingen . . . nowhere have I found such a pure scientific sense as here and nowhere are the students as industrious as here. I put specifically a great deal of weight on the last point. We chemists achieve very little ourselves, and only that which we achieve through our students is actually valuable."[23] Anyway, no one really expected that Beilstein would seriously abandon his alma mater.[24] That assumption, it turned out, could not be further off the mark.

## Beilstein, Editor: The *Zeitschrift für Chemie*

Before continuing with Beilstein's pedagogical peregrinations, however, we need to pause and explore a literary project that grew up alongside Beilstein's teaching style in the culture of Göttingen sociability: his joint editorship of the *Zeitschrift für Chemie* with Hübner and Fittig. The *Zeitschrift* is important to the history of the *Handbuch* both because Beilstein's editorship involved the correlation and processing of diverse chemical material (thus providing necessary experience for his later magnum opus), and because the *Zeitschrift* functioned for roughly a decade as the *only* regular publication outlet for Russian chemists—albeit in German—and thus Beilstein's role on the editorial board facilitated the constitution of a Russo-German chemical community, the eventual breakdown of which would leave traces in the *Handbuch*.

Like Beilstein himself, the *Zeitschrift für Chemie* had a confusing set of appellations. In 1858, a Heidelberg quartet centered on August Kekulé founded the *Kritische Zeitschrift für Chemie, Physik und Mathematik*, which consisted largely of book reviews in the associated three disciplines. In 1859, one of the four, Gustav Lewinstein, assumed control of the journal with former pharmacist, now chemist, Emil Erlenmeyer, who had just become a *Privatdozent* at Heidelberg, and the journal was renamed the cumbersome *Kritische Zeitschrift für Chemie und die verwandten Wissenschaften und Disciplinen als Pharmacie, Technologie, Agriculturchemie, Physik und Mineralogie*. In 1860, as Erlenmeyer assumed more and more control, the editors changed the name yet again to the *Zeitschrift für Chemie und Pharmacie*, which it would remain until 1865. From 1861 to 1864, Erlenmeyer was the sole editor.[25]

The *Zeitschrift* quickly became more than just a synopsis of the state of the field, as Erlenmeyer increasingly imposed himself on the content of the journal. When excerpting pieces for publication, he would frequently append lengthy and highly critical "Bemerkungen," and he turned the journal into a one-sided vehicle for the promotion of various chemical reforms. Erlenmeyer's frequently ad hominem and aggressive style alienated readers more than it attracted them.[26] One of the saving graces of his journal was the frequency with which Russian chemists, particularly those who had studied in Erlenmeyer's laboratory at Heidelberg, published original research in the *Zeitschrift*, including Aleksandr Butlerov's fundamental developments in structure theory. Erlenmeyer's alienation from the community, however, made it difficult to sustain high submission rates or circulation, and he began to offer it to other possible editors in 1864. He explicitly suggested that Butlerov adopt the journal as a German-language Russian chemical organ, an idea possibly suggested to him by Beilstein.[27] Butlerov turned it down.

Beilstein did not. In 1865 Erlenmeyer turned the journal over to Hübner at Göttingen, who began to edit it in collaboration with Beilstein and Fittig. The journal—once so tied to Erlenmeyer's name that antipathy toward the one translated into antipathy toward the other—became so identified with the triumvirate that a major obituary for Erlenmeyer neglected to mention that he had ever had any connection to the *Zeitschrift.*[28] Beilstein et al. turned the now-renamed *Zeitschrift für Chemie* into a fairly successful journal. They maintained the old format, and Russian contributions remained high—even after the creation of a Russian-language chemical journal under the auspices of the Russian Chemical Society in 1869. Beilstein was particularly insistent on developing the Russian connection, as he wrote to Butlerov on January 29, 1865:

> I will in closing stress again that the "Zeitschrift" has in my person a warm advocate of Russia's interests. I wish that Russian chemists won't just slave away laboriously on a Russian [*sic:* German—MG] edition of their works (for you, who writes German so expertly, this is obviously not necessary!). But many can thus postpone the publication of works, and thus I ask that they send me only the *Russian* articles. I will worry about getting a correct translation. . . . Chemists speak only *one* language and thus one should also know in Germany what is appearing in Russia.[29]

In several publications of this period, Beilstein stressed the important contributions made by Russians in their own language and the lamentable ignorance of these findings on the part of Western chemists. Beilstein's favorable review of Butlerov's Russian textbook in organic chemistry similarly attempted to include Russians into a broader community of chemists.[30]

The happily advancing juggernaut of the *Zeitschrift* ground to a complete halt in 1871, when it faced staggering competition for its small market niche from the *Berichte der Deutschen Chemischen Gesellschaft,* founded in 1868. The Berlin *Berichte* achieved much higher circulation after 1871 by virtue of its becoming the official organ of the unified German Chemical Society. Beilstein wrote to Erlenmeyer from St. Petersburg on April 26, 1871: "There remains no doubt: the *Zeitschrift für Chemie* can no longer be conducted *the way* that it is now. Through the successful appearance of the Berliner Berichte one of the chief tasks of the Zeitschrift—to publish quickly—is essentially lost."[31] Beilstein then took the somewhat perverse move of attempting to foist the journal *back* to Erlenmeyer. The latter, now affiliated with the *Annalen* in Munich, rebuffed these overtures, and the journal folded by the end of the year.

The central problem that the *Zeitschrift* had been created to ameliorate remained unresolved, however. There was still a proliferation of disjointed chemical knowledge among multiple subspecialties, compounded by increasingly nationalistic attitudes towards domestic organs. This linguistic nationalism was most evident among the Russians, a fact Beilstein lamented on several occasions.[32] He complained to Erlenmeyer in April 1871: "What is to become of us, when each city produces its own journal, where one must seek out one's bit of chemistry under dust, garbage, and mouse droppings."[33] The knowledge needed to be systematized or it would be lost.

**Beilstein, Russian: The St. Petersburg Technological Institute**

In 1864 Beilstein had joked in a letter to Kekulé: "You won't believe how much you have risen in value here [since you left Germany for Belgium]. Look at [Hermann] Kolbe, the poor devil, how he must struggle through with a lot of trouble, and on the other hand, what a big shot [August] Hofmann is now. . . . What kind of big shot will you be, if one draws you back to Germany. I constantly wish to become a professor of chemistry in Peking or in the Sahara desert. Then I should make it difficult for Hofmann to compete with me!"[34] He was soon to make a transition perhaps even more dramatic than the Sahara: in 1866, a year after he turned down St. Petersburg University, Beilstein took a position at the less prestigious Technological Institute in the same city.[35] The move caught his Göttingen patrons by surprise. As Wöhler wrote to Liebig in November 1866: "The Prussian ministry in Hannover has asked why we let Beilstein go. He has taken a call from St. Petersburg with 2,500 thalers and is already gone."[36] The immediate cause was personal. Beilstein's father had died somewhat suddenly at the age of 56 in April 1865, and by 1866 the family needed him back in St. Petersburg. Beilstein was officially appointed a professor at the Technological Institute on 24 September 1866, and by June 1867 he became a subject of the Tsar.[37]

The Technological Institute was far from ideal. Created under Tsar Nicholas I as a way to train large numbers of civilian engineers, the Institute featured high teaching loads and students who were less interested in the pure sciences.[38] Whereas St. Petersburg University was located right next to the Academy of Sciences, the Technological Institute, south of the city center, was marginalized from daily academic interaction. Furthermore, Alexander II's revised university statute of 1863 had expanded faculty posts and generated a renaissance of natural scientific work at the University. This was part of the reason why D. I. Mendeleev, later famous for his 1869 formulation of the periodic system of chemical elements, gradually moved his base of operations from the Technological Institute to the University. Although Mendeleev became a full professor at the University in 1867 and thereafter devoted essentially no attention to the Institute, he did not officially give up his post (or salary) at the Institute until 1871. Beilstein was hired to teach analytic and organic chemistry in a demanding institution that had been long neglected. Hardly a second Göttingen.

Beilstein expected to be in a tough position when he arrived, since his close interactions in Germany with Russian academics made him aware of their onerous duties. As he recalled in 1893 about A. A. Voskresenskii, Mendeleev's predecessor at St. Petersburg University:

With the paper on theobromine Voskresenskii's scientific activity stopped. . . . Teaching activities totally occupied him. Remember, that in those years the position of the scientist in Russia was unenviable. The meager salary for the position of professor far from protected one from material needs. It was necessary to seek out supplemental lessons for a living, for the preservation of a family, and simply for the needs of scientific activity. But even lessons paid poorly, so that a large part of working time went into teaching and there was no energy left for scientific works. This explains the verdict, which in the past was often heard among foreign scientists, that every year talented Russian scientists left Russia, worked passionately and successfully, but they stopped research as soon as they returned to their homeland.[39]

Those dire days had passed, but Beilstein faced the additional problem of repairing the consequences of what he would characterize as Mendeleev's neglect. Beilstein stayed here for 30 years and lectured several hours a week (usually three hours inorganic/analytic and four hours organic). Enrollments of engineering students continued to grow throughout the 1870s, with little hope for Beilstein to get either scientifically minded assistants or requisite compensation.[40] Despite his overt despair, Beilstein was remembered as a gifted lecturer by contemporaries: "Lively, in constant motion, quick with words and actions, even perhaps sharp, Beilstein was markedly exceptional by his order in lectures. He spoke well, very lively, sometimes stepping off the subject, but these were not digressions into other areas of knowledge, like Mendeleev did; if Beilstein digressed, then they always concerned chemistry, and it seemed that he made these digressions in order to invigorate perhaps the exposition of the subject."[41] In 1891, after 25 years of service, he became emeritus, but he continued reading analytical chemistry lectures. He also read weekly chemistry lectures at Nicholas Military Engineering Academy.

We get an excellent sense of the frustrations surrounding Beilstein's initial position at the Institute in an especially revealing letter to Butlerov (then at Kazan) of November 1866:

Perhaps the tidings have not yet reached you in the far East that I have now decided to move to Petersburg. I am Mendeleev's successor at the *Technological Institute* and am busying myself dealing with my imposed duties. That is no small affair, when I tell you that my predecessor—who, as you know, is not really a practical chemist—never bothered with the work of *Praktikanten* and went at most for a few minutes into the laboratory every 1/4 of the year. He was in such a rush to get Chancel and Gerhardt translated, without bothering to consider progress in analytic chemistry in the least. This book was shoved into the hands of each *Praktikant* and then he was discharged with a blessing. You can easily imagine in what kind of dilapidated circumstances I have encountered almost everything here. How hard it is for me to introduce discipline and order will be clear to you immediately when I add that in one crowded room there are presently—175—yes, yes—175 men working! [as opposed to 80–85 at Göttingen —MG]. . . . What kind of nonsense there is here under these conditions! I had to give up being an academic (*Gelehrter*) and am a schoolmaster (*Schulmeister*) in the harshest meaning of the word. How painfully I feel this biting contrast I don't need to tell you. Although I live here in the circle of my family, I still feel the greatest homesickness for my old laboratory and my old students. I don't have any time at all for my own work. All my free time goes to the preparation of my lectures, that there is so much left here for me to accomplish.[42]

There are several points of interest here. First, there is Beilstein's hostility to Mendeleev, the rising star of the Petersburg chemical scene. Second, Beilstein was alienated from his students, whose lack of preparation and interest forced him to abandon the collegial teaching style he had developed at Göttingen. With his two assistants, he organized his laboratory "*militarisch*," using the explicit model of Löwig's Breslau laboratory (which he had hated while there).[43] Given that he was still editing the *Zeitschrift* at a remove from Göttingen, and since he did not have equally qualified colleagues, he was forced into a cookie-cutter, chemistry-by-numbers approach to deal with his limited resources.

The third point of interest concerns the lack of a proper textbook for teaching analytical chemistry, which brings us to Beilstein's *Anleitung zur qualitativen Analyse* (*Instructions for Qualitative Analysis*), published in the late 1860s in response to the inadequate Mendeleev-generated Russian translation of Charles Gerhardt's book. Beilstein, who had distinguished analytical skills, is still widely known for his "Beilstein test," whereby a flame is used to detect the presence of halogens in an organic substance.[44] Beilstein's book was an attempt to integrate recent innovations in analytic chemistry in a format that would be useful to unskilled students in the laboratory. Its absence of a formal framework and emphasis on building transferable and widely applicable skills made it uniquely adaptable across major theoretical and experimental divides. The text was published simultaneously in Russian and German, going through six editions and translated into Dutch, English, and French.[45]

Analytic chemistry was widely recognized as the most basic skill set for chemists. You could not do higher levels of organic, inorganic, mineralogical, physiological, or any other kind of chemistry unless you could properly analyze substances in a laboratory.[46] Erlenmeyer, for example, was quite specific about the need for proper training in analytical chemistry: "Before [the beginner] can go on, he must first have gotten to know exactly the bodies that he is supposed to find from now on by their properties and chemical relations. His first order of business must therefore be to state without a doubt, that he has experimentally studied the properties and transformations of metals, their salts, oxides, chloride, sulfides, etc."[47] This further implied that not only was analytic chemistry the *first* branch of chemistry to be taught to students, and the most *widely* taught among the various other natural sciences, but it was also the most *laboratory-intensive* of the sciences. This visibility and importance of the field also made book reviews of analytic chemistry manuals some of the most contentious of the decade.[48] Beilstein felt his own book had to emphasize laboratory work and provide a sense of unity to the discipline, much as the *Zeitschrift* had tried to do for the chemical literature. He accomplished his pedagogical unification of chemistry by emphasizing the standardization of *skills*. His *Handbuch* would go even further by standardizing chemical concepts.[49]

Beilstein structured his book rather unusually. A very slim volume, the *Anleitung* is organized as a set of instructions, not as a conceptual organization of available techniques and information. The first twenty pages or so, "Examples of Practice in Analysis," were designed to calibrate the student's laboratory. That is, Beilstein walked the student through the basic procedures of titration, heating, etc.: what happens when you carbonize a substance, you moisten it, you put it to a flame, you dissolve it, etc. This part of the book ascertains that all students are using the same equipment in the same way—*militarisch* indeed. The other thirty pages of the book, the "Systematic Course of Analysis," show in an even more explicit step-by-step form what the student is to do when presented with an unknown substance, assuming the set of skills developed earlier. First you take part of the substance and add water. If it dissolves, then move to step 5, if it does not, try sulfuric acid, then move to step 7, and so on. In this fashion, the student should be able to identify the substance qualitatively at the end of the series.

Meanwhile, Beilstein was also one of the most visible members of the Petersburg chemical community. Although Beilstein was awarded the Lomonosov prize of the Academy of Sciences in 1876 on the recommendation of A. M. Butlerov and N. N. Zinin, served as consultant for the Ministry of Finances on patent questions since 1867, and was president of the chemical section of the Russian Technical Society, he still felt scientifically isolated in Petersburg. He wrote to Butlerov late in 1866: "I am here in the big city more than ever in the isolation room of the sciences and each stimulation from the outside would be newly cherished."[50] He continued in April 1867: "I am until now completely isolated in Petersburg. Zinin, the only really thinking and active natural scientist, is through his high-aristocracy relationships almost unapproachable. The other chemists are either not chemists or look only with indignation or incomprehension on the progress of science."[51] A few months after he moved to Petersburg he began to agitate for Butlerov to be brought there from Kazan as an intellectual companion, and despite some stonewalling by Mendeleev, Butlerov was eventually offered a post at St. Petersburg University and the chair of chemistry at the Academy of Sciences. Relations between Beilstein and Butlerov were exceptionally good throughout the 1860s, when

Beilstein helped to publish the German translation of Butlerov's structure-theoretical textbook and assisted in the propagandizing of his chemical ideas.[52]

The situation soured when it came to D. I. Mendeleev, long Beilstein's bête noire. Mendeleev recalled in his 1861 diary his first meeting with Beilstein, then visiting his family in Russia. At a party in chemist L. N. Shishkov's house, Mendeleev remembered the hubbub about this "dry German," who left him with a very unpleasant impression.[53] While Beilstein was still in Göttingen, he could paper over their mutual dislike. He wrote a rather flattering review of Mendeleev's *Organic Chemistry* (1861) in the *Zeitschrift für Chemie*, where he lauded its creative—if somewhat outdated—use of Gerhardt's type theory.[54] Right before his move to St. Petersburg, Beilstein wrote to Mendeleev to thank him for "all of your kindness [in] naming me your successor at the Institute, which you already promised me last year." He also warned: "I will at first often make you sick of me. Moving to a new circle of activity, I will be obliged to often ask for your advice and help."[55] After Beilstein had begun to clean up Mendeleev's mess there, however, the goodwill dissipated. Beilstein would have liked to ignore him, but that was becoming less and less feasible.

It became utterly impossible on November 11, 1880, when Mendeleev was denied the chair in technology by the Physical-Mathematical Division of the Imperial Academy of Sciences by a single vote. Butlerov had tried once before to get Mendeleev a post at the Academy (then an adjunct chair in physics), but was rebuffed by the Permanent Secretary, Konstantin Veselovskii. This more public rejection in 1880 sparked a massive outcry from Russian chemists and from newspaper reporters. N. A. Menshutkin, secretary of the Russian Physico-Chemical Society, asked chemists to sign a protest of the Academy's behavior, which he would then publish in a local newspaper. On the grounds that a newspaper was an improper forum, Beilstein was the *only* chemist who refused to sign, proposing instead an honorary address at their next professional meeting.[56] Beilstein's "desertion" of the cause was attributed to his being a "German," and thus an affiliate of the supposed "German party" in the Academy of Sciences that was widely (and erroneously) believed to have orchestrated Mendeleev's rejection.[57] In the midst of the newspaper campaign in support of Mendeleev, only one substantial article, published in the capital's German paper, attacked the support for Mendeleev on the grounds (mostly correct) that the Russian was not a technological chemist. Many Petersburgers attributed authorship of the anonymous article to Beilstein.[58]

Injury was added to insult when *Beilstein himself* was awarded the chair in technology in 1882. Drawing on their critique of Mendeleev, anti-Butlerov academicians (Helmersen, Schrenk, Savich, Wild, and Gadolin) proffered Beilstein for the post as a teacher at the *Technological* Institute with a decided emphasis on his contributions to the chemistry of the oil industry, and as an officer of the Russian *Technological* Society. Anticipating objections from the so-called "Russian party," they added in their December 22, 1881 nomination: "We also remind you, that F. F. Beilstein is a Russian subject, a native of St. Petersburg, where he received his education, and that he commands the Russian language fully."[59] Butlerov was hard-pressed to offer an adequate riposte. On January 19, 1882, the Physical-Mathematical Division voted on Beilstein's candidacy. Of the 16 present, Beilstein received 12 votes, one more than the necessary two-thirds majority.[60] Butlerov decided to blackball Beilstein at the General Assembly meeting of March 5, 1882, since a two-thirds vote there was necessary to confirm the January vote. Of the 27 academicians present, Beilstein received 17, one vote shy of the two-thirds majority. He would not receive his chair in technology until after Butlerov's death in 1886. Beilstein was thus marked as German, not Russian, and he began to behave accordingly.[61]

After 1880, Mendeleev's fame in the wake of his rejection made the Russian capital uniquely inhospitable for the displaced Göttingener. Before 1882, Beilstein had published 92 articles; afterwards only 21.[62] One explanation for this is the increasing demands the *Handbuch* placed on his time after its first edition began to appear in 1881. But this itself was a consequence of his increasing isolation, which plunged him more deliberately into his compilatory work and disinclined him from publishing for his local peers. Consider Beilstein's increasing disengagement with local Russian chemists at the Russian Chemical Society. He was one of the Society's founding members in 1868, but among all Petersburg chemists of distinction, he was unique in not being recognized by the Society in any substantial administrative or honorary capacity. In 1903, under the new charter of the Academy, Beilstein lost the presidential election to Aleksandr Zaitsev, a Kazan chemist who was never able to attend a single meeting.[63] Mendeleev, on the other hand, became the Society's third honorary member and was elected honorary president for life in the 1890s. Beilstein got the message.

## Beilstein Bound: The *Handbuch der Organischen Chemie*

In an often-quoted letter, Beilstein drew the central connection between Russia, pedagogical institutions, and the *Handbuch*: "Truly, I could only have written my Handbuch in Russia, and thus I have deferred calls back to Germany. At a Russian Polytechnicum professors don't have to be scientifically active, because the students don't give any reason for it, but in Germany they would have looked at me disapprovingly."[64] This case, however, is purely negative—pedagogy in Petersburg provided the context for the *Handbuch* because of an *absence* of demands. I contend for a more positive claim, that

Beilstein's book emerged out of an initial attempt to write another organic chemistry textbook, but upon facing isolation from his peers and the inability to generalize the model provided by his *Anleitung,* he defaulted to the form of the *Handbuch.*

Beilstein did not approach his material hoping to write a comprehensive reference work. Ever since he was a *Privatdozent* at Göttingen, he had been gathering material on organic compounds and checking them for accuracy. This was specifically for teaching. It was only after he realized the need for an updated organic textbook in Russia that he began to work the material into a *textbook.* As he wrote to Erlenmeyer on February 22, 1878:

I have gone now earnestly about carrying out a plan that I have had in mind for a long time: I am actually writing an organic chemistry. Now I am in a lot of trouble. As I have all the material *completely gathered* before me, so I hope that I can be done with the writing in about 2 years. Now I am already in the 2nd year of work & have only gotten to glycerin.

I have, you see, made it a rule while writing to monitor *all citations* myself & that is what has brought me slowly to such despair. But one can't get around this kind of work if the results are supposed to be reliable. What I have written so far is actually more a *catalog* of organic chemistry rather than a textbook. Since I want to cite *everything,* that is actually *everything,* it seems to me to appear in a purely organic classification (acids with $O_2$: $C_nH_{2n+2}O_2$ . . . acids with $O_3$ . . .). For the purpose of looking things up it is entirely excellent—and I have now had enough opportunity to convince myself satisfactorily of that. But with *reading* it is something else. The story comes out too dry. Thus I miss throughout the relevant pages precisely *your* textbook in which one can look things up, but can also *read.* Time is looming upon me, however: if I want to make everything also nice and easy to digest, I will be done in 5–6 years & that is too much for me. I don't have that much of my life to spend.[65]

The revision process of writing a textbook was forcing him to order the material purely by empirical formula.[66] He did not attempt any organization that would help students assimilate the material: "I can say clearly what my vision is: I want to put together in *one* volume the *complete* material of org. chemistry ordered completely and clearly with exact information about the literature. Pretty speeches, charming comparisons, lively pictures and the rest of it are as good as completely absent."[67] No wonder the textbook idea failed.

Nevertheless, Beilstein continually referred to the reference book as essentially rooted in its educational context. He accentuated these textbook origins—origins that *failed* to produce a textbook—before the Russian Chemical Society in 1893:

The material of organic chemistry and particularly of the aromatic series is growing by horrific measures. When I began to gather this material 33 years ago for its special study and for the goal of teaching, it was possible to follow the successes of organic chemistry easily; now this seems to be an impossible task. I began to read Liebig's *Annalen* correctly from its 101st volume. Then I had to reread all 100 volumes. Basing myself on Gerhardt's famous handbook (*rukovodstvo*) to organic chemistry and with constant corrections in Gmelin's classic handbook, I slowly gathered all the material in a form best suited for my goals. Then I had to look through the entire *Jahresberichte der Chemie* in order to convince myself that nothing had been missed. It stands to reason that it was necessary to take notes from the current periodical literature on all new facts and put them into the collection. Upon the appearance of each new guide to organic chemistry I compared its content with my notes, and in the event of a disagreement—which happened rather often—I had to check with the original articles. Thus I accumulated rather reliable material, but in a form unsuitable for a handbook. I had to rework the factual part, but when this was done, organic chemistry had again moved so far forward that it was necessary to redo everything over again. After 17 years of preparatory work, I went to publish it, and this explains why the entire work appeared in one brief period. . . . When my work was finished, I was sure that it was necessary to write everything over again. And this was done, but first it was necessary to redo everything a third time, so that now it is already clear that it is impossible to keep this arrangement any more. It is inconvenient to divide organic chemistry into a fatty series and an aromatic; an empirical division of camphors should be set up, sugary substances should also be grouped differently, ring-form compounds would be better united in different groups, etc. But I don't have time or energy for this.[68]

(The "Beilstein System," the organizing principle behind the present-day *Handbuch*, was only instituted in 1909, three years after Beilstein's death.)

What on earth did Beilstein mean by repeatedly referring to students in his autobiographical statements about the *Handbuch*? If we think of pedagogy too narrowly as what happens in the classroom, there is no connection between it and the *Handbuch*, a text so patently unusable in such contexts; but if we take a broader view of standardizing practitioners into the chemical community, then a clearer connection emerges. The *Handbuch*, more perhaps than any other text, standardized what it *meant* to be a practitioner of organic chemistry: you worked on the kinds of materials that were in the *Handbuch*. Similarly, the editorial abstracting of the *Zeitschrift* standardized what the field of "chemistry" meant by correlating the disparate strands of chemical publishing into one focused site. Circular, perhaps, but satisfying. A closer look at the first edition of the *Handbuch* bears this out. Beilstein had to create the demand for the kind of information he was providing among those less familiar with structure theory, and therefore distilled into a small introduction a presentation of the basic principles derived from Erlenmeyer's, Gmelin's, Kekulé's, Gerhardt's, and Kolbe's textbooks, and those textbooks alone. For example, on the first page of the introduction he offered the standard history of organic chemistry dating from Wöhler's synthesis of urea in 1828, and then continued to provide a framework to interpret the data included in the compilation, beginning with quantitative organic analysis—determining the composition of organic molecules—before moving on to how their structure might influence their properties.[69] This development from analysis to higher levels of chemical thinking bears strong marks of his strategy at the Technological Institute: first you teach the basics of measurement by rote, and then you construct theories of limited abstraction on this foundation. The stages of reasoning were clearly separated in St. Petersburg as they had not been at Göttingen. Beilstein's whirlwind survey of organic chemistry then moved through a series of topics in increasing specificity: "The Determination of Vapor Density," "Rational Formulas—Isomerism," "The Structure of Carbon Compounds," "Radicals," "Substitution," "Homology," and "The Physical Properties of Compounds." He presented a system of organic chemistry, although still a necessarily incomplete one: "As we look over the system of organic chemistry as it presently appears, we note still many holes. The filling of these holes is only a question of time."[70] This was the incentive to use the *Handbuch*: to ground oneself in basic principles and then become motivated to develop the science in defined directions—to fill in the holes.

For all the similarity to a textbook in motivation, genealogy, and program, the foreword exposed the unique nature of the Beilstein venture:

> In the present work I have made the attempt to present together as clearly as possible the completely analyzed organic compounds. I have refrained from an exhaustive characterization of the compounds; one individual doesn't have the strength to do that. Cursory, superficial remarks, imprecisely researched compounds and reactions, etc., have been omitted here or only partially included if there exist no other data for the nature of a body. Thus, as I have included references to the literature as completely as possible, it would be easy for the reader to look up anything that might be missing. Everything that contributes to an exact knowledge of the substances, such as melting point, boiling point, specific weight, solubility, etc., as well as precisely determined transformations and reactions of the bodies, are provided in their entirety.[71]

Beilstein wore on his sleeve the amount of personal labor and *individual* effort involved in the compilation of the handbook: "All comments were—as far as they were accessible to me—taken by the author from the *original articles*. In *the entire book there are no citations that I have not looked up before the writing*. All errors are from copying or printing."[72]

In a letter to Zincke, Beilstein noted the role the attack on his "Germanness" had played in the creation of the *Handbuch*, and the reciprocal effect of the *Handbuch* on the way local chemists identified his nationality:

> It will not be unknown to you, that since the political successes of Germans in Russia a hostile mood against Germany and Germans has been spreading ever wider. The systematic smearing of the newspapers was not without results. If I until now remained almost entirely unbothered by Germanophobia (*Deutschenhaß*), I have recently also had to suffer from this evil. The circumstance, that on the occasion of the elections to the chemistry [*sic*] post at the Academy of Sciences here some academicians also wanted to put my name on the list of candidates, has called forth a big storm and drawn a whole array of spiteful comments towards me. . . . Even the fact that my large handbook of organic chemistry was then appearing in *German* (there would be a dearth of buyers for a Russian work), has drawn the censure of the patriots to me.[73]

The process of revising the *Handbuch* for a second edition shows to what extent Beilstein turned his back on the Russian chemical community that had spurned him. Beilstein directed his solicitations for comments from readers almost entirely to German chemists.[74] Beilstein put out a second edition by himself by 1889, again in German, but his revision of the third edition pushed him to think about a successor. In 1895 he wrote to Paul Jacobson, who would eventually agree to become a co-editor of the *Handbuch*: "I am becoming old and have no co-worker. Now I worry only with all my powers that the third edition be finished—the heavens can worry about everything else."[75]

Beilstein wanted the *Handbuch* to be continued as a cooperative venture, run as the special branch of a scientific society, and he petitioned the German Chemical Society in Berlin, snubbing his local peers entirely in favor of an "international" (read: German) community. This was quite a transformation for the chemist who in 1873 at the Vienna international exposition had been quite sanguine about the prospects for Russia to become a full-fledged member of the international scientific community.[76] Opinion in the German chemical community was divided on whether the German Chemical Society should take on the task, with Erlenmeyer in support of the idea as "not bad," but Volhard opposing since he thought the matter was "better left to private industry."[77] It was eventually adopted in 1896, and after the third edition's appearance and Beilstein's death in 1906, it moved there entirely.[78] As Beilstein wrote to M. M. Richter: "Now I can peacefully retreat from the scene, since I am now sure that my work will be continued in the best manner, and good and complete handbooks will be available at a cheap price."[79]

Reviews of the *Handbuch* were ecstatic from the start. Richard Meyer wrote to Erlenmeyer in 1882: "I use Beilstein's book daily; it became immediately indispensable; but there is a mass of errors in it! One has to go to the original articles every time; but what is good is that you can find them quickly through the book."[80] Even the normally gruff V. V. Markovnikov was enthusiastic about "Beilstein's wonderful reference book," and P. P. Alekseev in Kiev commented in a letter to Butlerov in 1880: "I am now studying Beilstein and Kekulé. Beilstein's book is really a capital production. It is a pity, though, that the generalizations are rather short."[81] Perhaps the most personally satisfying review for Beilstein was the letter sent by Henry E. Armstrong, president of the Chemical Society of London, to congratulate the Russian Chemical Society on its twenty-fifth anniversary on November 6, 1893. In a statement addressed to Mendeleev, he wrote: "Our Society is proud to have enrolled your name in its liste [*sic*] of foreign members and to have welcomed you as one of its Faraday lectures [*sic*]; and the roll also includes the name of Beilstein—which, however, is no longer the mere name of an individual but a household word and one which cannot be mentioned without the feeling of gratitude arising in the chemists [*sic*] mind."[82] The ever-sarcastic Beilstein must have smiled internally from his seat on the same dais to hear Mendeleev intone such an endorsement. Beilstein's *Handbuch*, the rejection of both his Petersburg pedagogy and his Petersburg peers, was extolled in their midst.

## Conclusion: Beilstein Abandoned

Beilstein adapted his pedagogical styles to different contexts without changing certain principles about the necessity of theory for proper experimentation and the uselessness of theory once it moved beyond the limits experiment set for it. Thus, at Göttingen, he stressed the importance of collegial relations in the laboratory in order to initiate young chemists into a community he perceived as standardized by just such a lack of hierarchy—an attitude present in his collegial editing of the *Zeitschrift für Chemie* with Hübner and Fittig. When he moved to the Technological Institute in St. Petersburg, his less-prepared students compelled him to move to a yet more fundamental level of standardization—that of analytical chemical practice, as encoded in his *Anleitung*. Finally, after his hope for an integrated international community of practitioners was crushed by the polarizing of the scientific community into nationalized camps after the 1880 Mendeleev affair, when he was classed as a *German* chemist somewhat against his will, he moved once again to a more fundamental level and sought to standardize for practitioners of organic chemistry the very objects of their study. In each case, Beilstein maneuvered among the multiple instantiations of pedagogy as a category for history of science in his attempt simultaneously to define a chemical community and the means for stabilizing it.

One of Beilstein's last public actions at the Russian Physico-Chemical Society was to argue for the Geneva compromise on organic nomenclature. Already in 1890 Beilstein announced the existence of both the German and French commissions which were working to standardize organic terminology on an *international* footing, a sorely needed reform. As the local representative of these committees to St. Petersburg—mediated through his frequent vacations in Germany—Beilstein asked for any suggestions or comments from Russians.[83] Apparently, no one volunteered any suggestions, although debates about a *Russian* nomenclature in the early 1870s had sparked a good deal of interest. When the Geneva compromise system finally emerged, Beilstein was the only one to support it publicly. In the face of opposition from (among others) Beketov and Mendeleev—whose objections were remarkably insubstantial—Beilstein noted that the Geneva system was already being introduced into the *Handbuch* and would most likely become the standard, so it would behoove the Russians to adopt it.[84] Nikolai Menshutkin, editor of the Russian chemical journal, translated the nomenclature's French rules into Russian. When Beilstein correctly identified errors in the translation which would cause confusion, Menshutkin grudgingly admitted his mistake while belittling Beilstein's suggestions for repairing the Russian error.[85] Even in areas like organic classification, where Beilstein was clearly the resident expert, his advice was rejected.

A similar snub can be seen after Beilstein's death. It is a rule of thumb that a chemist's contemporary stature can be fairly well gauged by the kind of obituary he or she received. In the German *Berichte,* Beilstein's obituary was a full and judicious account penned by another Russian "foreigner," Helsinki chemist Edvard Hjelt. In the Russian *Zhurnal,* Beilstein received slightly over a page of impromptu comments by Beketov, his colleague at the Academy of Sciences (an institution not mentioned in the obituary), at a time when even minor chemists received at least ten pages and individuals of Beilstein's international reputation often received over 100 pages of memorial. Beilstein's death was quickly papered over to discuss a looming budget crisis.[86] Little attention was paid to his pedagogical importance, either in the Russian context or through his *Handbuch,* which received a passing mention.

It is important to remember that the constitution of a *scientific* community was far from the only kind of community these chemists were in. In particular, the formation of national communities of scientists who *thought of themselves* as Russian or German chemists, and not chemists in Russia or Germany, was an important causal factor in why such seemingly neutral texts as the *Handbuch* took on the form they did.[87] Even opting to publish it in German—and not, for example, simultaneously in Russian, as he had for the *Anleitung*—was a statement by Beilstein about linguistic dominance in the sciences, and such questions about language are central to both pedagogy and nation formation.[88] After both world wars, for example, the *Handbuch* was held up by German chemists as a testament of the possibility for goodness in German culture; Beilstein's systematization of chemical knowledge atoned for other sins.[89] These signs of the times continue: since 1981, all 230 volumes of "Beilstein" were published exclusively in English, as is today's on-line version.[90]

Finally, I would like to stress the contingency of seemingly universal, immobile, and "indispensable" reference works, such as Beilstein's *Handbuch.* Had his father's death not prompted him to abandon his beloved Göttingen for Petersburg, Beilstein himself believed he would not have even embarked on the project. And had the rejection of Mendeleev by the Academy of Sciences not taken on its nationalist dimensions, that *Handbuch* would most likely, I argue, have looked more like a textbook with a (very) long appendix of data on compounds. In a letter to Paul Jacobson shortly before his death, Beilstein commented on the contingency of the *Handbuch*: "If I look now, in all calmness, on the achievements of my last 40 years, it seems to me a lucky stroke of fate from the heavens that I was born and lived in a time, when such an undertaking as my Chemistry could be completed. Here all the necessary conditions happened to come together. Only a few years different and all would be in vain."[91] Beilstein's point was purely about the chemistry: had he been earlier, Kekulé's theory would not have existed as a basis; had he been later, there would have been too many compounds for one man to organize. But the contingency, like the *Handbuch* itself, can be applied much more widely than its author intended.

## Acknowledgments

I would like to thank Matthew Jones, David Kaiser, Robert Kohler, and Mary Jo Nye for helpful comments on earlier versions of this essay.

## Notes

The follow abbreviations are used in the notes: ADIM: Arkhiv-Muzei D. I. Mendeleeva (D. I. Mendeleev Archive-Museum), St. Petersburg, Russia; *Ber.*: *Berichte der Deutschen Chemischen Gesellschaft*; PFARAN: Peterburgskii Filial Arkhiva Rossiiskoi Akademii Nauk (Petersburg Division of the Archive of the Russian Academy of Sciences), St. Petersburg, Russia; RGIA: Rossiiskii Gosudarstvennyi Istoricheskii Arkhiv (Russian State Historical Archive), St. Petersburg, Russia; TsGIASPb: Tsentral'nyi Gosudarstvennyi Istoricheskii Arkhiv Sankt-Peterburga (Central State Historical Archive of St. Petersburg), St. Petersburg, Russia; *ZfC*: *Zeitschrift für Chemie*; *ZhRFKhO*: *Zhurnal Russkago Fiziko-Khimicheskago Obshchestva* (*Journal of the Russian Physico-Chemical Society*).

All dates in German texts or in correspondence to or from Germany are given in the new-style Gregorian calendar. Dates that occur unambiguously in Russia are given according to the old-style Julian calendar, which lagged 12 days behind the Western calendar in the nineteenth century, 13 in the twentieth. Transliterations from Russian follow a modification of the standard Library of Congress format. All unattributed translations are mine.

1. Edv. Hjelt, "Friedrich Konrad Beilstein," *Ber.* 40 (1907): 5041–5078, on 5041.

2. Highlighted in orange in *How to Use Beilstein: Beilstein Handbook of Organic Chemistry* (Beilstein Institute, 1979), 8.

3. F. Richter, "K. F. Beilstein, sein Werk und seine Zeit: Zur Erinnerung an die 100. Wiederkehr seines Geburtstages," *Ber.* 71A (1938): 35–71, on 46. Richter was then editor of the *Handbuch*. Chemical handbooks have, of course, a long history, beginning with French efforts in the eighteenth century and continuing, most importantly, in Leopold Gmelin's *Handbuch der Chemie*. In Gmelin's initial handbook, organic materials took up barely half the space. In the fourth edition (1841–1870), they occupied about two-thirds of the ten volumes. See Leopold Gmelin, *Handbuch der Chemie*, fourth edition (Karl Winter, 1843). Beilstein's organic classification is heavily based on the Laurent method used by Gmelin. On Gmelin, see E. Pietsch with E. Beyer, "Leopold Gmelin—der Mensch, sein Werk und seine Zeit," *Ber.* 72A (1939): 5–51. For a superficial survey of handbook history, see Heinz Götze, "Das wissenschaftliche Handbuch," in *Einhundert Jahre Beilsteins Handbuch der Organischen Chemie* (H. Stürtz, 1981), 83–98.

4. The list of works explicating the so-called "Beilstein system" is large. See, for example, B. Prager, D. Stern, and K. Ilberg, *System der organischen Verbindungen: Ein Leitfaden für die Benutzung von Beilsteins Handbuch der Organischen Chemie* (Julius Springer, 1929); Oskar Weissbach, *The Beilstein Guide: A Manual for the Use of Beilsteins Handbuch der Organischen Chemie* (Springer-Verlag, 1976); *How to Use Beilstein*; Friedo Giese, *Beilstein's Index: Trivial Names in Systematic Nomenclature of Organic Chemistry* (Springer-Verlag, 1986); Ernest Hamlin Huntress, *A Brief Introduction to the Use of Beilstein's Handbuch der Organischen Chemie* (Wiley, 1930); Friedrich Richter, *Kurze Anleitung zur Orientierung in Beilsteins Handbuch der Organischen Chemie* (Julius Springer, 1936). Some of these, such as Huntress's text, include problem sets for practice.

5. The only book-length biography of Beilstein in any language remains L. A. Shmulevich and Iu. S. Musabekov, *Fedor Fedorovich Beil'shtein, 1838–1906* (Nauka, 1971), which goes to some effort to dismiss Beilstein's real conflicts in Russia. The best overall works are the collection of essays on the *Handbuch* by Friedrich Richter: "How Beilstein Is Made," tr. Ralph E. Oesper, *Journal of Chemical Education* 15 (1938): 310–316; "Beilsteins Handbuch—75 Jahre organisch-chemischer Dokumentation," *Angewandte Chemie* 70 (1958): 279–284; "Friedrich Beilstein, Gedanken zur hundertsten Wiederkehr seines Geburtstages," *Angewandte Chemie* 51, no. 7 (1938): 101–107; "K. F. Beilstein, sein Werk und seine Zeit." See also *75 Jahre Beilsteins Handbuch der Organischen Chemie: Aufsätze und Reden*, ed. Richter (Springer-Verlag, 1957). For obituaries, see Hjelt, "Friedrich Konrad Beilstein"; Otto N. Witt, "Friedrich Konrad Beilstein," *Journal of the Chemical Society* 99 (1911): 1646–1649. Also valuable are Ernest H. Huntress, "1938: The one hundredth anniversary of the birth of Friedrich Konrad Beilstein (1838–1906)," *Journal of Chemical Education* 15 (1938): 303–309; Iu. S. Musabekov and L. A. Shmulevich, "Akademik F. F. Beil'shtein i ego vklad v khimiiu," *Voprosy Istorii Estestvoznaniia i Tekhniki*, no. 3 (28) (1969): 61–66; M. Gerchinov, "Beil'shteinovskie daty," *Khimiia i sotsialisticheskoe khoziaistvo*, no. 7 (1931): 142. Almost all the information on Beilstein on p. 15 of John Turkevich, *Chemistry in the Soviet Union* (Van Nostrand, 1965), down to the chemist's first name and his publications, is incorrect.

6. Sent into the Krause Album and preserved at the Deutsches Museum in Munich. Reproduced in *Beilstein-Erlenmeyer: Briefe zur Geschichte der chemischen Dokumentation und des chemischen Zeitschriftenwesens*, ed. Otto Krätz (Werner Fritsch, 1972), p. 8. Beilstein wrote a similar letter for his file in St. Petersburg, dated July 1, 1867. TsGIASPb, f. 492, op. 2, d. 2073, ll. 1–1ob. This version displays some grammatical lapses in the Russian.

7. Of course, man does not live by pedagogy alone, and I will pass over Beilstein's important work on industrial chemistry and especially the chemistry of Russian oil. See F. Beilstein, *Die chemische Grossindustrie auf der Weltausstellung zu Wien im Jahre 1873* (Quandt and Handel, 1873); F. Beilstein and A. Kurbatow, "Ueber die Natur des kaukasischen Petroleums," *Ber.* 13 (1880): 1818–1821; idem, "II. Ueber kaukasischen Petroleum," *Ber.* 14 (1881): 1620–1622; idem, "Ueber die Kohlenwasserstoffe des amerikanischen Petroleums," *Ber.* 13 (1880): 2028–2029. For an opposing contemporary view on Russian oil, see W. Markownikoff and J. Spady, "Zur Constitution der Kohlenwasserstoffe, $C_nH_{2n}$, des kaukasischen Petroleums," *Ber.* 20 (1887): 1850–1853. Markovnikov's citation of Beilstein's work was always positive and respectful. See, for example, Markovnikov and V. N. Ogloblin, "Issledovanie kavkazskoi nefti," in V. V. Markovnikov, *Izbrannye trudy. Klassiki nauki*, ed. A. F. Plate and G. V. Bykov (Izd. AN SSSR, 1955), 331–332. While some Soviet commentators ignore Beilstein's work on oil entirely in their surveys (such as S. R. Sergienko, *Ocherk razvitiia khimii i pererabotki nefti* [Izd. AN SSSR, 1955]), leaping directly from Mendeleev to Markovnikov, it is clear that Beilstein, not Mendeleev, was responsible for scientific interest in the composition of Baku oil. See V. I. Kuznetsov, *Vozniknovenie khimii alitsiklicheskikh soedinenii* (Izd. AN SSSR, 1961), 68.

8. This biographical outline is drawn from the sources in note 5.

9. Travel was crucial in this period for unifying various scientific cultures. See, for example, the account of Bunsen's travels in Fritz Krafft, "Das Reisen ist des Chemikers Lust—auf den Spuren Robert Bunsens: Zu Robert Wilhelm Bunsens 100. Todestag," *Berichte zur Wissenschaftsgeschichte* 22 (1999): 217–238.

10. For his early work, see F. Beilstein, "Ueber die Diffusion von Flüssigkeiten," *Annalen der Chemie und Pharmacie* 99 (1856): 165–197; idem, "Ueber das Murexid," *Annalen der Chemie und Pharmacie* 107 (1858): 176–191.

11. Huntress, "1938," 304.

12. The classic work on this transformation remains Peter Borscheid, *Naturwissenschaft, Staat und Industrie in Baden (1848–1914)* (Ernst Klett, 1976). On the laboratory at Göttingen, see the letter from Beilstein to Kekulé, June 3, 1860, reproduced in Huntress, "1938," 305.

13. Beilstein to Butlerov, 14/2 December 1862, reprinted in G. W. Bykow and L. M. Bekassowa, "Beiträge zur Geschichte der Chemie der 60-er Jahre des XIX. Jahrhunderts: II. F. Beilsteins Briefe an A. M. Butlerow," *Physis* 8 (1966): 267–285, on 268.

14. Quoted in Richter, "How Beilstein is made," 311. For the original researches, see F. Beilstein, "Ueber die Identität des Chlorbenzols mit dem gechlorten Chlorbenzyl (Bichlortoluol)," *Annalen der Chemie und Pharmacie* 116 (1860): 336–356; E. Reichenbach and F. Beilstein, "Ueber die Natur der sogenannten Salylsäure," *Annalen der Chemie und Pharmacie* 132 (1864): 309–321.

15. Quoted in Huntress, "1938," 305.

16. On Beilstein's two associates, see the obituaries by F. Beilstein, "Hans (Julius Anton Edward) Hübner," *Ber.* 17 (1884): 763–776; R. M., "Rudolph Fittig," *Journal of the Chemical Society* 99 (1911): 1651–1653.

17. Fittig's diary entry of Sunday, November 24, 1860, quoted in Fr. Fichter, "Rudolph Fittig," *Ber.* 44 (1911): 1339–1401, on 1352. As Fittig continued in 1860: "Everything that he [Beilstein] says is original and funny, he looks at everything with a sharp understanding and a very healthy judgment. His judgments and criticisms of others are most delightful. First he praises a lot, then follows a 'However,' by which the praise is supposed to be qualified a bit, only as a rule nothing of the praise remains at all." Quoted in Richter, "Friedrich Beilstein, Gedanken zur hundertsten Wiederkehr seines Geburtstages,"102.

18. Editor's introduction in Krätz, *Beilstein-Erlenmeyer,* 11.

19. Fichter, "Rudolph Fittig," 1364; Hjelt, "Friedrich Konrad Beilstein," 5046.

20. "In order to not be entirely hindered by false preconceived notions while working, I have bid farewell to all speculations and am in the best sense a fierce enemy of all theories." Quoted in Richter, "K. F. Beilstein, sein Werk und seine Zeit," 41. On Beilstein's support of structure theory, see Beilstein to Butlerov, 29/17 January 1865, in Bykow and Bekassowa, "II. F. Beilsteins Briefe an A. M. Butlerow," 270–271; F. Beilstein, Review of Marcellin Berthelot's *Chimie organique fondée sur la synthèse, Göttingische gelehrte Anzeigen,* no. 1 (1861): 542–560, on 560.

21. Quoted in Richter, "K. F. Beilstein, sein Werk und seine Zeit," 39–40.

22. TsGIASPb, f. 14, op. 1, d. 6203, l. 4, quoted in Shmulevich and Musabekov, *Fedor Fedorovich Beil'shtein,* 33–34.

23. Quoted in Richter, "K. F. Beilstein, sein Werk und seine Zeit," 41–42.

24. See, for example, the assessment by the typically astute V. V. Markovnikov to his mentor Butlerov, August 7 and September 13 [1865], reproduced in G. V. Bykov, ed., *Pis'ma russkikh khimikov k A. M. Butlerovu, Nauchnoe Nasledstvo,* v. 4 (Izd. AN SSSR, 1961), 216–217.

25. The best secondary article on the *Zeitschrift* remains G. V. Bykov and Z. I. Sheptunova, "Nemetskii 'Zhurnal khimii' (1858–1871) i russkie khimiki (K istorii khimicheskoi periodiki)," *Trudy Istorii Estestvoznaniia i Tekhniki* 30 (1960): 97–110. This episode, as well as the major institutions of German organic chemistry in this period, has been well treated in the seminal study by Alan J. Rocke, *The Quiet Revolution: Hermann Kolbe and the Science of Organic Chemistry* (University of California Press, 1993).

26. Rita Meyer, Emil Erlenmeyer (1825–1909) als Chemietheoretiker und sein Beitrag zur Entwicklung der Strukturchemie (dissertation, Medical Faculty of Ludwig-Maximilians Universität, Munich, 1984).

27. See Erlenmeyer to Butlerov, March 25, 1864, reproduced in G. W. Bykow and L. M. Bekassowa, "Beiträge zur Geschichte der Chemie der 60-er Jahre des XIX. Jahrhunderts: I. Briefwechsel zwischen E. Erlenmeyer und A. M. . Butlerow (von 1862 bis 1876)," *Physis* 8 (1966): 185–198, on 191. Beilstein later somewhat exaggerated his role in publicizing the work of Russian chemists through the *Zeitschrift*: F. F. Beil'shtein [Beilstein], "O rabotakh chlenov Russkago Fiziko-Khimicheskago Obshchestva po aromaticheskomu riadu," in *Russkoe Khimicheskoe Obshchestvo. XXV (1868–1893). Otdelenie khimii Russkago fiziko-khimicheskago obshchestva* (V. Demakov, 1894), 39–56, on 47–48.

Many of these Russians became acquainted with Erlenmeyer personally in Heidelberg. On the broader Russian colony in Heidelberg, see Willy Birkenmaier, *Das russische Heidelberg: Zur Geschichte der deutsch-russischen Beziehungen im 19. Jahrhundert* (Wunderhorn, 1995). Beilstein is never considered by Birkenmaier or other researchers as a true member of the Russian colony. He is mentioned, however, in the appendix to *Das russische Heidelberg*, on 175.

28. W. H. Perkin, "Emil Erlenmeyer," *Journal of the Chemical Society* 99 (1911): 1651–1653.

29. Reprinted in Bykow and Bekassowa, "II. F. Beilsteins Briefe an A. M. Butlerow," on 271, with ellipses. Beilstein's efforts in translation were marked by occasionally significant errors. When Mendeleev gave Beilstein the original article on the periodic system to be published in the *Zeitschrift*, Beilstein had a student, A. A. Ferman, undertake the translation, who erroneously replaced "*periodicheskii*" with "*stufenweise*," instead of "*periodische*," where Mendeleev described the nature of the changes in properties with increasing atomic weight. While not Beilstein's fault, Mendeleev resented the error, which later complicated the priority dispute with Lothar Meyer. K. Bening, *D. I. Mendeleev i L. Meier* (Tsentral'naia tip., 1911), i–iii.

30. F. Beilstein, Review of A. Butlerow's *Einleitung in das Studium der organischen Chemie*, *ZfC*, N.S. 1 (1865): 727–730. See also F. Beilstein, Review of Marcellin Berthelot's *Chimie organique fondée sur la synthèse*, *Göttingische gelehrte Anzeigen*, no. 1 (1861): 542–560, on 553; F. Beilstein, Review of August Kekulé's *Lehrbuch der organischen Chemie oder der Chemie der Kohlenstoffverbindungen*, *Göttingische gelehrte Anzeigen*, no. 1 (1863): 493–507, on 500: "He [Nikolai Beketov] placed his observations in a dissertation which was published in April 1853 in Russian and surely because of the latter circumstance remains unknown by the majority of chemists."

31. Reproduced in Krätz, *Beilstein-Erlenmeyer*, letter 1, 16–17.

32. Beilstein to Erlenmeyer, May 11, 1872: "Now the Russians however have become great patriots: they don't want to publish their articles any more in foreign languages. Only a few, e.g. [Nikolai] Menshutkin, are so kindly as to worry about taking care of a translation themselves. Thus it is predictable that many useful works will be lost" (Krätz, *Beilstein-Erlenmeyer*, 26). And again in Beilstein to Erlenmeyer, October 5, 1873: "My patriotic friends will make a stink if I don't provide the fatherland's journal with original articles" (ibid., 41).

33. Krätz, *Beilstein-Erlenmeyer*, 15.

34. Quoted in Richter, "Friedrich Beilstein, Gedanken zur hundertsten Wiederkehr seines Geburtstages," 102

35. Beilstein's appointment by the Technological Institute was a stunning 16–1 vote, far ahead of the other two (Russian) candidates. "Zhurnal uchebnago komiteta S. Peterburgskago Prakticheskago Tekhnologicheskago Instituta," September 24, 1866, RGIA, f. 733, op. 159, d. 12, l. 1ob.

36. Quoted in Reiner Luckenbach, "Der Beilstein: Geschichte, Gegenwart, und Zukunft," in *Einhundert Jahre Beilsteins Handbuch der Organischen Chemie*, 36.

37. This was necessary to become a professor under the Institute's statute. His official term of service only began with his renunciation of German citizenship. Director of the Technological

Institute to the Department of Trade and Manufactures, October 11, 1869, RGIA, f. 733, op. 159, d. 12, l. 3. The German papers attesting to his nationality as well as his academic transcripts are attached to this document. In his farewell letter to v. Warnsedt on October 18, 1866, Beilstein commented that he would prefer to keep his German citizenship as long as possible. See the excerpt in Richter, "K. F. Beilstein, sein Werk und seine Zeit," 53.

38. On the Technological Institute, see *Piatidesiatiletnii iubilei S.-Peterburgskago Prakticheskago Tekhnologicheskago Instituta: 28-go noiabria 1878 g.* (A. M. Kotomin, 1879).

39. Beilstein, "O rabotakh chlenov Russkago Fiziko-Khimicheskago Obshchestva po aromaticheskomu riadu," 40–41.

40. Beilstein to Erlenmeyer, October 2, 1871: "Our Institute counts at the moment about 1300 students, I have in one lecture course over 550 attending, whose honoraria here apparently end up landing in the State's money-bag." Krätz, *Beilstein-Erlenmeyer*, 25.

41. V. V. Kurilov, "Tri korifeia russkoi khimii. (F. F. Beil'shtein, D. I. Mendeleev i N. A. Menshutkin)," in *Sbornik statei Ekaterinoslavskago nauchnago obshchestva*, ed. I. F. Aldyrev, t. 7 (Tip. Arteli Ekaterin. Raboch. Pechatnago dela, 1907), 53–61, on 60. For a more favorable assessment of Mendeleev's style, see V. P. Veinberg, *Iz vospominanii o D. I. Mendeleeve kak lektor* (Tip. Gubernskago Upravleniia, 1910).

42. Letter of November 6, 1868, reprinted in Bykow and Bekassowa, "II. F. Beilsteins Briefe an A. M. Butlerow," 278–279. Beilstein continued to agitate for renovation of the laboratory at the Institute up to his full retirement in 1896. Shortly before he left the institution, he managed to wrangle funds for a complete overhaul of the facilities. See his report at RGIA, f. 741, op. 1, d. 186, ll. 8–9ob.

43. Hjelt, "Friedrich Konrad Beilstein," 5051. Beilstein developed a distaste for his Russian students: "He sympathized little with Russian students in general, who obviously belonged to an entirely different type than their German comrades" (ibid., 5053). Beilstein was on good terms with a few of his students at the Institute, such as A. Kurbatov and L. Jawein, with whom he coauthored pieces and whom he sponsored for membership in the Russian Chemical Society.

44. F. Beilstein, "Ueber den Nachweis von Chlor, Brom und Jod in organischen Substanzen," *Ber.* 5 (1872): 620–621. The test was simultaneously published in Russian as F. Beil'shtein, "Ob otkrytii khlora, broma i ioda v organicheskikh soedineniiakh," *ZhRFKhO* 4 (1872), no. 9: 358–359.

45. F. Beilstein, *Anleitung zur qualitativen chemischen Analyse*, second edition (Quandt and Handel, 1870; fifth edition, 1877; sixth edition, 1887). The first English translation was done by William Ramsay, later the famous discoverer of the noble gases. In his preface, he praises the book: "The translation of the present work has been undertaken with a view to furnish laboratory students with a manual, which should contain the principal methods of Qualitative Chemical Analysis. It is well known and extensively used in Germany, and the name of its author cannot fail to be a guarantee of its excellence." F. Beilstein, *A Manual of Qualitative Chemical Analysis*, tr. William Ramsay (G. P. Putnam's Sons, 1873), 5. For more on its popularity, see the later American translations and adaptations: W. S. Christopher, *Chemical Experiments for Medical Students Arranged after*

*Beilstein* (Robert Clarke, 1888), 3; Charles O. Curtman, *Dr. F. Beilstein's Lessons in Qualitative Chemical Analysis, Arranged on the Basis of the Fifth German Edition*, second edition (Druggist Publishing, 1886), v. Curtman indicates that Beilstein authorized the translation.

46. G. Lewinstein, Review of W. Stein's *Anleitung zur qualitativen Analyse und zu den wichtigsten Gehaltsprüfungen*, *ZfC* 3 (1860): 78–80, on 78.

47. E. Erlenmeyer, "Zur qualitativen Analyse," *ZfC* 3 (1861): 29–32, on 30. Beilstein agreed, as seen in his review of the *Zeitschrift für analytische Chemie*, *Göttingische gelehrte Anzeigen*, no. 2 (1863): 940–945, on 940.

48. Louis Ernst, Review of C. F. Rammelsberg's *Leitfaden für die qualitative chemische Analyse*, *ZfC* 3 (1861): 159–160.

49. On the importance of skills in transcending theoretical divides, see H. M. Collins, *Changing Order: Replication and Induction in Scientific Practice* (Sage, 1985); Peter L. Galison, *Image and Logic: A Material Culture of Microphysics* (University of Chicago Press, 1997).

50. Letter of November 6, 1866, reprinted in Bykow and Bekassowa, "II. F. Beilsteins Briefe an A. M. Butlerow," 280. We know from Beilstein's students that he rarely sought out companions in the capital and worked long hours alone. See Richter, "Beilsteins Handbuch," 280.

51. Reprinted in Bykow and Bekassowa, "II. F. Beilsteins Briefe an A. M. Butlerow," 282.

52. See Bykov, *Pis'ma russkikh khimikov k A. M. Butlerovu*, 51. Beilstein also wrote Erlenmeyer mockingly about Butlerov's increasing fascination with Spiritualism. Krätz, *Beilstein-Erlenmeyer*, 19. For more on Spiritualism and Butlerov, see Michael D. Gordin, *A Well-Ordered Thing: Dmitrii Mendeleev and the Shadow of the Periodic Table* (Basic Books, 2004), 85–87.

53. Diary entry of August 16, 1861, in D. I. Mendeleev, "Dnevnik 1861 g.," *Nauchnoe Nasledstvo* 2 (1951): 111–212, on 163. Mendeleev's impression did not improve when he went to a party in honor of Beilstein at Fritzsche's house on August 16 (ibid., 164). He also met Beilstein among a group of chemists on October 10 (ibid., 188). Shmulevich and Musabekov fudge somewhat in trying to explain away the blatant hostility between the two. Shmulevich and Musabekov, *Fedor Fedorovich Beil'shtein*, 52.

54. F. Beilstein, Review of Mendelejeff's *Organische Chemie*, *ZfC* 5 (1862): 271–276, on 271. On this textbook and its role in the creation of the periodic system, see Michael D. Gordin, "The organic roots of Mendeleev's periodic law," *Historical Studies in the Physical and Biological Sciences* 32 (2002): 263–290.

55. F. Beilstein to D. Mendeleev, September 27, 1866, Göttingen, ADIM I-V-39-1-46.

56. These events are chronicled in detail in Gordin, *A Well-Ordered Thing*, chapter 5, including a translation of Beilstein's response to Menshutkin on 122. For an alternative, Butlerov-centric, interpretation of these events, see I. S. Dmitriev, "Skuchnaia istoriia (o neizbranii D. I. Mendeleeva v Imperatorskuiu akademiiu nauk v 1880 g.)," *Voprosy Istorii Estestvoznaniia i Tekhniki*, no. 2 (2002): 231–280.

57. Beilstein ironically had denounced the silliness of the notion of German and Russian parties in a letter to Butlerov on October 15, 1867, a full 13 years before it would resurface. Bykow and Bekassowa, "II. F. Beilsteins Briefe an A. M. Butlerow," 284. As a further irony, it was Butlerov who resurrected the German party case most strongly in a popular article published in the right-wing journal *Rus'*. See A. M. Butlerov, *Sochineniia*, 3 v. (Izd. AN SSSR, 1953), III, 118.

58. "Zur Nichtwahl Mendelejew's," *St. Petersburger Zeitung*, December 24, 1880, 359: 2.

59. "Predlozhenie i balotirovanie professora F. F. Beil'shteina v ordinarnye akademiki po tekhnologii i khimii, prisposoblennoi k iskusstvam i remeslam," *Zapiski Imperatorskoi Akademii nauk* 41 (1882), no. 1: 84–167, on 86.

60. "Predlozhenie i balotirovanie professora F. F. Beil'shteina," 125.

61. To be specific about what I mean about Beilstein's alienation qua German: he was *not* alienated from the civil service or the Russian bureaucracy, and he attained the level of privy councilor—also Mendeleev's rank as Director of the Chief Bureau of Weights and Measures—in 1895, and he was elected honorary member of several institutions like Kiev University. But among *Petersburg chemists* he felt himself persona non grata. On his honors, see Hjelt, "Friedrich Konrad Beilstein," 5068.

62. Shmulevich and Musabekov, *Fedor Fedorovich Beil'shtein*, 44. Beilstein's complete bibliography of non-Russian articles shows a marked decline into the 1880s. See Hjelt, "Friedrich Konrad Beilstein," 5078.

63. Protocols of Chemical Division of the Russian Physico-Chemical Society, December 4, 1903, *ZhRFKhO* 35 (1903), no. 9: 1265.

64. Hjelt, "Friedrich Konrad Beilstein," 5054.

65. Krätz, *Beilstein-Erlenmeyer*, 60.

66. Beilstein continued to Erlenmeyer on March 17, 1878: "My worry is, that I can't finish revising my materials in 2 years, as I had thought at first. I beg you to consider, that I have before me *the totally gathered complete materials*. While writing I take the original in hand each time. Now I am only asking each Christian individual to gather the 'material'! Woe is the unfortunate one who has gotten it into his head to be able to complete the entire Jahresberichte alone. What I possess are excerpts crafted *only from the originals*. Those I bring now all prettily in order nicely and clearly arranged *by empirical formula*. . . ." Krätz, *Beilstein-Erlenmeyer*, 64.

67. Beilstein to Erlenmeyer, March 17, 1878, in Krätz, *Beilstein-Erlenmeyer*, 64. This may be a reference to Mendeleev's famously encyclopedic *Principles of Chemistry*.

68. Beilstein, "O rabotakh chlenov Russkago Fiziko-Khimicheskago Obshchestva po aromaticheskomu riadu," 52–53.

69. F. Beilstein, *Handbuch der Organischen Chemie*, 2 v. (Leopold Voss, 1883), I, 2–3. The front matter and format did not change until the post-Beilstein *Handbuch*. Of course, the material in the catalog was heavily expanded, corrected, and updated.

70. Beilstein, *Handbuch der Organischen Chemie,* I, 35.

71. Beilstein, *Handbuch der Organischen Chemie,* I, v.

72. Beilstein, *Handbuch der Organischen Chemie,* I, v. This was, in one sense, an attempt to remedy the faults of previous textbooks, like that of August Kekulé. As Beilstein commented in his review of that work: "The material is selected with skill and everywhere is pointed to by citations to the original articles. By these citations the author takes into account primarily Kopp's *Jahresbericht* and Liebig's *Annalen,* which however is unfair to many works." That is, by not looking at *all* journals, Kekulé risked undermining his own synthesis. F. Beilstein, Review of Kekulé's *Lehrbuch der organischen Chemie,* 493.

73. Quoted in Richter, "Friedrich Beilstein, Gedanken zur hundertsten Wiederkehr seines Geburtstages," 103.

74. Beilstein, *Handbuch der Organischen Chemie,* I, vii–viii. See also his request in the minutes of the March 10, 1884 meeting of the German Chemical Society, printed in *Ber.* 17 (1884): 489. Krätz reproduces one of Beilstein's letters to an anonymous colleague dated August 17, 1883, soliciting help on the *Handbuch.* Krätz, *Beilstein-Erlenmeyer,* 81–82.

75. Quoted in Hjelt, "Friedrich Konrad Beilstein," 5064.

76. Beilstein, *Die chemische Grossindustrie auf der Weltausstellung zu Wien im Jahre 1873,* 55.

77. Erlenmeyer to an unknown recipient June 13, 1895, and Volhard to Erlenmeyer, March 27, 1896, both in Krätz, *Beilstein-Erlenmeyer,* 83.

78. For the official acceptance, see "Rundschrieben," from the minutes of the February 4, 1896 meeting of the German Chemical Society, *Ber.* 29 (1896): 321–324. On the later history of the *Handbuch* after this transformation, see Richter, "Beilsteins Handbuch," 280–281.

79. Quoted in Hjelt, "Friedrich Konrad Beilstein," 5065.

80. Letter of November 5, 1882, reproduced in Meyer, Emil Erlenmeyer (1825–1909) als Chemietheoretiker, 391.

81. Markovnikov quotation from "Nafteny i ikh proizvodnye v obshchei sisteme organicheskikh soedinenii (1902)," in Markovnikov, *Izbrannye trudy,* 516. The Alekseev letter is reproduced in Bykov, *Pis'ma russkikh khimikov k A. M. Butlerovu,* 25. Alekseev had studied under Beilstein for a year in Göttingen. See P. P. Alekseev to A. M. Butlerov, December 8, 1863, in ibid., 13. See also K. M. Zaitsev to Butlerov, May 16, 1862, in ibid., 145. Beilstein maintained contact with Alekseev through the 1880s after the latter moved to Kiev (ibid., 20, 26).

82. *Russkoe Khimicheskoe Obshchestvo. XXV (1868–1893),* 4.

83. Protocols of Chemical Division of the Russian Physico-Chemical Society, September 13, 1890, *ZhRFKhO* 22 (1890), no. 7: 480.

84. Protocols of Chemical Division of the Russian Physico-Chemical Society, October 8, 1892, *ZhRFKhO* 24 (1892), no. 8: 542–543.

85. N. Menshutkin, "K voprosu o khimicheskoi nomenklature: Sostavlenie nazvanii organicheskikh kislot," *ZhRFKhO* 25 (1893), no. 1: 10.

86. N. N. Beketov, "Pamiati Fed. Fed. Beil'shteina," in Protocols of Chemical Division of the Russian Physico-Chemical Society, November 2, 1906, *ZhRFKhO* 38 (1906), no. 9: 1279–1280.

87. Consider the way Beilstein's "homeland" (*Heimat*) has been treated. *Heimat* is one of those notoriously tricky German words that serve as political touchstones. Biographers have tended to declare that Beilstein's birthplace—Russia—was his *Heimat*, as in Hjelt, "Friedrich Konrad Beilstein," 5047, while German commentators in the twentieth century have selected Germany for the honor of being "always his spiritual *Heimat*." Luckenbach, "Der Beilstein," 36. Beilstein himself sided with the latter, as he declared in a letter to Erlenmeyer on October 5, 1873: "I have lived alone 12 years in Germany and consider it always as my *scientific* homeland (*Heimath*), for which I always have and will have the greatest respect." Krätz, *Beilstein-Erlenmeyer*, 43. Although Beilstein was fluent in Russian (and French and English, and proficient in both Italian and Swedish), he strongly preferred to speak German at home and among friends.

88. William Coleman, "Prussian Pedagogy: Purkyne at Breslau, 1823–1839," in *The Investigative Enterprise: Experimental Physiology in Nineteenth-Century Medicine*, ed. William Coleman and Frederic L. Holmes (University of California Press, 1988), 15–64, on 35.

89. F. Richter, "75 Jahre Beilsteins Handbuch der Organischen Chemie: Ein Jubiläum der Wissenschaft," *Frankfurter Allgemeine Zeitung*, December 13, 1956, reprinted in Richter, *75 Jahre Beilsteins Handbuch der Organischen Chemie* (1957), 5–10, on 5; idem, "Zur Feierstunde am 14. Dezember 1956," in ibid., 19–25, on 19, for World War II; Paul Jacobson, "Beilsteins Handbuch der Organischen Chemie, ein Spiegel ihrer Entwicklung," in ibid., 85–93, on 93.

90. Dermot A. O'Sullivan, "Germany's Beilstein will change to English," *Chemical and Engineering News* 59 (1981), May 18: 21–22, reprinted in *Einhundert Jahre Beilsteins Handbuch der Organischen Chemie*, 123–127.

91. Letter of October 7, 1906, quoted in Hjelt, "Friedrich Konrad Beilstein," 5066.

# 2 Making Tools Travel: Pedagogy and the Transfer of Skills in Postwar Theoretical Physics

**David Kaiser**

Feynman diagrams revolutionized nearly every aspect of theoretical physics during the second half of the twentieth century. The young American theorist Richard Feynman introduced his diagrams in the late 1940s as a bookkeeping device for simplifying lengthy calculations in one area of physics—quantum electrodynamics, physicists' quantum-mechanical description of electromagnetic forces. Soon the diagrams gained adherents throughout the fields of nuclear and particle physics. Not long thereafter, other theorists adopted—and subtly adapted—Feynman diagrams for many-body applications in solid-state theory. By the end of the 1960s, a few physicists even wielded the simple line drawings for calculations in gravitational physics. With the diagrams' aid, entire new calculational vistas opened for physicists; theorists learned to calculate things that many had barely dreamed possible before World War II. With the list of diagrammatic applications growing ever longer, Feynman diagrams helped to transform the way physicists saw the world, and their place within it.

With few exceptions, historians, philosophers, and sociologists of science have overlooked the crafting and appropriation of theoretical tools such as Feynman diagrams. Instead, research in theoretical sciences has been analyzed as abstract thought, wholly separated from anything like labor, activity, or skill. Worldviews or paradigms seemed to be the appropriate unit of analysis, and the challenge became charting the birth and conceptual development of particular ideas. In short: more "night thoughts" than desk work; more *Weltbild* than *Fingerspitzengefühl*.[1] Yet since at least the middle of the twentieth century—and, arguably, during earlier periods as well—most theorists have not spent their days (nor, indeed, their nights) in some philosopher's dreamworld, weighing one cluster of disembodied concepts against another, picking and choosing among so many paradigms. Rather, their main task has been to *calculate*. They have tinkered with models and estimated effects, always trying to reduce the inchoate confusion of "out there"—an "out there" increasingly percolated through factory-sized apparatus and computer-triggered detectors—into tractable representations. They have accomplished such translations by fashioning theoretical tools and performing calculations. Theorists use calculational tools, in other words, to mediate between various kinds of representations of the world. These tools provide the currency of everyday work.

Since the late 1940s, generations of physicists have turned more and more often to Feynman diagrams as their tool of choice. For this reason, I follow Feynman diagrams around, focusing on how physicists fashioned—and constantly re-fashioned—the diagrams into a calculational tool, a theoretical practice. Once we begin to examine the tools of theory, we must also study the tools' users—a shift in emphasis from the isolated thoughts of Nobel laureates to the pedagogical work involved in training large numbers of researchers to approach physical questions in similar ways. After all, tools such as Feynman diagrams never apply themselves; physicists must be trained to use them, and to interpret and evaluate the results in certain ways. A link therefore always exists between research practices and the scientific practitioners who put them to work. This link, more often than not, involves some kind of pedagogical activity, such as advisors mentoring graduate students or postdocs working closely together. In this essay, I use the example of Feynman diagrams to disaggregate some of the different types of pedagogical activities involved in crafting theoretical tools and making them travel.[2]

My project is organized around three main questions: How did the diagrams spread so quickly? For what were they used during the late 1940s and throughout the 1950s and the 1960s? Given this variety of distinct applications, why did the diagrams "stick"? Resolving each of these questions clarifies the role of a specific pedagogical process for training young theorists during the decades after World War II. In pursuing these questions, it is helpful to consider two distinct meanings of the word "dispersion." One cluster of meanings is especially pertinent for the first question, regarding how the diagrams spread so quickly: "To distribute from a main source or centre . . . ; to put into circulation." A second meaning of "dispersion" is helpful for navigating through the ever-expanding, competing uses and interpretations given to the diagrams: "To cause to separate in different directions . . . ; to spread in scattered order."[3] "Dispersion" thus captures at once the work required to make theoretical tools travel and the plasticity of those tools when they travel.

## An Introduction in the Poconos

Feynman introduced his diagrams in a private, by-invitation-only meeting in the spring of 1948. Twenty-eight theorists gathered in the Pocono Manor Inn, in rural Pennsylvania, for several days of intense discussions. Most of the young theorists were preoccupied with the problems of quantum electrodynamics (QED), the physicists' description of how electrons interact with light. Physicists had known since the early 1930s that QED produced unphysical infinities, rather than finite answers, when pushed beyond its lowest-order, simplest approximations. Feynman began doodling his diagrams in the context of working on the problems of QED. Even apart from the divergence difficulties, calculations within QED had long been infamously unwieldy—often single terms within a calculation could stretch over four or five lines of algebra, and it was all too easy to conflate or (worse) omit terms within the algebraic morass. As Feynman explained in his talk at the Pocono Manor Inn, his new diagrams could serve as convenient guides for marching through the thickets of QED calculations.

As one of his first examples, Feynman considered the problem of electron-electron scattering. He drew a simple diagram on the blackboard, similar to the one later reproduced in his first article on the new diagrammatic techniques.[4] (See figure 2.1.) Feynman explained how the diagram provided a shorthand for a uniquely associated mathematical description: an electron had a certain likelihood, which Feynman called $K(5,1)$, to move as a free particle from the spacetime point $x_1$ to $x_5$; the other incoming electron moved freely—with likelihood $K(6,2)$—from spacetime point $x_2$ to $x_6$. This second electron could then emit a virtual photon at $x_6$, which would move—with likelihood $\delta(s_{56}^2)$—to $x_5$, where the first electron would absorb it. (Here $s_{56}$ represented the

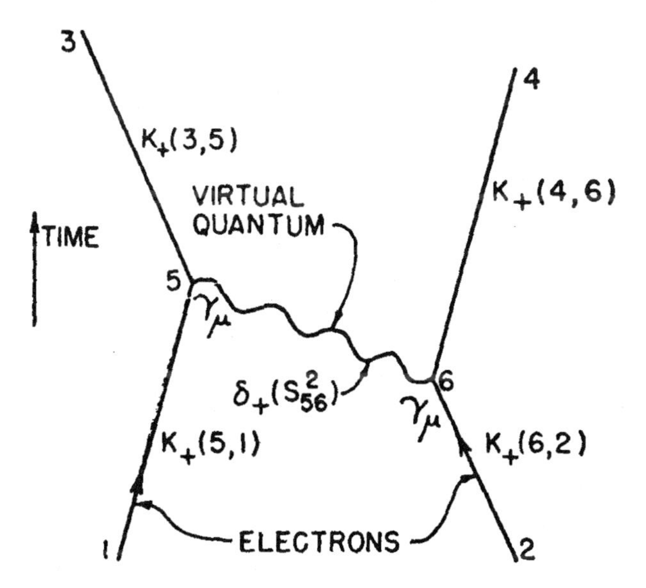


**Figure 2.1**
The simplest Feynman diagram for electron-electron scattering. Source: Richard Feynman, "Space-time approach to quantum electrodynamics," *Physical Review* 76 (1949): 769–789, on 772.

distance in space and time that the photon traveled.) The likelihood that an electron would emit or absorb a photon was $e\gamma_\mu$, where $e$ was the electron's charge and $\gamma_\mu$ was a vector of Dirac matrices. Having given up some of its energy and momentum, the electron on the right would then move from $x_6$ to $x_4$. The electron on the left, upon absorbing the photon and hence gaining some additional energy and momentum, would scatter from $x_5$ to $x_3$. In Feynman's hands, then, this simple diagram stood in for the following mathematical expression (itself written in terms of the abbreviations $K$ and $\delta$)[5]:

$$-ie^2 \iint d^4x_5\, d^4x_6\, K(3,5)\, K(4,6)\, \gamma_\mu\, \gamma_\mu\, \delta(s_{56}{}^2)\, K(5,1)\, K(6,2).$$

In this simplest process, the two electrons traded just one photon between them; the straight electron lines intersected with the wavy photon line in two places, called "vertices." The associated mathematical term therefore contained two factors of the electron's charge, $e$—one for each vertex. When squared, this expression gave a fairly good estimate for the probability that two electrons would scatter. Yet both Feynman and his listeners knew that this was only the start of the calculation. In principle, the two electrons could trade any number of photons back and forth—two, seven, forty-five, one million; there was an infinite number of distinct ways the two electrons could interact, and each of these possibilities had to be included. These additional possibilities, involving more and more interactions and hence more factors of $e$, should have been small compared with the lowest-order approximation, since $e^2 \sim 1/137$. That is, the additional terms should have been mere "perturbations" to the basic, starting calculation involving the lone, single photon. Feynman used his new diagrams to delineate the various possibilities. For example, there were nine different ways that the electrons could trade two photons back and forth, each of which would involve four vertices (and hence their associated mathematical expressions would contain $e^4$ instead of $e^2$; see figure 2.2). As in the simplest case (involving only one photon), Feynman could walk through the associated mathematical contribution from each of these diagrams, plugging in $K$s and $\delta$s for each electron and photon line and connecting them at the vertices with factors of $e\gamma_\mu$.

The main difference from the single-photon case was that most of the corresponding integrals for the diagrams in figure 2.2 blew up to infinity, rather than providing finite answers—just as physicists had found with their non-diagrammatic calculations since the late 1920s. Feynman next showed how one could remove some of the troublesome infinities by a combination of calculational tricks, some of his own design and others borrowed.[6] (The removal of infinities from QED calculations was dubbed "renormalization.") What is most important for our purposes is to consider Feynman's order of operations: *start* with the diagrams as a mnemonic aid in order to write down the relevant integrals, and only later alter these integrals, one at a time, to remove the infinities.

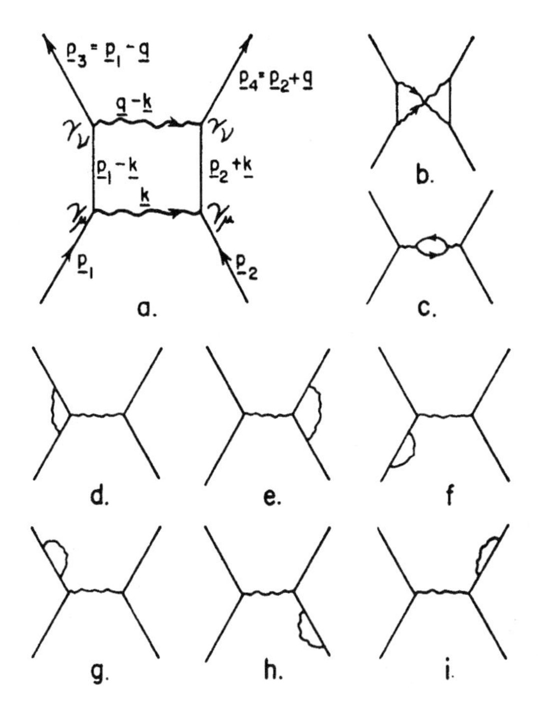


**Figure 2.2**
Feynman diagrams for electron-electron scattering correction terms; these diagrams play the role of "bookkeepers" for the perturbative analysis. Source: Richard Feynman, "Space-time approach to quantum electrodynamics," *Physical Review* 76 (1949): 769–789, on 787.

Diagrams in hand, Feynman had thus solved a puzzle that had stymied the world's best theoretical physicists for two decades. We might expect the reception from his colleagues at the Pocono Manor Inn to have been appreciative, to say the least. Yet things had not gone well at the Pocono meeting. For one thing, the odds were stacked against Feynman: his presentation followed a day-long lecture by Harvard's wunderkind, Julian Schwinger. Schwinger had arrived at a different method of removing the infinities from QED calculations, and the audience sat glued to their seats—pausing only briefly for lunch—as Schwinger unveiled his derivation. Coming late in the day, in contrast, Feynman's blackboard presentation was rushed and unfocused. No one seemed to be able to follow what Feynman was doing. He suffered frequent interruptions from Niels Bohr, Paul Dirac, and Edward Teller, each of whom pressed Feynman on how his new doodles fit in with the established principles of quantum physics. Others asked more generally, in exasperation, what *rules* governed the diagrams' use. By all accounts, Feynman left the meeting disappointed, even depressed.[7]

Feynman's frustration with the Pocono presentation has been noted often. Overlooked in these accounts, however, is the crucial fact that this confusion lingered long after the diagrams' inauspicious introduction. Even some of Feynman's closest friends and colleagues had difficulty following where his diagrams came from or how they were to be used. For example, Feynman had often discussed his new diagrammatic approach with Hans Bethe, both before and after the Pocono presentation. Bethe had been Feynman's boss at wartime Los Alamos and was at the time his senior colleague at Cornell; he was also a leading expert on QED and its problems. Yet Bethe wrote to Feynman while vacationing in England that summer that he tried to use Feynman's diagrams and kept getting stuck; Feynman had to coach his boss through the calculations through the mail.[8] Another theorist who had attended the Pocono meeting, Robert Marshak, remained flummoxed when trying to apply the new techniques. In a paper completed in December 1948, Marshak thanked Feynman in a footnote for completing a diagrammatic calculation at his request, since Marshak had been unable to undertake the calculation himself.[9]

Two years after the Pocono meeting, young physicists still struggled to make sense of the diagrams. "That great care must be taken" when calculating with "the different graphs," wrote a graduate student to two postdocs in February 1950, "is shown by the fact that we have obtained between us three different answers" for what was supposed to have been the same diagrammatic calculation. There followed a six-page enclosure that detailed everything from how to draw the various diagrams (which lines should be dashed or wavy, which lines should contain arrows); to which kinds of lines could and could not be inserted within a given diagram; to which kinds of diagrams contributed to which physical processes—all of which came before questions of how to translate the pages upon pages of tiny squiggles into mathematical expressions.[10]

Still the uncertainty lingered. During August 1952, the great architect of quantum theory, Wolfgang Pauli, admitted to a younger colleague that he was "not enough of an expert in 'graphs' to be able to check all the details" of a diagram-filled dissertation he had just received in the mail.[11] As late as May 1953—fully five years after Feynman had unveiled his new technique at the Pocono meeting—Stanford's Leonard Schiff wrote in a letter of recommendation for a recent graduate that his student *did* understand the diagrammatic techniques, and had used them in his dissertation.[12] The need to single out such hard-won skills for praise illustrates the larger point: as late as 1953, graduate students could not be assumed to understand or to be well practiced with Feynman's diagrams. The new techniques were neither automatic nor obvious for many physicists—the diagrams did not spread on their own.

### Evidence of Dispersion

And yet the diagrams did spread, and spread quickly, starting just months after Feynman's private presentation at Pocono. Articles that made use of Feynman's diagrams began to stream into the *Physical Review*. Eight such articles were published in 1949 alone, followed by an exponential rise, doubling every 2.2 years. By 1952, the biweekly journal carried, on average, one article making use of Feynman diagrams per issue. Within a few years, the articles were coming in from physicists on the East Coast, on the West Coast, and in the Midwest. All these articles—submitted by as many as 114 different authors by 1954—had been written before any textbooks on the new techniques had been published.[13] Somehow, Feynman's new techniques, introduced to little fanfare (if not outright hostility) before a small, private gathering, had made their way onto the scratch pads and blackboards of more than a hundred geographically dispersed physicists within just a few short years.

It wasn't just any physicists who picked up the diagrams and used them in print. Some actively resisted the diagrams—the most famous was Julian Schwinger, who scoffed years later that Feynman diagrams had "brought computation to the masses" but were at best a matter of "pedagogy, not physics." Schwinger's students at Harvard never encountered the new techniques in their advisor's polished lectures, and they avoided using the diagrams in their dissertations.[14] Other physicists likewise continued to pursue their calculations within QED without the aid of Feynman diagrams, though their relative numbers began to dwindle during the early and mid 1950s.

Scrutinizing those physicists who did begin to use the diagrams provides important clues as to how the new techniques spread. The authors of these diagrammatic articles shared three traits: they were *theorists*, they were *young*, and they were *in personal contact* with one another. By the 1960s, Feynman diagrams had become as routine for experimentalists as for theorists. In the early years, however, all the diagrammatic articles were written by theoretical physicists. Moreover, more than 80 percent of the authors were still in the midst of their training when they submitted their first diagrammatic articles, either as graduate students or as postdocs. Most of the others began using Feynman diagrams in print while young instructors or assistant professors, not more than seven years past their doctorates.[15] Older physicists simply did not "re-tool." Clearly something pedagogical was going on.

For the earliest users of Feynman diagrams, personal contact with other users of the diagrams proved critical. The acknowledgments to the articles in the *Physical Review* that included Feynman diagrams reveal a remarkably closed set. The names that recur most frequently in the acknowledgments—Freeman Dyson, Hans Bethe, Richard Feynman, Norman Kroll, Abraham Klein, Abraham Pais, Fritz Rohrlich—were the same people who published the greatest number of diagrammatic articles during this period. In fact, the 17 authors thanked most often in the acknowledgments contributed more than half of the diagrammatic articles published in the *Physical Review* between 1949 and 1954. Often the members of this core set thanked each other. Still more often, however, it was *other* young physicists—especially those making use of the diagrams for the first time—who thanked these authors. By and large, diagram users who published in the *Physical Review* were in personal contact with other diagram users. Feynman diagrams did not spread by texts alone. Rather, a particular pedagogical pattern put the diagrams in circulation.

## Dispersing the Diagrams: Dyson and the Postdoc Cascade

The spread of the diagrams was due mainly to the efforts of Feynman's younger associate Freeman Dyson. Dyson studied mathematics in Cambridge, England before traveling to the United States on a Commonwealth Fellowship to pursue graduate studies in theoretical physics. He arrived at Cornell in the fall of 1947, to study with Hans Bethe. Over the course of that year, he also began meeting with Feynman, just at the time that Feynman was working out his new approach to QED. Dyson and Feynman talked often during the spring of 1948 about Feynman's diagrams and how they could be used—conversations that continued in close quarters when the two drove from Ohio to Albuquerque together that summer, just a few months after Feynman's Pocono presentation.[16] Later that summer, Dyson attended the summer school on theoretical physics at the University of Michigan, which featured detailed lectures by Julian Schwinger on his non-diagrammatic approach to renormalization. The summer school offered Dyson the opportunity to talk informally and at length with Schwinger in much the same way that he had already been talking with Feynman. Thus, by September 1948, Dyson—and Dyson alone—had spent intense, concentrated time talking directly with both Feynman and Schwinger about their new techniques. At the end of the summer, Dyson took up residence at the Institute for Advanced Study in Princeton, where he spent the second and final year of his Commonwealth Fellowship.[17]

### Dyson as Diagrammatic Ambassador

Soon after arriving in Princeton, Dyson submitted an article to the *Physical Review* that compared Feynman's and Schwinger's methods. (Dyson also analyzed the methods of the Japanese theorist Tomonaga Sin-itiro, who had worked on the problem during and after the war; soon after the war, Schwinger arrived independently at an approach very similar to Tomonaga's.) More than just compare, Dyson demonstrated the mathematical equivalence of all three approaches—all this before Feynman had written a single article on his new diagrams. Dyson's early article, and a lengthy follow-up article submitted that winter, were both published months in advance of Feynman's own papers. Even years after Feynman's now-famous articles appeared in print, Dyson's pair of articles were cited more often than Feynman's.[18]

In these early papers, Dyson derived *rules* for the diagrams' use—precisely what Feynman's frustrated auditors at the Pocono meeting had found lacking. Dyson's articles offered a "how-to" guide, including step-by-step instructions for how the diagrams should be drawn and how they were to be translated one-for-one into their associated mathematical expressions. These Dysonian rules were captured a few years later by some of his earliest recruits at the Institute, Josef Jauch and Fritz Rohrlich, in their 1955 textbook *The Theory of Photons and Electrons*. (See figure 2.3.)

The correspondence between diagrams and $S$-matrix elements in momentum space

| Component of Diagram | Factor in $S$-Matrix Element | |
|---|---|---|
| Internal photon line $\nu$ - - - → - - - $\lambda$ | $g_{\nu\lambda}\dfrac{1}{k^2 - i\mu}$ | photon propagation function |
| Internal electron line •——→——• | $\dfrac{i\not{p} - m}{p^2 + m^2 - i\mu}$ | electron propagation function |
| Corner ($p$, $p'$, $\nu$, $k$) | $\gamma^\nu\delta(p - p' - k)$ | |
| External photon lines $\mu$, $k$   $k$, $\mu$ | $\dfrac{1}{\sqrt{2\omega}}e_\mu(\mathbf{k}),\ \dfrac{1}{\sqrt{2\omega}}e_\mu(\mathbf{k})$ | ingoing and outgoing photons |
| External negaton lines $p$   $p$ | $\sqrt{\dfrac{m}{\epsilon}}\,u(\mathbf{p}),\ \sqrt{\dfrac{m}{\epsilon}}\,\bar{u}(\mathbf{p})$ | ingoing and outgoing negatons |
| External positon lines $p$   $p$ | $\sqrt{\dfrac{m}{\epsilon}}\,\bar{v}(\mathbf{p}),\ \sqrt{\dfrac{m}{\epsilon}}\,v(\mathbf{p})$ | ingoing and outgoing positons |

**Figure 2.3**

The "Feynman rules" in momentum space, following Dyson's prescriptions. Source: Jauch and Rohrlich, *Theory of Photons and Electrons* (Addison-Wesley, 1955), 154.

In addition to systematizing Feynman's diagrams, Dyson *derived* the form and use of the diagrams from first principles, something that Feynman had not broached at all.[19] From the mathematics governing QED, for example, Dyson showed why each Feynman diagram had to have exactly two electron lines meet one photon line at each vertex. Dyson also included tiny arrows in the diagrams to distinguish particles (such as electrons) from antiparticles (such as positrons). With the aid of the arrows, Dyson could be sure to distinguish diagrams that differed only by the interchange of electrons and positrons—a distinction that had been lost on Feynman during some of his earliest diagrammatic calculations.[20] Beyond all these clarifications and derivations, Dyson—diagrams in hand—went on to demonstrate how the troubling infinities within QED could be removed systematically from any calculation, no matter how complicated. Until that time, Tomonaga, Schwinger, and Feynman had worked only with the first round of perturbative correction terms, and only in the context of a few specific problems. Building on the topology of the diagrams, Dyson generalized from these worked examples to offer a proof of renormalizability.

At first, Dyson's new boss at the Institute for Advanced Study, J. Robert Oppenheimer, remained underwhelmed by Feynman's diagrams and Dyson's use of them. Throughout the fall of 1948, Dyson gave a series of seminars on his diagrammatic work at the Institute. At every one, Oppenheimer behaved in his usual way, interrupting the talk with scathing criticisms or sarcastic dismissals. Unable to get a word in or to match Oppenheimer's renowned rhetorical skills face to face, Dyson sat down at his typewriter in exasperation one October evening to write a memorandum to Oppenheimer; this seemed to be the only way Dyson could say his peace. He argued that the diagrams were "considerably easier to *use, understand,* and *teach*" than the other approaches. Even the memorandum proved insufficient: only after his former advisor, Hans Bethe, intervened with Oppenheimer did Dyson get a fair hearing. In November 1948, after a new round of talks at the Institute, Oppenheimer left a simple note saying "Nolo contendere" in Dyson's mailbox.[21]

### Creating a Factory of Feynman Diagrams

From that moment on, Dyson converted the Institute for Advanced Study into a factory of Feynman diagrams. To understand how, we must first step back and consider changes in physicists' postdoctoral training during this period. Before the war, not all physicists who completed PhDs in the United States went on for additional postdoctoral training; it was still common to take a job with either industry or academia directly from one's PhD. Theoretical physicists were still a small minority among physicists within the United States before the war, and those who did pursue postdoctoral training usually traveled to the established European centers for their postdoctoral study. Only in Cambridge, in Copenhagen, in Göttingen, or in Zurich could these young American theorists learn the music (in I. I. Rabi's famous words) and not just the libretto of research in physics. Upon their return, many of these same American physicists—including Edwin Kemble, John Van Vleck, and John Slater, as well as Rabi and Oppenheimer—endeavored to build up domestic postdoctoral training grounds for young theorists.[22]

Soon after the war, the Institute for Advanced Study, newly under Oppenheimer's direction, became one of the most important centers for young theorists completing postdoctoral work. Having achieved worldwide fame for his role as director of the wartime Los Alamos laboratory, Oppenheimer was in constant demand after the war. He left his Berkeley post in 1947 to become director of the Princeton Institute in part to have a closer perch to his newfound consulting duties in Washington, D.C. He made it a condition of his accepting the position that he be allowed to increase the numbers of young, temporary members within the physics staff—that is, to turn the Institute into a center for theoretical physicists' postdoctoral training. The Institute quickly became a common stopping-ground for young theorists, who circulated through what Oppenheimer called his "intellectual hotel" for two-year postdoctoral stays.[23] The Institute quickly became one of the most commonly visited sites for young theorists upon completion of their PhDs. Oppenheimer captured something of what life was like at the Institute in a letter to Wolfgang Pauli in February 1952. In the midst of trying to lure Pauli to return to the Institute as a permanent senior member, Oppenheimer explained that the Institute "is not a school in the sense that even the younger people are not listening to lectures or working for doctor's degrees; but it is a school in the sense that almost everyone who comes learns of parts of physics . . . which are new to him. It is a very fertile group."[24]

The focused yet informal nature of the Institute's postdoctoral "school" proved to be crucial for spreading Feynman diagrams around. When Dyson arrived at the Institute in the fall of 1948—just a year after Oppenheimer became director and began to implement his vision of the Institute as a center for theorists' postdoctoral study—he joined eleven other junior theorists. One of the new buildings at the Institute, which was supposed to contain offices for the new visitors, had not been completed on time, so the entire crew of theory postdocs spent much of that fall semester huddled around desks in a single office. The close quarters bred immediate collaborations among the postdocs.[25] Very quickly, Dyson emerged as a kind of ringleader, training his peers in the new diagrammatic techniques and coordinating a series of collaborative calculations involving the diagrams.

The most famous of the diagrammatic collaborations was a paper by Dyson's fellow postdocs Robert Karplus and Norman Kroll on the fourth-order corrections to an electron's magnetic moment. Their paper, submitted to the *Physical Review* during the fall of 1949, announced that their purpose was "to demonstrate in a complete calculation of a particular example the feasibility of Dyson's program," in the prosecution of which "Dyson's methods have been followed quite closely."[26] Similar acknowledgments of Dyson appear in the other diagrammatic papers submitted by Institute postdocs at the time. Dyson's efforts with his fellow postdocs had become so effective that the ever-observant (and ever-sarcastic) Pauli wrote to another of the Institute's young theorists, Abraham Pais, in May 1949, asking what Dyson and the rest of "the 'Feynman-school'" were working on.[27]

Next these postdocs, having been tutored in the niceties of diagrammatic calculations by Dyson, left the Institute to take teaching jobs elsewhere. More than 80 percent of the *Physical Review* articles that used Feynman diagrams between 1949 and 1954 were submitted by these Institute postdocs or by graduate students (and other colleagues) whom they trained upon arriving at their new jobs.[28] The great majority of the 114 authors who made use of the diagrams in the *Physical Review* during the period 1949–1954 did so because they had been trained in the new techniques by Dyson or by one of his newly minted Institute apprentices. (All but two of the remaining authors interacted directly with Feynman.) The acknowledgments in graduate students' dissertations from Berkeley, Rochester, Chicago, Iowa City, Bloomington, Madison, Urbana, and Ithaca—and places in between—confirm the role of the Institute postdocs in taking the new techniques with them and teaching their own recruits how to use them. In this way, Feynman diagrams spread throughout the United States by means of a postdoc cascade, emanating from the Institute for Advanced Study. Personal mentoring and the postdocs' circulation thus dispersed the diagrams, distributing them from a main center and putting them into circulation.[29]

### Dispersion in Form, Use, and Meaning: Local Schools

Even as the diagrams began to disperse in this first sense, thanks to the postdoc cascade, they also underwent a second kind of dispersion, becoming more and more differentiated or "spread in scattered order." Physicists appropriated Feynman diagrams for many different kinds of calculations, often modifying the pictorial form of the diagrams they and their students drew to better suit their new purposes. Consider, for example, the diagrams from 1949–1954 reproduced here as figure 2.4, each of which was labeled a "Feynman diagram" by its author. Clearly Feynman's original diagram (figure 2.1) was not the only model on offer at the time.

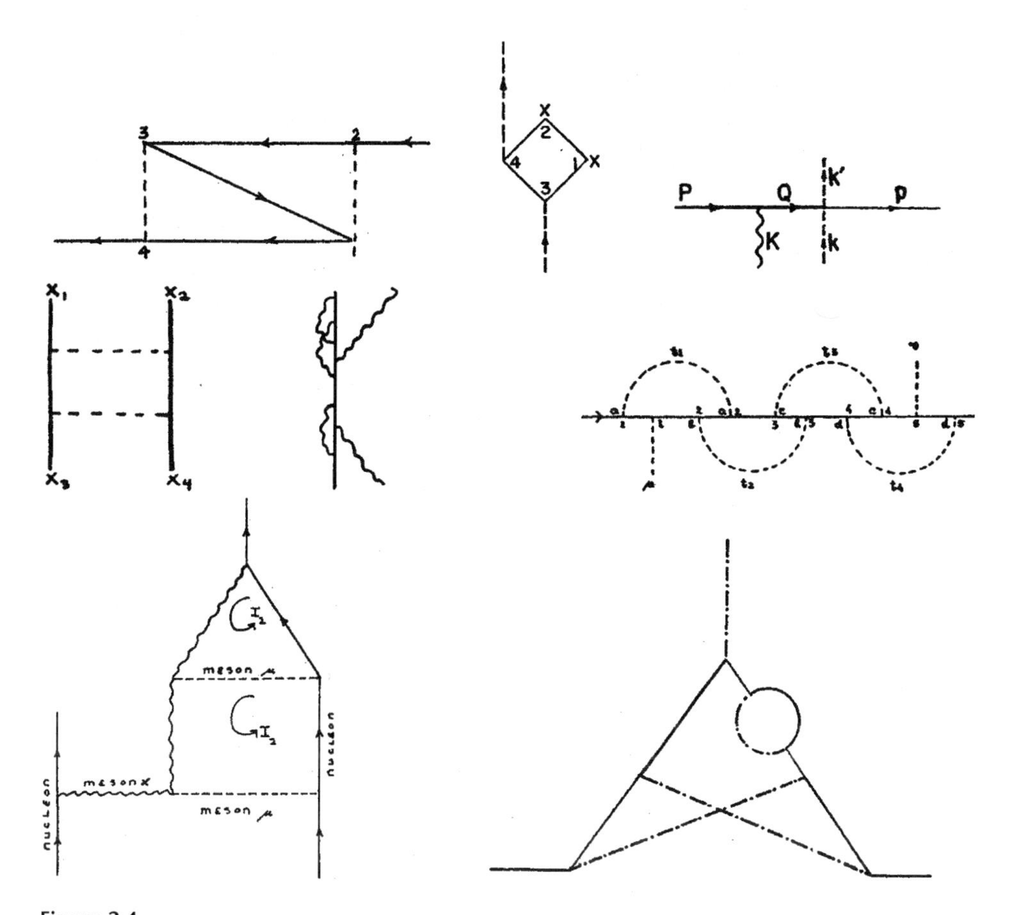


**Figure 2.4**

Feynman diagrams differentiate. Sources: Top row, left to right: Felix Villars, "Quantum electrodynamics," unpublished lecture notes from a July 1951 MIT course, 65; F. Rohrlich and R. Gluckstern "Forward scattering of light by a Coulomb field," *Physical Review* 86 (1952): 1–9, on 2; A. Lenard, "Inner Bremsstrahlung in μ-meson decay," *Physical Review* 90 (1953): 968–973, on 971. Middle row, left to right: M. Gell-Mann and F. Low, "Bound states in quantum field theory," *Physical Review* 84 (1951): 350–354, on 352; F. Low, "Natural line shape," *Physical Review* 88 (1952): 53–57, on 55; A. Salam, "Renormalized S-matrix for scalar electrodynamics," *Physical Review* 86 (1952): 731–744, on 735. Bottom row, left to right: J. Steinberger, "On the use of subtraction fields and the lifetimes of some types of meson decay," *Physical Review* 76 (1949): 1180–1186, on 1182; N. M. Kroll and M. A. Ruderman, "A theorem on photomeson production near threshold and the suppression of pairs in pseudoscalar meson theory," *Physical Review* 93 (1954): 233–238, on 235.

Some order may be brought to the random-looking display of figure 2.4 by considering more closely the *pedagogical* links between young instructors and their students. In each local setting, young physicists adapted the diagrams to better bring out aspects deemed most important for the new kinds of calculations. Thus the diagrams drawn by young Cornell physicists began to look different—and to be used in subtly distinct ways—from those drawn by their peers at Rochester, Columbia, Urbana, and so on. (With practice, one can actually "predict" where a physicist was trained based on the kinds of diagrams he or she drew and the kinds of calculations in which the diagrams were enrolled.) Consider the examples in figure 2.5, taken from *Physical Review* articles during the period 1949–1954: in each pair, the diagram on the left comes from an advisor in one of the major training centers, and the example on the right from someone

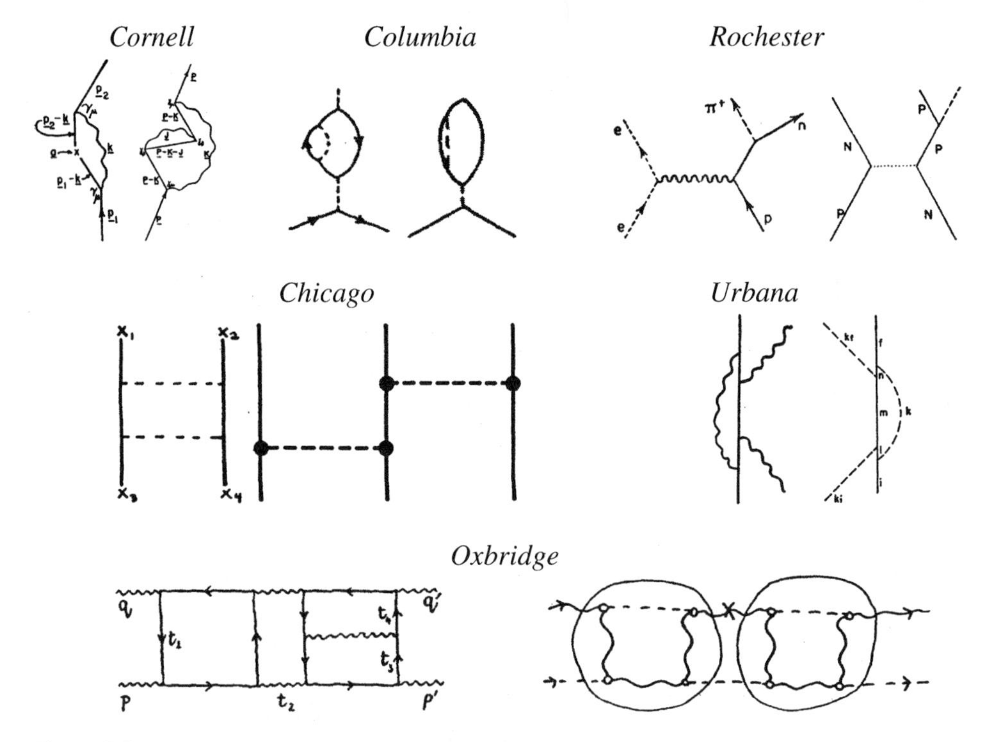


**Figure 2.5**
"Family Resemblances": mentors and students crafted diagrams for new purposes. Note the differences in which items were labeled, which lines received arrows, which lines were inclined at an angle, and so on. These pictorial differences were intimately tied to the different calculational roles the diagrams were meant to perform, and to more subtle differences in how the diagrams were interpreted. (Sources listed in note 30.)

he trained.[30] The pictorial similarity between advisors' and students' diagrams provides a second hint that something pedagogical was going on: students were clearly learning something from their supervisors above and beyond an abstract notion of what a Feynman diagram is or for what it should be used. They were practicing how to *apply* the semi-standard techniques to specific research questions—research questions that varied from place to place. Among the many distinct schools that emerged, consider two examples: young physicists at Columbia University and at Rochester University.

### Columbia: Kroll's Perturbative Bookkeepers

The Columbia students learned of the new diagrams from Norman Kroll, who took up a teaching position in the department in 1950 directly upon completing his postdoctoral training at the Institute. Kroll had been one of the earliest "converts" to Dyson's diagrammatic program; his famous article with Robert Karplus on fourth-order corrections to an electron's magnetic moment, submitted to the *Physical Review* from the Institute in October 1949, had been heralded immediately as a triumph both for Feynman diagrams and for QED. Karplus and Kroll followed Dyson's prescriptions to the letter, introducing the five distinct classes of Feynman diagrams involved in their unprecedented calculation. (See figure 2.6.) Note, for example, their careful application of Dyson's antiparticle arrows on the electron and positron lines. As the postdocs made explicit, these arrows carried mathematical bite: precisely because the arrows in the triangles of the two diagrams of class V ran in opposite directions, the contributions from these distinct diagrams cancelled exactly. What Feynman had at first overlooked in his energetic doodling, these two disciples of Dyson could discern with tiny arrows and clarify with a single sentence. Next came the laborious job of evaluating the integrals associated with each of the remaining diagrams—a painstaking calculation rendered feasible by using Feynman's diagrams in the strict manner that Dyson had specified.[31]

In Karplus and Kroll's hands, then, Feynman diagrams had been disciplined into trusty bookkeepers for perturbative calculations. After "much helpful discussion with F. J. Dyson," they had drawn the diagrams carefully to Dyson's specifications, and applied his methodical rules to translate each dashed and solid line uniquely into its corresponding mathematical term.[32] When the dust settled and each of these integrals had been evaluated and added together, Karplus and Kroll had demonstrated that an electron in an external electromagnetic field will behave as if it had a magnetic moment of 1.001147 (in appropriate units) instead of 1, as Dirac's equation would have it. More than this, they had shown how the second-order result, 1.001162, first calculated without diagrams by Schwinger and re-calculated diagrammatically by both Feynman and Dyson, was further modified by the fourth-order correction terms to yield their answer.

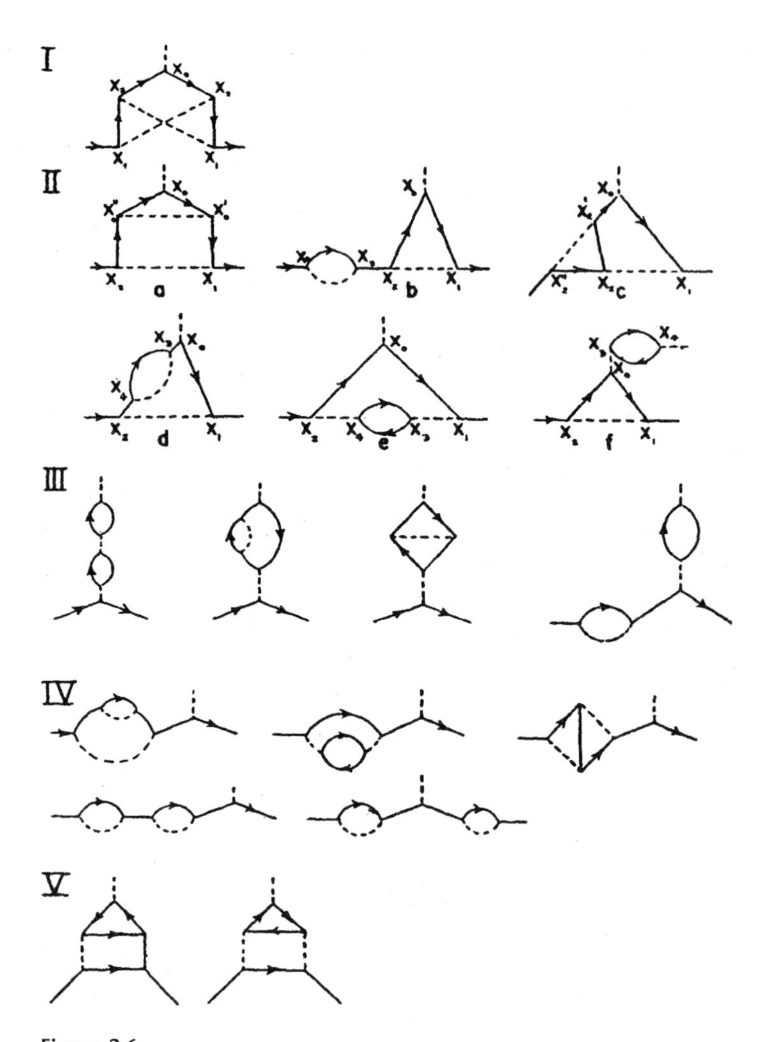


**Figure 2.6**

Fourth-order diagrams for the scattering of an electron in an external field. Source: R. Karplus and N. M. Kroll, "Fourth-order contributions in quantum electrodynamics and the magnetic moment of the electron," *Physical Review* 77 (1950): 536–549, on 537.

They had demonstrated, with this concrete example, how to put the diagrams to work for tracking the ever-tinier wisps of QED's correction terms.

Upon arriving at Columbia to begin teaching in the autumn of 1949, Kroll carried these hard-won skills with him, and soon a series of his own graduate students and postdocs submitted diagrammatic articles to the *Physical Review*. Their calculations followed Kroll's example closely, employing Dyson's rules to set out and methodically evaluate the fourth-order corrections to other important quantities within QED.[33] No one had been able to produce, let alone evaluate, each of the distinct, competing contributions that enter into a fourth-order perturbative QED calculation before Karplus and Kroll. Yet by treating the diagrams as handy mnemonic devices, as Dyson had taught them, the two postdocs—and soon Kroll's own students—could draw the five distinct classes of diagrams without confusing or conflating the various terms. Moreover, with the distinct diagrams written down, it became almost trivial to translate each of these, step by step, into its own integral expression.

The efficiency of calculating with Feynman diagrams in this manner was undeniable. Surely the rapidity and ease with which such labyrinthine correction terms could be clarified and evaluated would have convinced great numbers of theorists to pick up the diagrams and march along their own perturbative calculations. And surely this, in turn, would explain the diagrams' rapid dispersion to theorists throughout the world. And yet this simply wasn't so. Only a handful of other physicists followed Karplus and Kroll's famous perturbative calculation with similar ones, trotting out the diagrams as bookkeepers for the ever-smaller correction terms within Dyson's power-series expansion. In fact, fewer than twenty percent of the diagrammatic articles within the *Physical Review* between 1949 and 1954 made use of Feynman diagrams in this manner. Nearly all of these papers, in turn, were contributed by graduate students at Cornell (working with Feynman, Dyson, and/or Bethe) and by Kroll's students at Columbia.

### Rochester: Marshak's Meson Markers

Young physicists elsewhere rarely used the diagrams for perturbative calculations within QED. What captured most theorists' attention soon after the war was not electron physics, but rather the embarrassment of riches suddenly pouring forth from the new accelerators. A flood of new particles, similar to but in many ways distinct from the familiar electrons and photons, surprised physicists when they began to probe high-energy interactions with the aid of accelerators, rather than relying only upon cosmic rays. As quickly became clear, the new particles—dubbed "mesotrons" or "mesons," since the masses of many of them were intermediate between electrons and protons—interacted with each other differently than electrons and photons did. Most important, most of these new particles interacted *strongly*, with coupling constants $g^2$ between 7 and 57, unlike the weak electrodynamic interaction, governed by the electron's charge of $e^2 \sim 1/137$. If theorists tried to treat interactions among, for example, pions and protons in the same way as they treated electron-photon scattering, with a long series of more and more complicated Feynman diagrams, each containing more and more vertices, then each higher-order diagram would include extra factors of the large number $g^2$. In contrast with the situation in QED, then, these complicated diagrams, with many vertices and hence many factors of $g^2$, would overwhelm the lowest-order, more basic contributions. Perturbative approaches seemed impossible within meson physics.

If there could be no reliable perturbative expansions for mesonic calculations, then what place could there be for Feynman's diagrams, which had been introduced for the sole purpose of simplifying perturbative calculations? As it turned out, there would be plenty of room for them. In fact, more than half of the diagrammatic articles published in the *Physical Review* between 1949 and 1954 applied Feynman diagrams in one way or another to problems in meson physics, including the first four diagram-filled papers published after Dyson's and Feynman's own. One of the early groups to bring Feynman diagrams to bear on meson physics was Robert Marshak's group at the University of Rochester. Rochester was in the process of building its own cyclotron, and Marshak set his team to work preparing to be useful once their colleagues had the new machine up and running.[34]

Marshak had attended Feynman's original introduction of the diagrams at the Pocono meeting, and his group benefited from frequent visits by Feynman over the next two years. In addition, one of Marshak's younger collaborators, Julius Ashkin, was a close associate of Feynman's whom Feynman thanked several times in his articles about the diagrams.[35] Whereas Kroll and his Columbia students worked with the diagrams in Dysonian fashion, Marshak's group at Rochester learned to use the diagrams much more as Feynman himself did. They learned directly from Feynman that his line drawings could provide "intuitive" help far beyond the narrow dictates of perturbative QED.

In the spring of 1950, Marshak explained some of the differences between QED and meson physics to his students. Whereas theorists had long ago narrowed the range of options for the basic interaction between electrons and photons down to one in the case of QED, the options remained frustratingly open in the mesonic realm. For one thing, the pions' characteristics were still unclear: Were they spin-0, or spin-1? Under parity transformations were they even, or odd? The way in which they interacted with protons and neutrons remained equally unclear: Did they couple to the nucleons via scalar, pseudoscalar, vector, pseudovector, tensor, or pseudotensor interactions? The various possibilities, Marshak lectured, led to eight distinct choices for the basic interaction, each of which was still in the running for describing the nuclear domain.[36]

Awash in this sea of open-ended possibilities, and unable to calculate anything beyond the lowest-order in meson models because of the large size of $g^2$, Marshak and his students thus had different goals in mind when they picked up Feynman diagrams and began to calculate. They wanted to fashion a useful means of distinguishing the eight different possibilities for the meson-nucleon interaction. Marshak and his young team pursued this goal by comparing the lowest-order predictions of the eight contenders, scanning for qualitative differences between the various models' phenomenological predictions. Beginning in the spring of 1950, they fastened onto each model's theoretical predictions for the angular distribution of decay products—that is, how likely it would be for pions to be detected at various angles or directions as they careened away from the interaction region. The predicted angular distributions, Marshak's team soon realized, depended sharply upon the symmetry properties of the mesons and their interactions—symmetry properties that were encoded in the various interaction terms, and hence reflected in calculations that included only the simplest Feynman diagrams for a given process. In his first diagrammatic article, Marshak, together with one of his graduate students and Feynman's friend Ashkin, explained that working only with the lowest-order Feynman diagrams "is extremely crude, but it is thought that the qualitative features will persist in a more correct theory." Never dreaming of pushing any given calculation beyond $g^2$, they would nonetheless try to bring order to the nuclear realm.[37]

In the process, Marshak and his young collaborators refashioned the diagrams, designing them for better use toward their own goal—all within a year of Feynman's and Dyson's original articles on the diagrams. The Rochester group used the diagrams as illustrations of basic processes; then they could run down their list of eight contenders for the mesonic interaction and calculate each model's lowest-order contribution to the same physical process. Only five pages into his 1952 textbook on the new techniques (culled primarily from his 1950 lectures), Marshak's students found a full page of examples of these newly refurbished Feynman diagrams. (See figure 2.7.) Calculating a particular lowest-order contribution—for instance, the prediction for photon-proton scattering coming from the pseudoscalar interaction term—in itself was no trouble. Theorists had written down the analogous terms for QED for years without the aid of Feynman diagrams. The complicated task in Marshak's meson physics was navigating the maze of competing versions of each lowest-order calculation; Marshak and his students had to distinguish the different theoretical predictions for a given process. By doodling the lowest-order Feynman diagrams and labeling carefully the

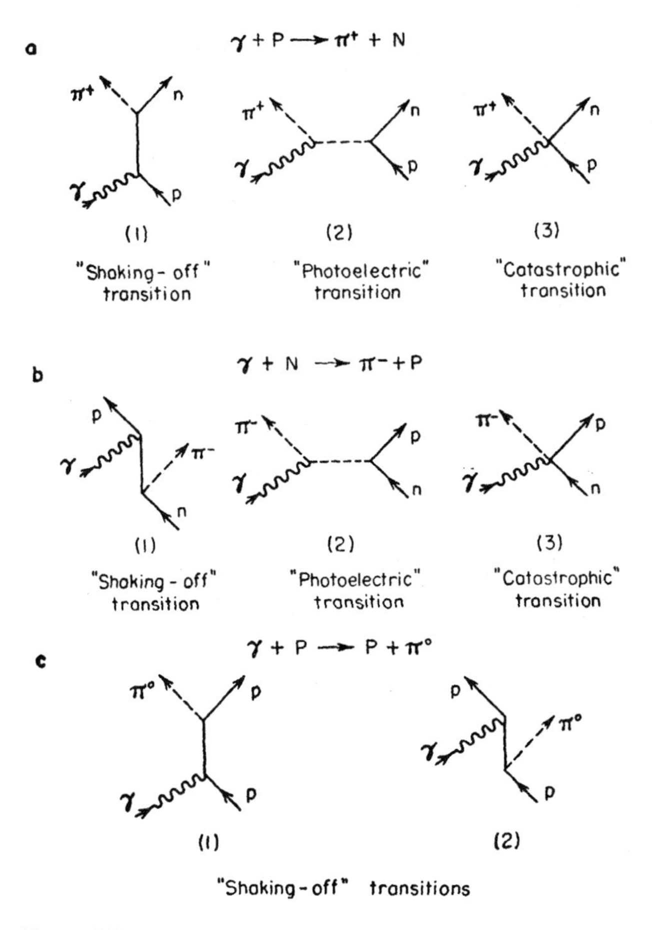


**Figure 2.7**
Feynman diagrams for the photoproduction of pions. Source: Robert Marshak, *Meson Physics* (McGraw-Hill, 1952), 6.

specific particles involved in each specific process, they could march through the contributions in turn, calculating and re-calculating each model's predictions.[38]

As Marshak and his students unglued Feynman diagrams from their stipulated Dysonian rules, the diagrams they drew thus changed as well. Features of the diagrams that had carried specific meanings when considered under their QED rubric could now be discarded without being missed. Particle labels replaced the coordinate-space or momentum-space labels that had been so important for cranking through perturbative calculations within QED. In Marshak's hands, moreover, upward-directed arrows appeared on all external lines—in other words, he did not make use of Dyson's antiparticle arrow convention for distinguishing particles from antiparticles. In the context of Kroll's perturbative-bookkeeping calculations, these antiparticle arrows had proven essential for correctly distinguishing between distinct closed-loop diagrams; to Kroll and his students, the tiny arrows carried specific meaning. In the Rochester meson work, by contrast, no one was calculating any closed-loop contributions, so there was hardly any need to distinguish between different kinds of closed loops; the antiparticle arrows meant nothing. In only a few months, the antiparticle arrow convention had become a difference that no longer made any difference. The diagrams' pictorial forms, calculational roles, and interpretations thus became intertwined. All three shifted as young theorists deployed the diagrams toward different ends.

The Rochester and Columbia approaches were clearly different from each other. How easy was it for members of one group—say, graduate students working with either Kroll or Marshak—to understand what the other group was doing? Some physicists at the time spoke as if it were impossible for members of such different "schools" to understand one another—thereby reaching a similar conclusion to that of sociologists such as Harry Collins, in his work on tacit knowledge and skills-transfer. Consider, for example, Wolfgang Pauli's amusing formulation, in a letter written in 1954, of how one could define various schools within theoretical physics:

> What is the definition of the 'Wigner School'? The question can be answered in the practical American way by an *operational* definition. In order to decide, whether or not a person is a member of the Wignerschool [*sic*], you give him a paper of Schwinger. . . . If he then says, that this is very obscure, and that he can't understand it, he is a member of the Wigner school. But he, who says that it is quite well understandable and clear will be excluded from the Wigner school. . . . Similar tests are certainly very popular at Harvard where the test object for membership of the *Schwinger* school is a paper of *Wigner*.[39]

Thus Pauli tied the social question of young physicists' training under senior colleagues directly to the epistemic question of understanding various calculational approaches.

But such ties were not always so tight in practice. In 1950, Ashkin, Marshak, and Marshak's graduate student Albert Simon published a long article in the Japanese journal *Progress of Theoretical Physics* in which they went through the motions of a fourth-order perturbative calculation within a specific meson model. With much help and personal coaching from Dyson, whom they thanked for "a lucid presentation of his method," Marshak and his group could perform calculations more like those of Kroll and the Columbia contingent.[40] They worked in terms of the parameter $g^2$, neglecting the fact that $g^2 >> 1$, in order to demonstrate that Dyson's systematic renormalization program for QED could also fix certain types of calculations among mesons and nucleons (even if the swollen size of $g^2$ prevented the Rochester physicists from comparing their calculations with quantitative experimental data). This appeared to be a one-time exercise: neither Marshak nor his students pursued mesonic calculations beyond the lowest order after this early paper. Students other than Simon appear not to have sweated such details; at least, no traces of such types of calculations were left in their dissertations or in their published articles. In Marshak's textbook, published in 1952, he made it clear that students were expected to have a background in non-relativistic quantum mechanics but need not ever have studied QED, let alone Dyson's diagrammatic renormalization program.[41]

The differences between Marshak's and Kroll's pedagogical programs thus point to local choices about what to work on and what to drill one's students in, rather than wholly incommensurable or incommunicable epistemic regimes. Both groups (and many others in between) made Feynman diagrams central to their calculations. Students in Rochester, just as much as those at Columbia, spent hours practicing how to draw the diagrams and put them to work in certain kinds of calculations, even though the diagrams they drew and the calculations for which they became central were clearly different from each other. Different research programs called for different pedagogical patterns. Students' Feynman diagrams bore the marks of their distinct training.

## Why Did They Stick? Global Aspects

Marshak's energetic doodling, in which he and his students loosened Feynman diagrams from Dyson's systematic rules for their use, was among the earliest examples of the diagrams' dispersion. Over the course of the 1950s, many more theorists, especially in the United States, grew ever more frustrated with their inability to use quantum field theory to get any phenomenological handle on the "zoo" of new strongly interacting particles. By the late 1950s and on into the 1960s, a few groups, most notably Geoffrey Chew's group in Berkeley, declared that quantum field theory itself was dead, at least for nuclear and high-energy physics. Chew announced, with increasing gusto, that Dyson's careful QED apparatus—Lagrangians, interaction Hamiltonians, perturbative series, even the exchange of virtual particles as carriers of force—was less than useless for studying what had come to dominate the attention of most high-energy theorists (not to mention experimentalists).[42]

One might expect, given that Feynman diagrams had been designed for perturbative calculations within weakly coupled theories like electrodynamics, that the alleged fall of that entire theoretical program would bring Feynman diagrams down with it. Instead, just the opposite happened: Chew and his students and postdocs extracted the diagrams from their original field-theory embedding and used them as a heuristic scaffolding for building what they hoped would become a rival theory. A decade after Marshak began tinkering with the diagrams for making meson calculations, Chew and company announced a clean break between the diagrams and their so-called derivation.

Why did the diagrams stick, even as many theorists tugged so doggedly at (and eventually discarded) their original theoretical embedding? The diagrams' curious tenacity points to a different pedagogical feature in the making of young theorists in the middle of the twentieth century: associations shared by Feynman diagrams and another, more widely adopted, diagrammatic practice.[43] In short, physicists during the late 1940s and throughout the 1950s often associated Feynman's diagrams with Minkowski's spacetime diagrams used for special relativity, which most physicists had practiced drawing and analyzing from their earliest studies in physics. Clearly the two types of diagrams were distinct—physicists at the time rarely proclaimed them to be conceptually equivalent, and some even argued out loud that the two types of diagrams should not be confused. All the same, most physicists continued to draw their Feynman diagrams according to certain specific learned pictorial conventions. Their occasional verbal denials to the contrary, the visual features of their Feynman diagrams betray a distinct pedagogical legacy.

As physics students around the world routinely learned by the late 1940s, Minkowski diagrams could be used for charting objects' propagation through space and time. They learned about Minkowski diagrams in the context of special relativity.[44] In the already-standardized diagrams, students were taught to place the time axis vertically and one spatial axis horizontally. (See figure 2.8.) In addition, students learned to scale the speed of light to 1, so that light would travel along 45° diagonals within their Minkowski diagrams.

Feynman consistently introduced his own diagrams only after including explicit Minkowski diagrams—in his 1949 articles introducing the diagrams, in his unpublished lecture notes from Cornell and Caltech, and in his later popular book on QED.[45] In Feynman's many presentations, Feynman diagrams came laden with talk of "worldlines," the order of "events" along particles' "trajectories," and "spacetime pictures"—all clear markers of "Minkowski-talk." This language—and, more important, these visual features—continued even after Feynman moved from coordinate space to momentum space. Still his diagrams displayed particles moving along tilted trajectories, with photon lines often moving at or near 45°—even though in momentum space, 45° diagonals bore no relation whatsoever with light-like travel. (Compare figure 2.1 with figure 2.9.)

The tacit, visual assimilability between Feynman's doodles and Minkowski diagrams had stirred Niels Bohr, that elder statesman of quantum physics, to object vigorously at

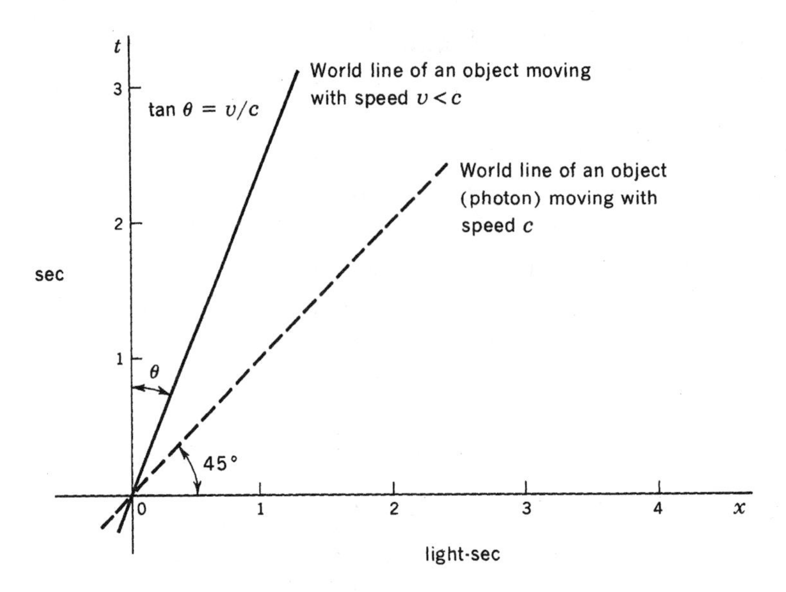


**Figure 2.8**
A Minkowski diagram. Source: N. David Mermin, *Space and Time in Special Relativity* (McGraw-Hill, 1968), 160.

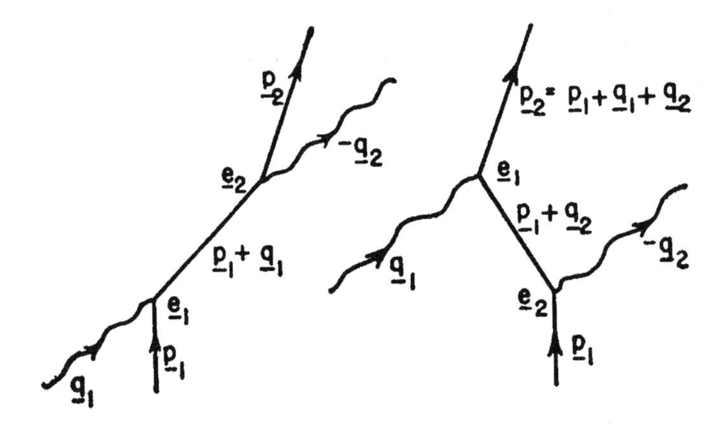

**Figure 2.9**
Feynman diagrams in momentum space, drawn with Minkowski-diagram conventions. Source: Richard Feynman, "Space-time approach to quantum electrodynamics," *Physical Review* 76 (1949): 769–789, on 775.

Feynman's original presentation at the 1948 Pocono meeting. Spacetime trajectories were one thing for macroscopic objects, Bohr reminded Feynman, but were ruled out of court for quantum mechanics: Heisenberg's uncertainty principle denied the possibility of knowing a quantum object's simultaneous position and momentum, the two ingredients needed to construct something like a spacetime trajectory. Feynman's flustered reply—that his diagrams were not meant to be read literally as spacetime trajectories, but rather as a convenient shorthand notation—had convinced few in the room.[46]

Despite Bohr's objections, Feynman's pictorial scheme caught on. As a small sampling of the enormous repetition of these pictorial conventions, consider the diagrams shown here in figure 2.10, all of which were published in the *Physical Review* between 1949 and 1954. The earliest diagrammatic textbooks, published in the mid 1950s, likewise used Minkowski-diagram imagery when introducing Feynman diagrams, sometimes even including explicit space and time axes as Feynman had done. (See figure 2.11.) These associations did not always remain tacit. Sometimes textbook authors actually tried to explain away (in lengthy footnotes) any conceptual links between the two types of diagrams, even as they continued to draw Feynman diagrams in this "stylized"

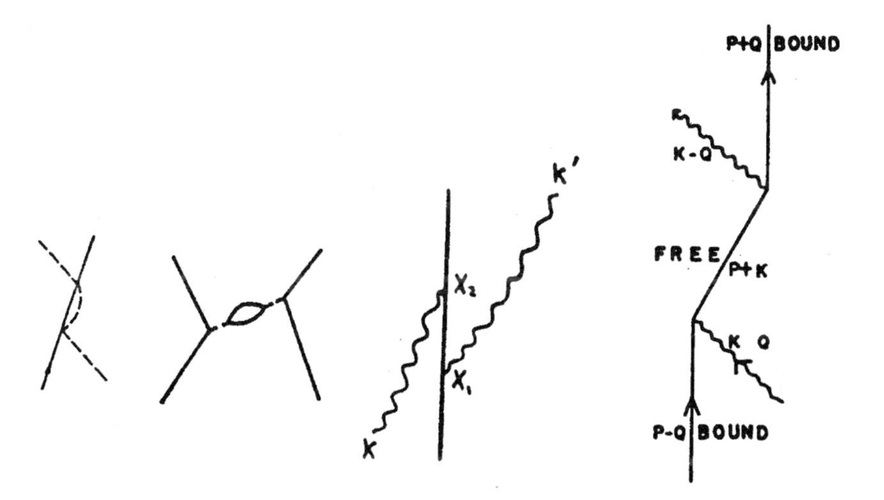


**Figure 2.10**

Repetition of Minkowski conventions in Feynman diagrams. Sources, left to right: F. Rohrlich, "Quantum electrodynamics of charged particles without spin," *Physical Review* 80 (1950): 666–687, on 671; J. L. Anderson, "Green's functions in quantum electrodynamics," *Physical Review* 94 (1954): 703–711, on 706; Francis Low, "Natural line shape," *Physical Review* 88 (1952): 53–57, on 54; and J. S. Levinger, "Small angle coherent scattering of gammas by bound electrons," *Physical Review* 87 (1952): 656–662, on 661.

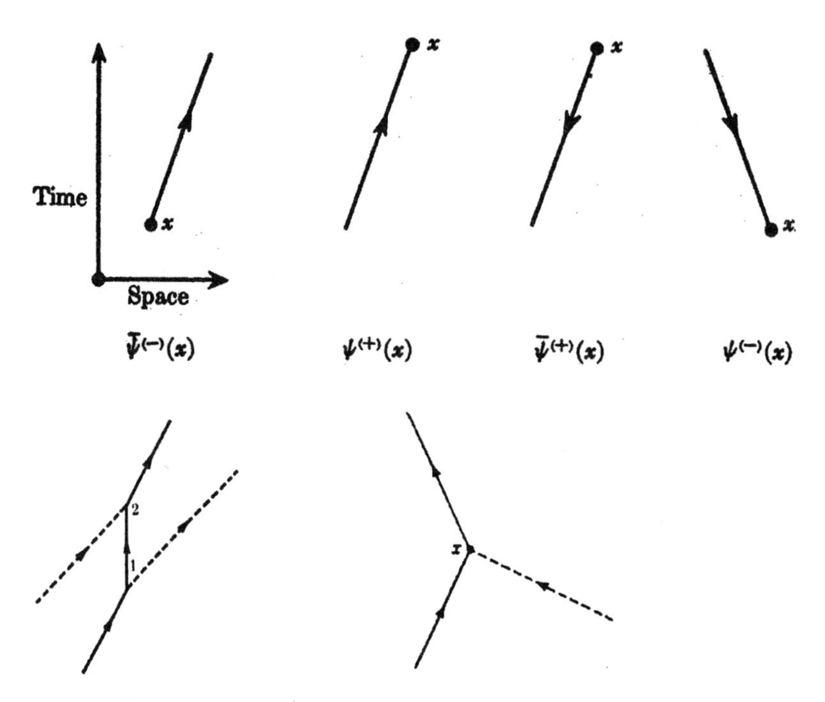


**Figure 2.11**
Feynman diagrams in early textbooks. Sources: (top) S. Schweber, H. Bethe, and F. de Hoffmann, *Mesons and Fields* (Row and Peterson, 1955), volume 1, 219; (bottom, left to right) J. Jauch and F. Rohrlich, *Theory of Photons and Electrons* (Addison-Wesley, 1955), 150; F. Mandl, *Introduction to Quantum Field Theory* (Interscience, 1959), 73.

way. Still other physicists trumpeted the associations as an especially good method for introducing Feynman diagrams to beginning physics students or to audiences of non-physicists.[47]

Thus, despite Feynman's denials to Niels Bohr that his diagrams were intended literally to picture particles' physical paths, they were consistently drawn *and taught* as being of a piece with the reigning pictorial standards for studying particle trajectories through space and time. The mnemonic device simply was not "innocent" of physicists' prior inculcation in the visual practice of depicting particle paths, regardless of the distinct meanings attributed in different contexts to the stick figures. This long, geographically dispersed pedagogical tradition carried tacit, visual "baggage" far more general than the specific functions Feynman diagrams had been designed to play in the context of perturbative QED calculations. Minkowski diagrams had already become second nature to young physicists in many parts of the world. In a broad, general way, they helped to pattern how theorists and their students set up problems and carved out solutions. Within this broader pedagogical space, physicists could improvise and tinker with Feynman diagrams without being unduly constrained by Feynman's, Dyson's, or anyone else's diagrammatic rules of the game.

### Conclusions: Three Pedagogical Functions amid the Dispersion

Robert Kohler's recent work on the history of genetics during the early decades of the twentieth century provides a telling example of how intertwined crafting research practices and creating scientific communities can become. As Kohler demonstrates, *Drosophila melanogaster* was never a research tool outside of a specific community of "drosophilists," and a specific set of social, political, and economic ties that these researchers forged and shared. In the case of fruit-fly genetics, it took a lot of work to domesticate a particular variation of the fly into a useful and interpretable tool. At the same time, it took a lot of work by these same drosophilists to domesticate the nascent community of fruit-fly investigators to share their stocks of mutant-fly varieties, communicate their findings, and regulate intellectual-property claims.[48] In the case of Feynman diagrams in postwar physics, the need to craft both the tool and its user was even more extreme, since the diagrams were never anything more than paper-and-pencil representations—representations that could therefore do absolutely nothing without an interpreter. In the drosophilists' case, at least the bugs were in some sense "out there," even if the artificially stabilized basis of study, *Drosophila melanogaster*, was not an independently existing tool simply found in nature. We are much harder pressed to point to any particular example that would be able to stand in, on its own, for the myriad ways in which theorists drew, calculated with, and interpreted Feynman diagrams.

Three lessons about pedagogy emerge from studying Feynman diagrams' dispersion during the middle decades of the twentieth century. The great majority of physicists who used the diagrams did so only after working closely with a member of the diagrammatic network: the diagrams were put into circulation largely by means of the postdoc cascade. Postdocs in theoretical physics circulated through the Institute for Advanced Study, participating in intense study sessions and collaborative calculations while there. Then they took jobs throughout the United States (and elsewhere) and drilled their own students in how to use the diagrams. Something like tacit knowledge seemed to be crucially important: for the most part, physicists outside of this rapidly expanding network did not pick up the diagrams for their research. It is therefore no accident that more than 80 percent of the physicists who first used Feynman diagrams in print (in the *Physical Review*) between 1949 and 1954 did so as either graduate students or postdocs. Personal contact and individual mentoring remained the diagrams' predominant means of circulation even years after explicit instructions for their use had been in print. Face-to-face mentoring, rather than the circulation of texts, provided the most robust means of inculcating skill with the new diagrams in expanding groups of users. In fact, the homework assignments these postdocs assigned to their students often stipulated little more than to draw the correct Feynman diagram for a given problem, not even to translate the diagrams into mathematical expressions.[49] These students learned early, in ways rarely shared by students outside this early network of dispersed postdocs, that calculations would now begin with Feynman diagrams.

When it comes to physicists' actual uses of the diagrams, however, the simple cascade model must be augmented. Here we see evidence of the second meaning of "dispersion": pictorially and calculationally, physicists' appropriations of the diagrams showed greater and greater differentiation. Local traditions emerged, within which "family resemblances" can be found. Young physicists at Cornell, Columbia, Rochester, Berkeley, and elsewhere practiced using the diagrams in distinct ways, toward distinct ends. These diagrammatic appropriations bore less and less resemblance to Dyson's original packaging for the diagrams. His first-principles derivation and his set of one-to-one translation rules guided students at Columbia, for example, but were deemed less salient for students at Rochester, and were all but dismissed by Geoffrey Chew's students at Berkeley. Students' mentors made choices about what to work on and what to train their students to do. Thus, as with any tool, we can understand physicists' uses of Feynman diagrams only by considering their local contexts of use.

Why did physicists continue to use the diagrams, often basing entire calculational and pedagogical programs on them, even as the problem for which they were invented faded from view? In part because of visual links with other learned practices, such as Minkowski diagrams. These links were largely shared across local groups and national borders; thanks to decades of systematic training, they had become second nature to generations of physicists long before Feynman began doodling his new diagrams. Thus, whereas much of the diagrams' pedagogical dispersion highlighted local groups and face-to-face communication, at least some of the diagrams' dispersion and staying power must be understood in terms of more broadly shared pedagogical resources.

Thus, in considering the diagrams' dispersion, it remains impossible to separate the research practices from how various scientific practitioners were trained. Within a generation, Feynman diagrams became the "theoretical technology" that undergirded calculations in everything from electrodynamics to nuclear and particle physics to solid-state physics. This was accomplished through much pedagogical work, postdoc-to-postdoc, mentor-to-disciples, and department-by-department. Feynman diagrams do not occur in nature, and theoretical physicists are not born, they are made. During the decades after World War II, the practices and practitioners were forged as part of the same pedagogical process.

## Notes

1. Research during the past decade has begun to challenge these older assumptions. See esp. Kathryn Olesko, *Physics as a Calling: Discipline and Practice in the Königsberg Seminar for Physics* (Cornell University Press, 1991); Martin Krieger, *Doing Physics: How Physicists Take Hold of the World* (Indiana University Press, 1992); Andrew Pickering, *The Mangle of Practice: Time, Agency, and Science* (University of Chicago Press, 1995), chapter 4; *The Cultures of Theory*, ed. Peter Galison and Andrew Warwick, published as *Studies in History and Philosophy of Modern Physics* 29 (1998): 287–434; Andrew Warwick, *Masters of Theory: Cambridge and the Rise of Mathematical Physics* (University of Chicago Press, 2003). See also Ursula Klein, "Techniques of modelling and paper-tools in classical chemistry," in *Models as Mediators*, ed. Mary Morgan and Margaret Morrison (Cambridge University Press, 1999), 146–167; idem, "Paper tools in experimental cultures," *Studies in History and Philosophy of Science* 32 (2001): 265–302. Cf. Eric Livingston, *The Ethnomethodological Foundations of Mathematics* (Routledge & Kegan Paul, 1986).

2. This essay is based on my book *Drawing Theories Apart: The Dispersion of Feynman Diagrams in Postwar Physics* (University of Chicago Press, 2005).

3. *Oxford English Dictionary*, ed. J. A. Simpson and E. S. C. Weiner, second edition (Oxford University Press, 1989), s.v. "disperse." For historiographical, rather than etymological, clarity I have combined the *OED*'s definitions 4a and 4b for the first meaning of "disperse" quoted here, and combined definitions 1a and 2b for the second meaning of "disperse." See also Kathleen Jordan and Michael Lynch, "The sociology of a genetic engineering technique: Ritual and rationality in the performance of the 'plasmid prep,'" in *The Right Tools for the Job: At Work in Twentieth-Century Life Sciences*, ed. Adele Clarke and Joan Fujimura (Princeton University Press, 1992), 77–114.

4. Richard Feynman, "Space-time approach to quantum electrodynamics," *Physical Review* 76 (1949): 769–789. For more on Feynman's own route to his diagrams, see Feynman, "The development of the space-time picture of quantum field theory," *Science* 153 (1966): 699–708; Silvan Schweber, "Feynman and the visualization of space-time processes," *Reviews of Modern Physics* 58 (1986): 449–508; Jagdish Mehra, *The Beat of a Different Drum: The Life and Science of Richard Feynman* (Oxford University Press, 1994), chapters 5, 6, 10–14; Silvan Schweber, *QED and the Men Who Made It: Dyson, Feynman, Schwinger, and Tomonaga* (Princeton University Press, 1994), chapter 8; Peter Galison, "Feynman's war: Modelling weapons, modelling nature," *Studies in History and Philosophy of Modern Physics* 29 (1998): 391–434.

5. Feynman, "Space-time approach," 771–773. The absolute square of this integral yielded an approximation, to lowest-order in $e$, for the probability that two incoming electrons will scatter into two outgoing electrons. Actually, this is one-half of the lowest-order contribution: because electrons are indistinguishable, Feynman next explained that one must include a similar amplitude which would describe the case in which the incoming electron on the left ended up, after

scattering, as the outgoing electron on the right, and vice versa. Feynman had already fixed upon these features of his approach before his presentation at the Pocono meeting, though other elements of this 1949 article included developments which Feynman only worked out after the meeting. See Schweber *QED*, chapter 8; Mehra, *Beat of a Different Drum*, chapters 12 and 13, for further details on the evolution of Feynman's work between 1947 and 1949.

6. Feynman, "Space-time approach"; Schweber, *QED*, chapter 8.

7. Schweber, *QED*, 318–334, 436–445; Abraham Pais, *Inward Bound: Of Matter and Forces in the Physical World* (Oxford University Press, 1986), 458–459; Robert Crease and Charles Mann, *The Second Creation: Makers of the Revolution in Twentieth-Century Physics* (Macmillan, 1986), 137–138; Gleick, *Genius*, 255–261. For more on Schwinger's approach, see esp. Schweber, *QED*, chapter 7.

8. Mehra, *Beat of a Different Drum*, 262–263.

9. L. L. Foldy and R. E. Marshak, "Production of $\pi^-$ mesons in nucleon-nucleon collisions," *Physical Review* 75 (1949): 1493–1499, on 1493.

10. Paul Matthews to Wolfgang Pauli, February 25, 1950, in Wolfgang Pauli, *Wissenschaftlicher Briefwechsel*, ed. Karl von Meyenn, volume 4, part I (Springer, 1996), 27–33, on 27. Matthews, a graduate student at Cambridge University, had been corresponding with Fritz Rohrlich and David Feldman, both postdocs at the time, as well as with Pauli. See also the other corresopndence between Matthews, Pauli, and Rohrlich from spring 1950, in ibid., 4–5, 46–47, 72–73, and 97–99.

11. Wolfgang Pauli to Gunnar Källén, August 19, 1952, in Pauli, *Wissenschaftlicher Briefwechsel*, volume 4, part I, 708; my translation. The dissertation in question was by C. A. Hurst, from Cambridge University.

12. Leonard Schiff to Edward Teller, May 26, 1953, in folder "Schiff, Leonard Isaac, 1915–1971," Raymond Thayer Birge, *Correspondence and Papers*, call number 73/79c, Bancroft Library, Berkeley, California.

13. Kaiser, *Drawing Theories Apart*, chapter 2. Two diagrammatic textbooks appeared in 1953 in Japanese and Russian, but these were not translated into English until 1956 and 1957, respectively, and had no impact on American physicists ca. 1949–1954.

14. Julian Schwinger, "Renormalization theory of quantum electrodynamics: An individual view," in *The Birth of Particle Physics*, ed. Laurie Brown and Lillian Hoddeson (Cambridge University Press, 1983), 329–353, on 343, 347. See also *Julian Schwinger: The Physicist, The Teacher, and the Man*, ed. Y. Jack Ng (Singapore: World Scientific, 1996). Although Schwinger's graduate students did not study Feynman diagrams in their formal coursework, some did learn about the diagrams by other means: see Kaiser, *Drawing Theories Apart*, chapter 3.

15. Details may be found in chapter 2 of Kaiser, *Drawing Theories Apart*.

16. In typical fashion, Dyson's letters home to his family in England during and after the trip speak of a Homeric affair, as he and Feynman climbed intellectual heights even as they battled floods and closed roads. Feynman's later reminiscences of the trip, in contrast, centered around bawdy stories of staying overnight in a brothel because there was no vacancy in the local hotels.

Compare Dyson to his parents, June 5, 1948, reprinted on pp. 327–330 of Freeman Dyson, *From Eros to Gaia* (Pantheon, 1992), with pp. 65–66 of Richard Feynman with Ralph Leighton, *"What Do You Care What Other People Think?"* (Norton, 1988).

17. Freeman Dyson, correspondence with his family, 1947–1960, in Professor Dyson's possession, Institute for Advanced Study, Princeton, New Jersey. See also Freeman Dyson, *Disturbing the Universe* (Basic Books, 1979), 47–68; Schweber, *QED*, chapter 9.

18. F. J. Dyson, "The radiation theories of Tomonaga, Schwinger, and Feynman," *Physical Review* 75 (1949): 486–502; idem, "The *S* matrix in quantum electrodynamics," *Physical Review* 75 (1949): 1736–1755; *Science Citation Index* (Institute for Scientific Information, 1961–), s.v. "Feynman" and "Dyson."

19. In his published article, Feynman avowed unapologetically that "Since the result was easier to understand than the derivation, it was thought best to publish the results first in this paper," noting that "in the interest of keeping simple things simple the derivation will appear in a separate paper"—a "separate paper" that was only submitted a full thirteen months later (Feynman, "Space-time approach," 770). The later paper was published as Feynman, "Mathematical formulation of the quantum theory of electromagnetic interaction," *Physical Review* 80 (1950): 440–457. Years later, Dyson recalled: "Nobody but Dick [Feynman] could use his theory, because he was always invoking his intuition to make up the rules of the game as he went along. Until the rules were codified and made mathematically precise, I could not call it a theory." (Dyson, *Disturbing the Universe*, 62)

20. During Dyson's and Cécile Morette's visit with Feynman at Cornell in October 1948, Feynman had calculated the probability for the scattering of light by an external electromagnetic potential, a result that had eluded physicists for nearly a decade. Yet Feynman wrote to Dyson the following week, explaining that in fact the effect vanished (at least at lowest-order in the electron's charge), because the contributions from two different diagrams exactly cancelled out: in one diagram, the virtual electrons circled around the closed loop in one direction, while in the second diagram, they circled in the opposite direction. Only if one clearly distinguished between these two separate diagrams could one thereby reproduce Furry's theorem. See Schweber, *QED*, 450. On Furry's theorem, see W. Furry, "A symmetry theorem in the positron theory," *Physical Review* 51 (1937): 125–129; Pais, *Inward Bound*, 381.

21. Dyson to J. Robert Oppenheimer, October 17, 1948, in Dyson papers; emphasis added. See also Dyson, *Disturbing the Universe*, 72–74; Schweber, *QED*, 520–527.

22. Rabi's famous quotation is reprinted in John Rigden, *Rabi: Scientist and Citizen* (Basic Books, 1987), 46. Theorists' need to travel to Europe had begun to change by the mid to late 1930s, by which time theorists could pursue postdocs at a handful of American institutions with the aid of National Research Council fellowships, most notably with Oppenheimer at Berkeley, John Van Vleck at Wisconsin, or Eugene Wigner at Princeton. See Alexi Assmus, "The creation of postdoctoral fellowships and the siting of American scientific research," *Minerva* 31 (1993): 151–183.

23. Beatrice M. Stern, *A History of the Institute for Advanced Study, 1930–1950* (unpublished typescript, 1961), chapter 11; Ed Regis, *Who Got Einstein's Office? Eccentricity and Genius at the Institute for Advanced Study* (Perseus, 1987), 137–140.

24. J. Robert Oppenheimer to Wolfgang Pauli, February 20, 1952, in Pauli, *Wissenschaftlicher Briefwechsel*, volume 4, part I, 553–554.

25. Freeman Dyson, interview with David Kaiser, January 8, 2001, Princeton.

26. Robert Karplus and Norman Kroll, "Fourth-order corrections to quantum electrodynamics and the magnetic moment of the electron," *Physical Review* 77 (1950): 536–549, on 536–537.

27. Wolfgang Pauli to Abraham Pais, May 26, 1949, in Pauli, *Wissenschaftlicher Briefwechsel*, volume 4, part I, 655.

28. Kaiser, *Drawing Theories Apart*, chapter 3.

29. The diagrams spread to physicists working in other countries largely by similar means. See Kaiser, *Drawing Theories Apart*, chapter 4.

30. The examples from Cornell, Columbia, and Rochester involve advisors and their graduate students. The cases of Chicago and Urbana involve older theorists who learned about the diagrams from their colleagues. In the final example, from Oxford/Cambridge, the putative "advisee," John Ward, had already made use of Feynman diagrams for different types of analyses, and talked extensively with his colleague, Abdus Salam, about how to use the diagrams for the types of calculations Salam had been working on; the diagrams Ward began to draw shifted accordingly. Cornell: Feynman, "Space-time approach," 775; R. M. Frank, "The fourth-order contribution to the self-energy of the electron," *Physical Review* 83 (1951): 1189–1193, on 1190. Columbia: R. Karplus and N. M. Kroll, "Fourth-order corrections in quantum electrodynamics and the magnetic moment of the electron," *Physical Review* 77 (1950): 536–549, on 537; J. Weneser, R. Bersohn, and N. M. Kroll, "Fourth-order radiative corrections to atomic energy levels," *Physical Review* 91 (1953): 1257–1262, on 1258. Rochester: Robert Marshak, *Meson Physics* (McGraw-Hill, 1952), 39; A. Simon, "Bremsstrahlung in high energy nucleon-nucleon collisions," *Physical Review* 79 (1950): 573–576, on 574. Chicago: Gell-Mann and Low, "Bound states," 351; G. Wentzel, "Three-nucleon interactions in Yukawa theory," *Physical Review* 89 (1953): 684–688, on 684. Urbana: Low, "Natural line shape," 55; G. F. Chew, "Renormalization of meson theory with a fixed extended source," *Physical Review* 94 (1954): 1748–1754, on 1749. Oxford and Cambridge: A. Salam, "Overlapping divergences and the *S*-matrix," *Physical Review* 82 (1951): 217–227, on 223; J. C. Ward, "Renormalization theory of the interactions of nucleons, mesons, and photons," *Physical Review* 84 (1951): 897–901, on 899.

31. Actually, the postdocs next clarified that for their stated problem—calculating corrections to an electron's magnetic moment in the presence of an external field—only the diagrams in Classes I and II would contribute. The other diagrams contributed instead to different types of radiative corrections, such as the photon and electron self-energies.

32. One of these terms, in fact, proved simply too lengthy to write out in full. After setting up the integral with the aid of Dyson's version of Feynman's diagrams, the two postdocs quoted their result, explaining in a footnote, "The details of two independent calculations which were performed so as to provide some check of the final result are available from the authors." Karplus and Kroll, "Fourth-order," 548, n. 23. They were describing the integral associated with diagram I in figure 2.6.

33. See in particular Norman Kroll and Franklin Pollock, "Second-order radiative corrections to hyperfine structure," *Physical Review* 86 (1952): 876–888; Weneser, Bersohn, and Kroll, "Fourth-order radiative corrections."

34. See, e.g., J. Ashkin, T. Auerbach, and R. Marshak, "Note on a possible annihilation process for negative protons," *Physical Review* 79 (1950): 266–271, on 266. See also Robert Marshak, interview with Charles Weiner (1970), part II, 63, and part IV, 1–11. Call number OH308, Niels Bohr Library, American Institute of Physics, College Park, Maryland.

35. Ashkin was an assistant professor in Rochester. Feynman thanked him in the acknowledgements to both "Space-time approach" and "Mathematical formulation."

36. Marshak, *Meson Physics*, chapter 1. As explained on x, this textbook was based largely on lectures Marshak had given at Rochester during spring 1950.

37. Ashkin, Auerbach, and Marshak, "Possible annihilation process," 266–267.

38. This approach was adopted in Ashkin, Auerback, and Marshak, "Possible annihilation process," as well as in papers by Marshak's graduate students: Albert Simon, "Bremsstrahlung in high energy nucleon-nucleon collisions," *Physical Review* 79 (1950): 573–576; Morton Kaplon, "The contribution of the Pauli moment to π-meson production by photons," *Physical Review* 83 (1951): 712–715.

39. This addendum was apparently intended to be included in a letter to Léon Rosenfeld; it continued themes that Pauli and Rosenfeld had been discussing in recent correspondence. The quotation is reprinted in Pauli, *Wissenschaftlicher Briefwechsel*, ed. Karl von Meyenn, volume 4, part II (Springer, 1999), 956.

40. J. Ashkin, A. Simon, and R. Marshak, "On the scattering of π-mesons by nucleons," *Progress of Theoretical Physics* 5 (1950): 634–668; quotation on 635.

41. Marshak, *Meson Physics*, preface.

42. See David Kaiser, "Nuclear democracy: Political engagement, pedagogical reform, and particle physics in postwar America," *Isis* 93 (2002), June: 229–268.

43. The argument in this section is condensed from David Kaiser, "Stick-figure realism: Conventions, reification, and the persistence of Feynman Diagrams, 1948–64," *Representations* 70 (2000): 49–86.

44. Peter Galison, "Minkowski's space-time: From visual thinking to the absolute world," *Historical Studies in the Physical Sciences* 10 (1979): 85–121; Stanley Goldberg, *Understanding Relativity: Origin and Impact of a Scientific Revolution* (Birkhäuser, 1984), esp. part III.

45. Feynman, "Space-time approach"; idem, Quantum Electrodynamics (unpublished lecture notes, Cornell, fall 1949); idem, Quantum Electrodynamics and Meson Theories (unpublished lecture notes, Caltech, February-March 1950); idem, Quantum Mechanics III (unpublished lecture notes, Caltech, fall 1953); idem, *QED: The Strange Theory of Light and Matter* (Princeton University Press, 1985). Copies of the 1949 and 1950 lecture notes are in the possession of Sam Schweber; a

copy of the 1953 lecture notes are in the possession of Elisha Huggins. My thanks to both for sharing the notes with me.

46. Pais, *Inward Bound*, 458–459; Schweber, *QED*, 444.

47. Jauch and Rohrlich, *Theory of Photons and Electrons*, 149; Ernest Henley and Walter Thirring, *Elementary Quantum Field Theory* (McGraw-Hill, 1962), 146. Cf. Richard Mattuck, *A Guide to Feynman Diagrams in the Many-Body Problem* (McGraw-Hill, 1967), chapters 2–3; Kenneth Ford, *The World of Elementary Particles* (Blaisdell, 1963), 191–201; M. Stanley Livingston, *Particle Physics: The High-Energy Frontier* (McGraw-Hill, 1968), 204–205; Feynman, *QED*, chapter 3.

48. Robert Kohler, *Lords of the Fly: Drosophila Genetics and the Experimental Life* (University of Chicago, 1994).

49. See chapter 7 of Kaiser, *Drawing Theories Apart*.

# 3 A Pedagogy of Diminishing Returns: Scientific Involution across Three Generations of Nuclear Weapons Science

**Hugh Gusterson**

In science studies we hear a lot about the construction of facts, the building of networks, and the growth of disciplines, but much less about the withering, decay, or arteriosclerosis of fields of knowledge (brought on either through natural exhaustion of a particular approach or through external constraints). We have a disciplinary bias toward scientific evolution and revolution rather than what I am here calling involution.[1] A generation of historians, sociologists, and anthropologists of science has learned from actor-network theory and the sociology of scientific knowledge (SSK) to focus on the building of scientific institutions and facts, and from Thomas Kuhn to expect a certain historical rhythm in the evolution of scientific fields of knowledge: first, a dynamic burst of creativity (the "revolution") as the foundational ideas of the new field are laid down; second, a period of "normal science" in which gaps are filled in as the new knowledge is institutionalized; and, finally, as puzzles emerge that cannot be fully explained by the established paradigm, a new burst of creativity as another generation redefines the fundamental precepts of the field.[2] In this essay, looking at three generations of nuclear weapons designers, I follow and then depart from the Kuhnian script. Although the first two generations of nuclear weapons scientists conformed perfectly to the Kuhnian storyline, the final story is not about the punctuated equilibrium of scientific revolution, but about a process of scientific involution as nuclear weapons science has simultaneously matured and withered in a way that is beautifully evoked in a blues ballad once sung for me by a group of weapons designers from the Lawrence Livermore National Laboratory:

Went down to Amarillo
Lookin' for my sweet '53[3]
It was laying on a long white table
Looked cold and hard to me

Let it go, let it go, retire it
No city scrapers do we need

Take a 61[4] and modify it.
Call it the mod 11-E

Now you can search this whole world over
From Frisco to Albuquerque
You can mentor anyone that you want to
But you'll never find designers like me

Now when I'm gone, just put me way down
In a hole off the old Orange Road.
'ttach a cable to my device can
So I can run those legacy codes (fading)
So I can run those legacy codes
So I can run those legacy codes.[5]

The great breakthroughs in nuclear weapons design were made in the early, heroic phase of American nuclear weapons design—roughly coinciding with the period of atmospheric nuclear testing—when the nuclear weapons laboratories at Los Alamos and Lawrence Livermore were smaller than today, the designers were younger, and the organization of weapons design much less formal. It was in this period that American nuclear weapons scientists learned how to control nuclear fission in the engineering of the first atomic bomb[6]; devised two-stage devices in which the energy unleashed by a fission bomb was used to create a larger, thermonuclear explosion[7]; shrunk these hydrogen bombs so they would fit atop missiles[8]; learned to use tritium to "boost" the yield of atomic weapons, thus creating fission-fusion hybrids; and designed narrow cylindrical weapons that would fit in artillery shells despite the need for a spherical implosion at the moment of detonation.[9] The names of the men behind many of these achievements include some of the great physicists of the twentieth century and are well known outside the parochial world of nuclear weapons science: J. Robert Oppenheimer, Edward Teller, Stan Ulam, John von Neumann, Hans Bethe, Richard Garwin, Herb York, and Ted Taylor.[10]

The 1970s and the 1980s, when nuclear testing moved underground, were a period of routinization: the institutional apparatus for nuclear weapons design and testing grew, its scientific achievements shrank, and the arteries of the weapons design bureaucracy hardened. Attempts to perfect a third-generation nuclear weapon—the x-ray laser—failed and were abandoned in an atmosphere of scandal and disgrace.[11] The art of weapons design progressed, but by increments rather than great leaps: weapons designers learned to squeeze greater yields out of smaller quantities of plutonium so that nuclear weapons could be made lighter and smaller, weapons were made safer through the addition of Permissive Action Links (PALS) and the substitution of

Insensitive High Explosive (IHE) for conventional explosives,[12] and the supercomputer codes used to model the behavior of nuclear weapons were gradually refined. The names of the men (and now women) behind these achievements are largely unknown outside the nuclear weapons bureaucracy, and in some cases their achievements are only partially known within the weapons laboratories, thanks to the compartmentalizing effects of official secrecy in the weapons complex.[13]

Nuclear tests were forbidden after the end of the Cold War, and the practice and pedagogy of nuclear weapons science shifted again. Forced to largely abandon their nuclear test site in Nevada—a place where the desert sands encroach on the old bowling alley and cinema, now disused, as tourist buses disgorge camera-laden voyeurs to gawk at the nuclear craters—many of the old-timers elected to retire. Those that stayed have regrouped their forces in the virtual world of simulated testing, where they are attempting to train a new generation of scientists to maintain devices they cannot test. In some ways the scientific challenges of nuclear weapons design have shrunk to microscopic proportions: new designs are not built or deployed, and even the decision to substitute a new epoxy in an aging weapon can send a tremor of fear through design teams unsure if their weapons will still work. In other ways, the scientific challenges are suddenly magnified: how to design implosion, shock wave, and laser fusion experiments that will shed light on the performance of aging nuclear weapons in the absence of nuclear testing? How to use the physics knowledge of today to understand test data, long buried in dusty filing cabinets, from the 1950s and the 1960s? And how to convert old two-dimensional codes designed for Cray supercomputers into three-dimensional codes that can run on massively parallel systems now being designed?

With a nod to Max Weber, I call two of the three periods I have described here the "charismatic" and the "routine"; I call the third (nothing to do with Weber) the "virtual."[14] Obviously the periodization I have sketched out here is schematic, not to say contrived, but it is a useful optical device through which to investigate the practice of knowledge creation and pedagogy in the nuclear weapons laboratories of Los Alamos and Livermore. My argument, summarized in the crudest terms, would be that, as nuclear weapons science has progressed, as the weapons laboratories have grown and matured as institutions, and as the laboratories lost their central experimental practice, nuclear testing, the training of new weapons designers has become more lengthy and formalized, the knowledge they acquire more carefully codified, the contributions of individuals more anonymous, and the incremental leaps in knowledge less substantial. We see, in other words, a scientific bureaucracy engaged in an increasingly involuted pedagogy of diminishing returns.

My claims here are based on more than 15 years of research into the organizational culture of nuclear weapons scientists. I have been investigating the strange closeted worlds of nuclear weapons designers since 1987, when I first arrived in Livermore as an anthropologist engaged in my own apprenticeship program as a graduate student, embarking on field research among nuclear weapons designers.[15] Since 1992 I have also been conducting fieldwork at Los Alamos and, to a lesser degree, the engineering support laboratory at Sandia (which has branches in both Albuquerque and Livermore), trying to understand how the end of nuclear testing has altered the practice of nuclear weapons design and the means by which a new generation of designers is trained. In my earlier work I focused on what, adapting a well-known phrase from Foucault, we might call the "pedagogy of the self": the processes through which new weapons scientists learned to internalize the appropriate emotional and ideological orientation toward their work from the everyday discourses and practices in which they were immersed. In this essay I look instead at a more orthodox kind of pedagogy: the means by which neophyte weapons designers learn and elaborate their life's craft.

## The Charismatic: The Atmospheric Testing Era

In reading accounts of the Manhattan Project and the early years of the Cold War arms race at the Los Alamos and Livermore laboratories, one is struck by the extraordinary youth of the scientists and the relative informality (especially by contemporary standards) of the laboratories as organizations. Wartime Los Alamos was a place where, dislocated from the hierarchies of university life to a remote desert mesa in the West, a community of scientists crystallized in which divisions between faculty and students or European émigrés and native-born Americans were subordinated to the collective goal of mapping out a new kind of weapons science and engineering in which none had prior expertise in the strict sense of the term.[16] At the outset of wartime work at Los Alamos, J. Robert Oppenheimer won a struggle with General Leslie Groves, the military overseer of the atomic bomb project, about the internal organization of the laboratory. Where Groves, concerned about the possibility of espionage at the laboratory, wanted to compartmentalize the laboratory and inhibit conversation among its scientists, Oppenheimer insisted that good scientific work could be done only in an environment in which open discussion and the sharing of ideas among all the leading scientists was possible.[17] New scientists arriving at Los Alamos were given a 24-page mimeographed document called The Los Alamos Primer (Edward Condon's notes on Robert Serber's introductory lectures explaining the basics of what was known about neutron cross-sections, critical mass, and so on).[18] In the meantime, the team of scientists Oppenheimer assembled to undertake this collaborative research project was extraordinarily young, the average age being 25. Oppenheimer, the director of the laboratory, was himself only in his late 30s, while his two leading theorists, Hans Bethe and Edward Teller, were 36 and 34 respectively. "At thirty-four," recalled Stan Ulam, "I was already one of the older people." Other scientists working on the project were, like Oppenheimer's students Philip Morrison and Robert Wilson, much younger. Many were on leave from graduate programs in physics while they worked at Los Alamos. Frederic de Hoffman, who would go on to help found General Atomic in the 1950s, was 19 when he arrived at Los Alamos.[19]

This pattern was repeated when the second nuclear weapons laboratory was established in the California town of Livermore in 1952. The scientists who ran the Livermore Laboratory were even younger than those in charge of the Manhattan Project, most being "newly minted PhDs from Berkeley" with no prior nuclear weapons design experience.[20] One of these new PhDs from Berkeley, Harold Brown (later Jimmy Carter's Secretary of Defense), was put in charge of the thermonuclear weapons design division, A Division, at the age of 24. John Foster, in charge of the design division for fission bombs, B Division, was 29. Herbert York, the man put in charge of the new laboratory by E. O. Lawrence and one of the few with previous experience in nuclear weapons science, was 31. Discussing the group of men running the Livermore Laboratory, York recalled in his memoirs that "Teller aside, the average age of its members was just thirty and, except for certain modest projects set up deeply in and supported solidly by the larger laboratory structure, none had ever directed or managed any very substantial free-standing enterprise."[21]

Remarking on "the relaxed and unstructured atmosphere" at the new Livermore Laboratory, Sybil Francis observed that "the relatively small Livermore staff contributed to informality and overlapping organizational identities."[22] Herbert York recalled that "Lawrence firmly believed that if a group of bright young men were simply sent off in the right direction with a reasonable level of support, they would end up in the right place. He did not believe that the goals needed to be spelled out in detail or that the leadership had to consist of persons already well known."[23] Although York was known to be the person Lawrence had put in charge of the new laboratory, he did not call himself "Director" for two years. "Whenever I wrote a letter to officials in Washington to propose some new element of the program, to arrange for the construction of a new facility, to ask for more money, or the like and whenever I wrote to officials at Los Alamos or Sandia to arrange for cooperation on some new project, I simply signed my name followed only by my address: UCRL, Livermore."[24]

Los Alamos and Livermore in the early years, then, were institutions where managerial titles and hierarchies were relatively unimportant. Despite the high wall of secrecy placed around the laboratories,[25] a group of young scientists was given great freedom to explore new ideas in an atmosphere of openness and informality. Since these young scientists were inventing (rather than reproducing and extending) a field of knowledge and practice and the most experienced were still relatively inexperienced, there was little in the way of formal pedagogy and those without formal qualifications could be recognized for their contributions in a way that would have been improbable in larger, more hierarchical institutions. Thus, for example, one of Livermore's senior scientists from this era, Carl Haussmann, was said not to have even finished his bachelor's degree.[26] Later Livermore would appoint John Nuckolls as its director and Roy Woodruff as its Associate Director for Weapons Development—both men lacking PhDs but respected for their contributions to weapons science.

Ted Taylor, perhaps Los Alamos' most prolific warhead designer, recalls the informality and lack of bureaucracy at Los Alamos in the 1950s thus:

> In my seven years at the laboratory I never had to participate in the writing of a single grant proposal. . . . In Los Alamos in the '50s someone would get an idea and go down the hall and get Preston Hammer to put it on the computer and six weeks later you get printouts and find out whether the guess was right. If the results came out interesting you go up and talk to Carson Mark [the head of division] and he often would find some flaw. Or he'd say something like, 'Well, I'll be damned,' and then you'd cut across the overpass to the middle of the laboratory, the head of the explosives division, and he'd say, 'That sounds great! We'll put it on the fission committee agenda.' A week, two weeks later we'd have a fission-weapon committee meeting, and sometimes flaws turned up there. When they didn't, OK, we'll put it on the list. List for what? List for testing, either in Nevada or Eniwetok, quite often in less than a year from the initial concept to the successful test.[27]

It was in this early period—roughly, the era ending in 1958 with the commencement of the first moratorium on nuclear testing—that the most important breakthroughs in American nuclear weapons science were made. (Thus, in 1959, Darol Froman, scientific advisor to the director of Los Alamos, arguing that the end of nuclear testing might not greatly matter, was able to write to a member of the Atomic Energy Commission that "we have milked the nuclear weapons business pretty dry . . . [while] they [the Soviets] have a few quarts yet to go."[28]) In these years American nuclear weapons scientists had developed reliable processes for the enrichment of uranium and the production of a new element: plutonium. They had devised two different ways of making effective fission devices: the gun-assembly weapon used on Hiroshima and the plutonium implosion bomb dropped on Nagasaki.[29] (This, in turn, required important advances in metallurgy, high-speed photography, the physics of shock waves, the engineering of high explosives, and so on). They had then learned how to make two-stage devices—hydrogen bombs—that used x-ray radiation from fission devices ("primaries") to heat and compress tritium and deuterium (stored as lithium deuteride) in "secondaries" so that nuclear fusion was achieved. They had also learned to use tritium and oralloy (enriched uranium) to boost the yield of fission devices, mastered the design of non-spherical implosion devices that had narrower diameters than spherical implosion bombs (and were therefore well suited to missile warheads and nuclear artillery shells), and had shrunk hydrogen bombs while enhancing yield-to-weight ratios enough that they could fit a one-megaton bomb into a missile warhead. Finally, they had learned how to make a bomb that maximized radiation output relative to blast (the neutron bomb), had developed sophisticated methodologies for measuring weapons effects beyond the mere explosive yield of the bombs, and had demonstrated the possibility of testing nuclear weapons underground with full radiation containment and accurate diagnostics. They also made bold attempts, finally cut off by the ban on atmospheric nuclear testing in 1963, to develop a nuclear-bomb-powered spaceship (the Orion Project) and to develop nuclear weapons optimized for the construction of harbors and canals, and even for mining and oil drilling.[30] Although Livermore managers argued as the testing moratorium went into effect in 1958 that, if only testing were to resume, further exciting advances were possible (atomic hand grenades, "clean bombs," and ultra-small nuclear bombs for recoilless rifles, for example[31])—nuclear weapons design from the 1960s onwards increasingly became, in the memorable phrase of one university physicist I interviewed, like "polishing turds."[32]

The period up to 1958 also saw the emergence and consolidation of a striking divergence in the organizational and design cultures of the two nuclear weapons laboratories. Los Alamos was founded by J. Robert Oppenheimer, possibly the premier theoretical physicist of his generation in the United States, and it bore both his imprint and that of the other great Los Alamos theorist, Hans Bethe, in that Los Alamos scientists relied heavily on theoretical calculations in their preparation of new designs. The heavy reliance on calculations also minimized the need for expensive nuclear tests and therefore conserved nuclear material which was scarce at the time Los Alamos first institutionalized this approach to nuclear weapons design during and immediately after World War II. The Livermore Laboratory, by contrast, was established by Ernest Lawrence, arguably the leading experimental physicist of his generation, and its scientists (the first of whom were largely Lawrence's students) relied heavily on experimental trial and error and, increasingly, on computer codes calibrated against test experience. Sybil Francis described Livermore thinking in her fine historical study of the Livermore Laboratory: "Weapons development, within limits, could be pursued without deepening fundamental scientific knowledge. This could be done by testing a variety of designs and selecting those that worked without complete knowledge of the reasons for their success."[33] Francis contrasted the Livermore style with Los Alamos' more theory-oriented approach:

> Livermore scientists were comfortable pursuing designs for which it was "hard to make advanced calculations of expected results," a consequence of their experimentally-oriented background. As a UCRL budget document explained, experiment provided Livermore scientists the means of studying designs "that might go unused because of difficulties in making calculations." Another useful technique was computer modeling, in which Livermore invested substantial effort. Designs difficult to calculate meant Livermore developed empirically based models, resulting in a more incremental approach to weapons design: computer modeling, tests, then more modeling in preparation for the next test. Livermore thus made its earliest contributions to the weapons program in areas especially difficult to calculate theoretically.[34]

Francis also argued that a sibling rivalry dynamic emerged in the relations between Livermore and Los Alamos, and that the Livermore Laboratory, always insecure about its right to exist as the second laboratory whose founding had been opposed by Los Alamos, consistently cast itself in the role of the "new ideas" laboratory in order to legitimate itself. (Dan Stober and Ian Hoffman report that, when the Livermore Laboratory was established, "the orders from its young lab director, Herbert York, were clear: 'Whatever you do, don't do it like Los Alamos.' For the researchers in the lab's Small Weapons program, that meant make 'a fission bomb that was anything but spherical.'")[35] Also, since two-thirds of the weapons entering the stockpile through the late 1950s were Los Alamos weapons, Livermore could afford to do speculative and exploratory design work, while Los Alamos was forced to devote its resources to validation of actual designs. In its preference for more ambitious and exploratory design work the Livermore Laboratory was not only responding to the structural exigencies of its relationship with Los Alamos; it was also expressing the persona of its other distinguished founding scientist, Edward Teller. Thus, while Los Alamos developed a conservative approach to weapons design, "the new laboratory worked on 'bolder' designs, less certain of success than those of Los Alamos."[36] Francis argued that Los Alamos, "more experienced, and with established ties to the military . . . gained responsibility for the highest priority and most urgent military requirements. Livermore found opportunities in nuclear systems that were more speculative, that did not yet have formal JCS [Joint Chiefs of Staff] authorization, or were low priority."[37]

Thus, after a rocky start in which the "new ideas" laboratory's first three nuclear tests were "fizzles" (the weapons designers' term for low-yield duds), Livermore consistently pushed the envelope of nuclear weapons design in a way that Los Alamos did not. It was Livermore that shrunk the diameter of fission weapons enough to make the low-diameter Davy Crockett artillery shell for the army in the 1950s.[38] It was Livermore that pushed yield-to-weight ratios and shrunk a hydrogen bomb enough to produce the first hydrogen bomb warhead for a strategic missile—the Polaris system—after Los Alamos scientists had scoffed at the Livermore proposal.[39] It was Livermore that designed the first MIRV warhead. It was Livermore that introduced the first two-dimensional computer code in the mid 1950s. And it was Livermore that conducted the first underground nuclear test in 1957, in the face of opposition from Los Alamos.[40]

But there was a dark side to Livermore's organizational culture of risk and creativity. Livermore's institutionalized propensity to cut corners and push forward with poorly understood design approaches led to periodic failures, even scandals, of a kind unknown at Los Alamos. Livermore's first two tests, which failed to even vaporize the towers from which the bombs were suspended, made the new laboratory the butt of jokes in Los Alamos, whose scientists eagerly photographed the still-standing towers.[41] Meanwhile, during the test moratorium, Livermore rushed Polaris warheads into the stockpile despite a poorly tested mechanical safing system that, it was subsequently discovered, turned many of the warheads into duds.[42] This risk-taking culture, essential to the younger laboratory's organizational persona, was to endure: in the 1980s Livermore scientists' optimistic and mistaken assurances that the x-ray laser—a critical component of President Ronald Reagan's Strategic Defense Initiative—was ready for the engineering phase would create a public relations disaster for the Laboratory when the weapon was finally cancelled amidst revelations that it never really worked and that the Associate Director for Weapons Design had secretly resigned in protest against his colleagues' representations of the weapon to the White House.[43] And in the 1990s Livermore's Associate Director for Lasers would resign after public revelations of cost overruns and wildly over-optimistic performance projections by Livermore's management regarding its huge new laser project: the National Ignition Facility (NIF).[44]

At the end of the Cold War, a Livermore weapons designer said to me, only half joking, "The Soviets are the competition, but Los Alamos is the enemy." The fierce rivalry between the two weapons laboratories encouraged each, in the complex competition for contracts from the Atomic Energy Commission and the three armed services, to develop partly opposed institutional personae: Los Alamos as a conservative organization of good pedigree making carefully understood incremental design changes as it developed higher and higher yield hydrogen bombs for air force bombers and improved tactical weapons for the army, Livermore as a bolder organization using riskier design ideas Los Alamos had shelved in order to miniaturize nuclear weapons for the navy and the army. The divergence of design cultures at the two laboratories was also a legacy of the approach to physics as craft of the charismatic scientists who established the research orientations of the two laboratories at the outset: Oppenheimer and Bethe, both theorists, at Los Alamos; Lawrence, an experimentalist, and Teller, a dreamer, at Livermore. In the words of the physicist Brian Dunne, "You've got the Livermore gang and you've got the Los Alamos gang, two different cultures. . . . Acolytes of Teller, disciples of Bethe. They are different tribes, warring tribes."[45] As the two laboratories matured and grew organizationally, these two design cultures were passed on to a new generation of weapons scientists who, instead of inventing a new field de novo as their forebears had, found themselves internalizing an established field of knowledge and making incremental refinements of it.

## Routine: The Era of Underground Testing

The period from the Limited Test Ban Treaty of 1963 to the last U.S. nuclear test (so far) in 1992 saw the weapons laboratories grow into larger, more complex, and, inevitably, more hierarchical organizations. The Livermore Laboratory, for example, added 2,000 to its staff of 3,000 between 1958 and 1963 and, by the time I arrived to do fieldwork in 1987, had added 3,000 more employees for a total staff of 8,000 and an annual budget of $1 billion. The Los Alamos National Laboratory was the same approximate size.[46] As the laboratories grew, the men who had pioneered the design of nuclear weapons trained a new generation of scientists to reproduce and refine their art. These were years when an increasing gulf between rank-and-file weapons scientists and their managers appeared. (One Livermore designer, lamenting the end of what I am calling the charismatic period of weapons science, told me: "Back in the early days there wasn't a separate group of managers looking out for themselves the way there is now. The scientists would just take it in turns to do managerial work, or they'd take decisions by consulting each other. Then there was a very small number of professional managers—financial people and so on, but they didn't dominate the whole thing. The lab was run by real scientists. Now, do you know how many managers there are at the lab? 105! 105 ADs (Associate Directors), PADs (Principal Associate Directors), DADs (Deputy Associate Directors), and DOODADS.")[47] Meanwhile, as weapons science became routinized, the weapons design process became increasingly conservative and bureaucratized and weapons design teams within each laboratory more specialized. The new generation of designers learned to further improve the yield-to-weight ratios of U.S. nuclear weapons and to make some safety improvements, but did not make any spectacular breakthroughs in the field of nuclear weapons design. Safety improvements included the development of Permissive Action Links (PALs)—devices that allow a weapon to be armed only when the correct authorization code is entered; the development of Insensitive High Explosive (IHE), making it less likely that a bomb would detonate if dropped by accident; and the development of Fire-Resistant Pits (FRPs) less likely to detonate if, for example, a plane carrying the weapon crashed and caught fire.[48]

Reflecting on the increasingly conservative design culture at his[49] laboratory, one younger Livermore designer, recalling Enrico Fermi's dictum that "you were no good if you didn't have some major surprises," lamented that "pushing the design physics" was no longer acceptable in an era when nuclear tests were fewer in number but greater in cost: "Before, when there were a lot more tests. . . then there was more chance taking—having someone pick a novel idea and trying to find out if it's doable. But in the short time that I've been at the lab, people are very careful and conservative. They try to get as much out of each test as possible."[50] An older designer echoed this: "These days there's tremendous pressure from Washington for each test to be a spectacular success. If the bomb doesn't go off or something, then there's all hell to pay in Washington."[51] One designer who joined the Livermore Laboratory in the early 1960s, only to quit in the 1970s out of, he says, boredom, remarked of a highly respected designer at Livermore: "He's spent his whole life designing the same bomb over and over again, shaving a few centimeters off here and there but never doing anything fundamentally new. It only seems like exciting physics because it's so secret."[52]

With these comments as prologue, let's now follow the process by which a "typical" young weapons designer was apprenticed in the arts of weapons design in this period of increased routinization and professionalization. Young physicists considering accepting a position in one of the design divisions would make the rounds of different design groups to see if they found what seemed to be a good fit anywhere. They might be drawn toward a specific area of weapons physics or toward a particular group of people. One designer recalls picking an assignment in Livermore's B Division[53] because he felt drawn toward a particular group of people whom he described as "very reasonable. I can see some unreasonable people in other groups. I feel that my group is wise, that's the word. The whole group is that way."[54]

New designers were encouraged to maintain broad interests and even to publish in the open literature as they were learning their corner of the art of weapons design. This was partly to ensure that young physicists—accustomed to the university environments where they did their graduate work and unsure if weapons design would hold their interest over the long run—did not become too disaffected from their new specialization and instead focused on the extraordinary opportunities offered by employment at a laboratory with the fastest supercomputers and the largest community of physicists in the country. The same designer quoted above recalled:

Once you step inside the gates [of the laboratory] it's a very open and university-like atmosphere. It's the closest that I've come to that kind of atmosphere outside a university. The Laboratory has tremendous resources to do work that would not be possible academically. Even now, I still believe that it's possible for someone to come up with an idea and do some interesting work, even if it's not related to the weapons program. They may not get a lot of recognition for it, but there is the possibility.[55]

Similarly, a senior manager in Livermore's other design division, A Division, said: "We like people to keep publishing. When we bring people in, we tell them that: we'll give you a certain percentage of your time, and we want you to keep to that because we want you to interact with people outside the lab. I have continued to publish—not a lot, but I have a colleague who works with me, and over the years we've published quite a few papers. It's low key, but I have an international reputation from that work. It's not my number one priority in life, but. . . ."[56]

Many of the younger weapons designers developed strong apprenticeship relationships with older designers who served as mentors in relationships that carried great intellectual and emotional force. They learned from these mentors not only the arcane and secret knowledge about weapons design of which they were custodians, but also how to navigate the bureaucratic shoals of laboratory life, how to carry one's expert judgment as a weapons scientist, and how to interpret ethical conundrums and geopolitical puzzles in the broader world. One Livermore designer, when I asked him which relationship in his life was most important to him, mentioned the relationship with the weapons designer to whom he had apprenticed himself, not his relationship to his wife of many years. Describing this relationship, he said his mentor "is totally honest and has a total lack of respect for authority. He will be straight with everyone, whether it's an Associate Director or a shop-floor machinist. And, because he's a man of integrity, he's disliked by managers. He's incredibly knowledgeable. If a company is producing a part, he often knows more about what's happening on the shop floor than the company's managers do. That's why they often dislike him! If he'd worked for Morton Thiokol, he would have stopped the *Challenger* launch. He's more responsible than anyone for the integrity of our weapons program."[57]

The heavy emphasis on mentorship and apprenticeship at the weapons laboratories was largely an adaptation to the small scale of the weapons design community and to the powerful effects of official secrecy on the exchange of ideas within that community. In the weapons laboratories of the 1960s, the 1970s, and the 1980s, intellectual life was both compartmentalized and informal. Although weapons designers sometimes told me that becoming a weapons designer was like doing a second PhD in physics (and, indeed, it took about the same length of time), unlike graduate programs in physics, the weapons laboratories provided neither formal classes nor textbooks. In this period weapons designers did not yet have their own classified journal in which to publish so, when knowledge was written up, it was often in a gray literature of shot reports and other documents that was not well inventoried and was opportunistically and eccentrically stored in the accumulated files of individual scientists building collections of documents over their careers. Moreover, the Department of Energy's classification system, segmenting knowledge and obstructing a scientist's access to knowledge in areas where he or she was deemed not to have a "need to know," meant that there was little in the way of a public roadmap of the entire topography of nuclear weapons knowledge. More important still, a number of scientists worried that some of the most important knowledge about nuclear weapons science at the laboratory was not written up at all, existing instead as tacit knowledge in the heads of the most experienced scientists. "Seymour's retirement will be a blow," one weapons designer told me, speaking of his legendary senior colleague Seymour Sack in Livermore's B Division. "He has such a great memory that he hasn't written down lots of important stuff. How will people know it?"[58] The emphasis on learning through personal relationships that permeated nuclear weapons pedagogy at the laboratories was an adaptation to this situation where knowledge was not formally codified in carefully evaluated and ranked reports of the kind that circulate in geographically dispersed academic communities unified by open literatures, and where some of the most important knowledge was not written down at all but could only be acquired, experimentally, through trial and error and, dialogically, through interaction with mentors. It is this feature of nuclear weapons pedagogy and expertise that has led Laura McNamara, an anthropologist who has recently written a fine PhD dissertation on weapons scientists at Los Alamos, to describe weapons design knowledge as "a form of situated action, located in a nexus of relationships that linked weapons experts to nuclear artifacts."[59]

To show exactly how this process of knowledge acquisition and elaboration can work, I reproduce here extended excerpts from an interview with a weapons designer who had gone on to become a manager in one of the weapons design divisions. The narrative, self-consciously rendering a neophyte's progress as an abstract ideal type almost as an ethnographer might, shows how a new recruit generates questions and ideas by working through the new field of weapons physics and by performing calculations sub-contracted by senior team members, learns to defend his or her ideas in the often brutal environment of review meetings, and works as part of a team to translate an abstract idea into a test. Recruits who excel are good at working collaboratively in inter-disciplinary teams and in defending their ideas and calculations in meetings. The narrative also underlines the vital importance of experiments, especially nuclear tests, as core instruments of pedagogy and apprenticeship. As Laura McNamara observes, "throughout the Cold War . . . there was always another test just on the horizon, another iteration of the design and test cycle reaching fruition; iteration after iteration, like a series of waves, slowly rising and building towards an end point, then breaking into memory to make space for the next event. There was a constant flow of work, a recycling of the same process, year in and year out, so that the activities involved in testing were constantly being exercised on the various experiments that were ongoing at any one point."[60] The weapons designer agrees:

You'd come in and you'd learn a lot of new physics which is not known outside the lab, which is very interesting in itself. And in the process of learning that physics you often get new ideas. It's our fresh people coming in that are really hot with new ideas. So what you'll do is you'll mess around with theory at your desk, and you don't get very far with that usually because we work with very complex physical processes. That's why we work with these big computer codes. They are to handle many types of physics simultaneously. You can't do that with equations on paper. And so we train all the people that come in to use these computer codes as tools to study the physics. So you'd mess around with computer codes to study your idea, and, if it looks good, then you'd go to your group leader. Everybody's in small groups of 4 or 5 people—that's what we find is most productive—and so you have this great idea and you want to work on it. We work as teams; if your team is about to have an experiment come up, particularly one in Nevada, then your group leader will say to you, "you can't work on that right now because we need you. Everybody's working together and people are depending on you to do predictions for this experiment." If you've just done an experiment, then he'll say "go for it!"

And so there you are with your idea. What you do then if your idea still looks good is you go to your division leader. Someone might come to me and say, "I want to propose an experiment."[61] At that point you are asking for $50 million, typically. So that what happens then is that you go through a very intensive peer review process. We start pulling together groups of people, and you'd brief your idea to maybe fifteen or twenty people, and they'd try very hard to find reasons why it doesn't work because this is a very expensive business and you do everything you can to find errors before you actually go do the experiment. So you keep doing that. These are not nice reviews. They're very critical. I've seen men all in tears.

Question: What percentage of the ideas would survive?

Answer: small percentage. The culture in that division—this is really interesting too—the big reward in our division is to do an experiment, to get your idea tested. It's highly competitive. For every twenty things people propose, maybe one is going to make it onto that shot schedule. If you have a really good idea. Let's say it's a REALLY good one, and it's recognized to be that way, then what we will do is put everything aside in the shot schedule. We have a lot of flexibility. And, just like magic, we'll matrix a team together. We can do that in a couple of days. The minute we put an experiment on the shot schedule that happens. So all of a sudden you find yourself with fifty people supporting you, all trying to help you research this idea.[62]

Of those fifty probably half of them do not have the skills to run a big program, and we will have found that out by having them run smaller projects, which is part of life there; and you can see as you go through that process who has the ability to successfully carry out a big project and who doesn't. . . . We have people, for example, who are forgetful of details, and forgetting certain details in these experiments can be disastrous. That's why we have these reviews. . . . A typical review for a new idea lasts about three hours.

So then we have this big team there, and they've got a goal they're aiming for and they're all starting to work together, and they're all depending on you to tell them, to make predictions for them using computers and theory and whatever else you have, to tell them what to measure. They might be building big spectrometers. They can cost millions of dollars and they get blown up in the experiment. So there's tremendous pressure to get the predictions met. So now you start down this track and the whole thing is just a very fast-moving process. We keep reviewing over and over again. . . . I spend probably fifty percent of my time reviewing things. We have a review on a project almost every day. Then we will go through a long discussion about which [experiments] look best scientifically, which ones have the most national importance, and I hope that what we'll end up doing is making a balance between the various needs we have. We will say, ok we'll spend a third of our resources on the fundamental physics issues, and a third of our resources on the direct national need.

[In preparing for the review] [y]ou do lots and lots of computer calculations. You try and anticipate every question that somebody's going to ask you. And you prepare computer graphs that will answer those things as best you can. You try and identify what are the weak parts of it. Usually there's physics in it that we don't know, otherwise we wouldn't be doing the experiment. You identify what you need to find out, need to measure. Oral briefings are the way we communicate. There's even a culture about the review, and people learn it by osmosis. You start out by presenting what's been done in the past that's relevant, and then you present your idea, everything you know about the physics, what you think the pitfalls are.

Question: How many hours a week does your average physicist work?

Answer: I'd say probably sixty. It varies. When you have a shot coming, a lot. The experimentalists are waiting for you to get them predictions so they can design the diagnostics to fit them, so then you'll be in there night and day keeping the computers going. You kind of ramp up and work and, after the experiment, usually people take a vacation.

As you get closer to shot-day, there are some critical times called "Freezes." First you freeze what the design of the experiment is like, and that's because you have to give people time to build, and so there's a frenzy of activity, and that's the last chance you have to decide what you want in the experiment. Then you'll freeze the diagnostics, the measurement devices, and that's another frenzy because that's when you've got to put it on the line and say "I'll measure 10 to the 22 gamma rays coming out at 30 degrees," and then they set their measuring devices around that number. As soon as they start building the diagnostic can you go out and crawl through it and you peer up at places where you're looking for anything that could possibly go wrong. You crawl all over it, frequently.

And then you reach the day where you go out to the test site, and there's this huge 200-foot canister filled with all this beautiful equipment and they're about to put it down. That's a real gut-wrencher.

That's it. You go through a period where you have a lot of doubts because the computer codes don't cover everything, and you have to use your judgment in a lot of places about what it's going to do physics-wise, and there's a lot of places where it could go wrong and you won't know. Most of our experiments don't come back with data as predicted. That's why we do the experiments. This is a hard period for the designer, especially the younger designers. If it's your first shot, it's really worrisome. Actually, if it's your second shot, it's really worrisome. I find that people believe the codes too much the first time. They always think that those big fancy computer codes can't be wrong.[63]

The narrative finishes with an invocation of one of the most important words in the weapons designer's lexicon: "judgment." The goal of apprenticeship, teamwork and experimental experience at the Laboratory is to enable weapons designers to acquire this elusive quality of judgment—what we might think of, to bend Evelyn Fox Keller's celebrated phrase, as a "feeling for the bomb." Judgment is demonstrated not simply in the ability to predict whether a design will work, which can be difficult enough if a weapon is designed to operate near what weapons designers call "the cliff"—the point at which a self-sustaining chain reaction fizzles. Judgment is also demonstrated in a more refined ability to predict the exact yield of a particular design and whether small changes in design or in fabrication materials will affect the yield of the weapon.

Donald MacKenzie and Graham Spinardi, in a brilliant and much-noted article on the possibility of uninventing nuclear weapons, speak of judgment as follows:

Judgment is the feel that experienced weapons designers have for what will work and what won't, for which aspects of the codes can be trusted and which can't, for the impact on the performance of a weapon of a host of contingencies, such as ambient temperature, aging of the weapon, and the vagaries of production processes. . . . According to our interviewees, the judgment goes beyond the explicit knowledge that is embodied in words, diagrams, equations, or computer programs. It rests upon knowledge that has not been, and perhaps could not be, codified. It is built up gradually, over the years, in constant engagement with theory, the codes, the practicalities of production, and the results of testing. Knowing what approximations to make when writing a code requires judgment, and some crucial phenomena simply cannot be expressed fully in the codes.[64]

The scientist quoted at length above believed that mature judgment could only be achieved by working on at least 15 nuclear tests. Another weapons scientist I asked about this, himself in the process of apprenticeship at the time, believed it would take 15 years. Donald MacKenzie and Graham Spinardi were told that it takes "five years to become useful" and ten years "to really train" a weapons designer.[65] The Los Alamos designer Jas Mercer-Smith told a journalist: "After five years you can do work without hurting yourself. After four or five shots a designer knows how to do his job. At fifty million a shot, that's a quarter of a billion dollars in training. . . . In the design group's apprenticeship-journeyman-master system, the hierarchy is based on years of successful testing. After eight years, I'd call myself a senior journeyman, or maybe a junior master. You don't get paid more or get a better office because you have brought off a test. You get the respect of your peers."[66]

A respected weapons designer's judgment was based on the patient accumulation through experience of tacit knowledge.[67] Tacit knowledge has been variously defined. Harry Collins, writing about the difficulties experienced by laser physicists trying to make their lasers work on the basis of formal descriptions and instructions alone, says "building up tacit understandings is not like learning items of information, but is more like learning a language or a skill."[68] Philippe Baumard likens tacit knowledge to the master chess player's instinct that enables him or her to know which moves to explore and describes this kind of knowledge as "something that we know but cannot express."[69] Similarly, Kathryn Henderson defines tacit knowledge as "knowledge that is not verbalized, sometimes because it is taken for granted but often because it is not verbalizable." This would include, for example, "a carpenter's knowledge of how to choose the appropriate nail for a particular kind of wood or the way humans normally recognize a face."[70] In their article on tacit knowledge and nuclear weapons design Donald MacKenzie and Graham Spinardi define tacit knowledge as "knowledge that has not been (and perhaps cannot be) formulated explicitly and, therefore, cannot effectively be stored or transferred entirely by impersonal means. Motor skills supply a set of paradigmatic examples of tacit knowledge in everyday life. Most of us, for example, know perfectly well how to ride a bicycle yet would find it impossible to put into words how we do so. There are (to our knowledge) no textbooks of bicycle riding, and when children are taught to ride, they are not given long lists of verbal or written instructions."[71]

Throughout the Cold War apprenticeship in the art of nuclear testing was the principal means through which this kind of tacit knowledge was transmitted and cultivated. In the period after the end of the Cold War, as nuclear testing slowed and then disappeared entirely, weapons designers and managers at the weapons laboratories have come to fear that, in Laura McNamara's words, "fifty years' worth of weapons-related knowledge might simply evaporate as experienced Cold Warriors retire from the laboratory."[72]

## Virtual Life: Pedagogy after Nuclear Testing

Immediately after the end of the Cold War, the pace of nuclear testing slowed and only tests for safety improvements to existing weapons were approved. Then, in September 1992, President George H. W. Bush signed into law a moratorium on nuclear testing that Senate supporters of a test ban had shrewdly attached to a bill funding the superconducting supercollider in Bush's home state of Texas—a state he believed would be crucial in that year's presidential election. At the time of writing the United States has conducted no nuclear test since that date, and it has signed but not ratified the Comprehensive Test Ban Treaty of 1996, negotiated by the Clinton Administration.

As I hope the previous section will have conveyed, the test ban ended a way of life at the weapons laboratories. Weapons designers came to understand nuclear weapons physics and engineering, trained the next generation of scientists, and made their reputations by conducting nuclear tests. In Laura McNamara's words: "The design and test cycle acted as an engine for the ongoing integration of expertise and the social reproduction of the weapons community; indeed, experimental activity was critical in organizing social relations among the hundreds of staff members involved in weapons work at Los Alamos."[73] Without nuclear tests the laboratories as organizations were in a very real sense adrift—unable to validate new weapons designs, unsure how to assure the continued reliability of old designs, and unclear how to train and test new weapons scientists. In these circumstances, morale plummeted at both weapons laboratories and many older scientists took early retirement. An early retirement drive in 1993, the third in four years, pushed out 9 percent of the staff at Livermore and 11 percent at Los Alamos. The total number of designers at Livermore and Los Alamos fell by about 50 percent in the decade after the end of the Cold War—a trend that left designers feeling that theirs was indeed a dying art.[74]

By the mid 1990s, a group of managers from both weapons laboratories, working together with senior officials from the Department of Energy and select members of Congress and their staffers, had devised a plan to replace nuclear testing at the Nevada Test Site, at least in the short to medium term, with a program of simulated testing called Science-Based Stockpile Stewardship distributed across the laboratories and other facilities in the nuclear complex and funded at almost $6 billion per year by 2002.[75] The central components of this program are the following:

**Subcritical tests** Underground tests at the Nevada Nuclear Test Site that use conventional explosives to shock small quantities of plutonium. The tests are called "subcritical" because the plutonium does not undergo a run-away chain reaction. These tests help scientists to refine their equations of state for plutonium and, in particular, understand its changing behavior as it ages.

**Dual-Axis Radiographic Hydrotest Facility (DARHT)** A Los Alamos facility to enable scientists to take x-ray snapshots of a nuclear primary made from a non-fissile surrogate for plutonium as it implodes, disintegrating at almost 6,000 miles per hour. These pictures enable scientists to peer inside the implosion of a pit to check its speed and evenness, to see how different material surfaces interact, and to trace the propagation of shock waves through the pit.

**ATLAS** A Los Alamos facility that discharges electricity stored in huge capacitor banks to create enormous magnetic fields that briefly create within the laboratory high energy-density regimes like those found inside stars. This helps scientists simulate the conditions inside a nuclear weapon as the imploding primary ignites fusion in the secondary.

**National Ignition Facility (NIF)** A 192-arm laser being built at the Livermore Laboratory. The laser will create within the laboratory temperatures and pressures higher than those in the sun by using the laser energy to fuse pellets of tritium and deuterium, thus enabling scientists to refine their modeling of fusion processes within hydrogen bombs.

**Accelerated Strategic Computing Initiative (ASCI)** A lavishly funded program to help the laboratories replace their old supercomputers with massively parallel computing systems with such power and speed that "all of the calculations used to develop the U.S. nuclear stockpile from the beginning could be completed in less than two minutes."[76] Codes run on the new computer systems will, finally, be three-dimensional, and will, in theory at least, enable scientists to integrate results from subcritical tests and experiments on NIF and DARHT with old nuclear test data.

There are three principal rationales for the Science-Based Stockpile Stewardship Program: to help diagnose problems in the aging nuclear arsenal and assure the efficacy of repairs; to refine the computer codes that model the behavior of nuclear weapons; and, finally, to provide a new means of nuclear pedagogy. Regarding the first rationale—diagnosing and fixing problems caused by aging—experiments at different facilities can shed light on performance changes as the materials in weapons age and are replaced. Subcritical tests, for example, can show whether the behavior of plutonium changes as helium bubbles form within the aging metal. Experiments on the DARHT can show whether the compressive behavior of conventional high explosive lenses changes as they age and whether substitution of one material in the lenses with another alters their ability to produce perfectly symmetrical implosion waves.

Second, in regard to the supercomputer codes, the plan is to transform the old two-dimensional codes used for weapons design in the 1980s into three-dimensional codes continually refined with data from subcritical tests and shots on DARHT, ATLAS, and the NIF that improve the modeling of particular phenomena within the explosion of a nuclear weapon. As one material scientist working on plutonium put it to me, "all my recent work has been an analysis of data published twenty to thirty years ago, applying correct chemical understanding and mechanisms, and modeling them on modern computers, showing, in fact, that we understand now what happens to chemical bases and then extending that."[77] The ultimate dream, as described by science journalist Dan Stober, is of a virtual reality Cave that allows a weapons designer to stand inside an exploding thermonuclear weapon and watch the explosion "in three dimensions on the walls, floors, and ceiling," with the ability to "zoom in on a piece of plutonium right down to the microscopic level. . . . A physicist might stand inside the cube wearing special 3-D glasses while images of his simulated weapon are projected on the walls of the cube. The view would change as the physicist turns his head or 'flies' with an electronic wand."[78]

The refinement of the codes (the largest of which are a million lines long)[79] and the development of much more powerful computer systems on which to run the codes will shift the balance of power between weapons designer and code, relocating some of the expert judgment thought to reside in the human designer's tacit knowledge to the formalized protocols of the computer codes. As with so many other expert systems, from airplane autopilots to accounting software, the result will be a partial deskilling of the human expert—here the weapons designer—and a fetishization of the authority of the code. If in the old days nuclear tests (whether they produced visible mushroom clouds or inky flickers on seismographs) were the embodiment of the efficacy and power of the U.S. nuclear deterrent, today it is the codes that, in a very real sense, signify the power and reliability of the U.S. nuclear deterrent. (It is only in this context, for example, that we can understand the extraordinary outburst of national hysteria in the United States over allegations in the late 1990s that the Los Alamos code developer Wen Ho Lee had transferred computer codes to China—as if he had somehow emailed the U.S. nuclear arsenal to a foreign power.)

There is, however, a limit to the transfer of a weapons designer's predictive ability to the codes, a limit that returns us to the issue of tacit knowledge and judgment. As one senior manager told Donald MacKenzie and Graham Spinardi, "the codes only explain 95% of physical phenomena at best, sometimes only 50 percent."[80] The codes are thought to be more reliable in predicting the behavior of "secondaries" than of "primaries" or "boosting" within a nuclear weapon, and they have limitations because of "fudge factors" or "knobs" deliberately introduced into the codes as a way of bridging the divide between the weapons scientists' imperfect theories and the empirical realities of nuclear testing. Dan Stober and Ian Hoffman explain it as follows: "In the desert a bomb might blow up with 20 percent less energy than the code had predicted. If the code writers or weapons designers understood the reason for the mistake, they would change the basic physics of the code. If not, they simply added a 'fudge factor' to the code, to make the computer prediction match the 'ground truth' of the actual nuclear explosion. The code jocks adjusted these 'knobs' in the software until the answers came out right, even if the underlying physics was not understood completely."[81] One designer told Laura McNamara: "You don't know exactly what's going on, but you've got a hunch—if you tweak a knob, the model fits the data better. You can't explain exactly why it fits—it's intuition."[82]

The inherent limitations of the codes underline the importance of the third rationale for stockpile stewardship: as a replacement for nuclear testing in the cultivation of judgment in a new generation of weapons scientists. Testifying to the Senate in 1996, Paul Robinson, director of Sandia, said: "Many of the systems in the stockpile will require replacement at about the same time at some point in the first half of the next century. The engineers and scientists who will do that work are probably entering kindergarten this year. No old-timers will be around in 2025 who have had actual experience in designing a warhead. We must find ways to qualify these people."[83] These new designers are being qualified through an adapted version of the old apprenticeship relationship and through newly formalized ways of archiving and transmitting nuclear weapons design knowledge. There is now a formal class in nuclear weapons design at Los Alamos called TITANS (Theoretical Institute for Thermonuclear and Nuclear Studies), described in the *Los Alamos Insider* as "a formalized training curriculum in nuclear weapons design and analysis"[84] in which experienced nuclear weapons designers take it in turns to lecture weapons designers in training. Meanwhile the nuclear laboratories have established a peer-reviewed classified journal to which young and old scientists are encouraged to submit articles. The laboratories are also asking older designers to archive their knowledge before they retire and are videotaping interviews with older designers in which they discuss everything from neutron cross-sections to the social customs at the Nevada Test Site.[85] Insofar as nuclear weapons design instruction is now more formalized, with knowledge transmission partially dislocated from apprenticeship relationships to classroom instruction and the consumption of archived articles and videos, pedagogy at the nuclear weapons laboratories increasingly resembles that practiced in the world of universities, with its heavily routinized conventions for knowledge consumption and production. At the same time, the old emphasis on intense dyadic apprenticeship relationships endures, albeit in a form adapted to a world without testing. A scientist at Los Alamos explained to me how this system now works:

An experienced designer would show the person he's mentoring where you start. That is, you've got some specs from DoD [the Department of Defense]. They've got a mission that needs an 80-kiloton warhead that can't weigh more than 23 kilograms or whatever. This could be an actual mission from real life or a hypothetical one. They could take a new guy through the design process that is already in place for one of the weapons in the stockpile, show them what the mission was and how they do the scoping study to see if the mission was possible; and if the scoping study showed that it wasn't possible, how they went back to the Department of Defense and said "Look, you're going to give us 25 kilograms, because we can't do it in 23.. . ." Now you've got the rough outline and you've got to fill in the details and take the guy through it. It takes a tremendous amount of time, basically as much time as he probably spent designing it in the first place to take the guy through. And the guy probably would have to go through a couple of times before he would be comfortable doing the same exercise on his own, without the mentor telling him "No, that's a blind alley. Don't do that, do this." It wouldn't be full time because the designer's got other things to do, but he'll be spending a significant fraction of his time for years or more mentoring a new person.[86]

One older scientist at Los Alamos worried that this kind of mentorship experience could not replicate the intensity of the old mentorship relations structured around nuclear testing. "There is nothing that has that level of intensity and excitement, nothing that you pour your heart and soul into. In terms of the intensity of a nuclear test, I'd literally worked days straight without sleep designing a lot of my devices and they had to pull the design out from under my hands in time to go manufacture it for the test. And literally you'd go design it over a period of time and you were lucky if you got six hours of sleep on any night for that entire period. But you had a sense of accomplishment when you were done."[87] New designers now are encouraged to propose subcritical experiments or shots on DARHT, ATLAS, or NIF that will enable them to make predictions and get experimental feedback. These experiments will not only enable them to develop judgment, but will also enable laboratory managers to determine which of the new designers have the best judgment when it comes to deciding in the future who can certify the continuing reliability of the U.S. nuclear stockpile. A senior manager at Livermore put it this way:

At a fundamental level the way you evaluate people is the same; that is, you set up situations where people can succeed or fail. In the past, those were things like a nuclear test. You go out and do a nuclear test and make some predictions. If it works, you're a hero; if it doesn't work, your career ends. . . . The same thing is occurring now but, instead of judging on the basis of nuclear tests, you're having to judge on the basis of other kinds of experiments. Ultimately you can only succeed or fail at an experiment. . . . And so, with the National Ignition Facility, for instance, you will be able to judge whether or not the people doing the experiments are good by virtue of whether they succeed.[88]

A number of older weapons designers have told me that the power of the U.S. nuclear deterrent resides not only in the thousands of nuclear weapons that have been built and stockpiled, ready to be used against an adversary with only a few moments' notice, but also in the perceived expert judgment of U.S. weapons designers who must guarantee that these weapons really work as advertised.[89] In the days of atmospheric and underground nuclear testing, the judgment of American weapons designers was displayed, both to the community of weapons scientists and to foreign adversaries, in elegantly designed and successfully executed nuclear tests. An intricate and intense program of apprenticeship structured around the preparation of nuclear tests enabled older designers to transmit their tacit knowledge accumulated over a lifetime—the elusive quality called "judgment"— to the younger scientists who would replace them. Now nuclear weapons scientists, young and old, worry whether it is really possible to cultivate this kind of judgment in the world of simulations, and wonder how the acquisition of this judgment might be advertised to foreign adversaries to signal that the American nuclear arsenal is still robust. Formulating the notion of "capability-based deterrence," Steve Younger, until recently Associate Director for Nuclear Weapons Design at Los Alamos, had suggested that the traditionally reclusive Los Alamos scientists, serving as human embodiments of the national nuclear deterrent, should be more active on the international conference and publication circuit, publicly displaying to other national expert communities their excellence in unclassified fields of knowledge related to weapons science—a suggestion he stopped making after the Wen Ho Lee case exacerbated public concerns about loose stewardship of classified information at Los Alamos.[90]

Other designers worry that their managers sold them out by consenting to the replacement of nuclear testing with Science-Based Stockpile Stewardship and that the gulf between simulated and real testing is too wide for pedagogy and the cultivation of judgment in a new generation of designers to be truly possible.[91] Bob Barker, a former designer and Associate Director at Livermore, testified as follows to Congress in 1997:

> As a nuclear weapons designer I learned the limitations of simulations and the humility that comes with the failure of a nuclear test. Computer calculations, regardless of how good or fast the computer is, are only as good as the data and models you give them and the knowledge and experience of the person doing the calculations. Even today no computers are big enough or fast enough to simulate all that goes on when a nuclear weapon explodes. The true knowledge of and experience with the limitations of calculations came from understanding the differences between calculations and experiments, including nuclear tests.[92]

Another older designer told me: "Good judgment comes from experience and experience comes from bad judgment. Now we're going to have people out there with no experience."[93] Another older designer told me that novice weapons designers relying on their codes reminded him of drunk drivers: the drunker they were, the better they thought they were driving.

## Conclusion

The world of TITANS classes, billion-dollar lasers, and teraflop computer simulations in which nuclear weapons designers can physically stand as they seek to inherit half a century of nuclear weapons science is far removed from the 1940s and 1950s world of men in their twenties and thirties in charge of design divisions trying to understand for the first time the fundamental processes of atomic and hydrogen bombs. At the outset of this essay I referred to a "pedagogy of diminishing returns" at the weapons laboratories. Paradoxically, the more elaborate, drawn-out, and formalized nuclear weapons pedagogy has become over the years, and the more extensive the laboratories' understanding of nuclear weapons science has become, the more problematic pedagogy has become. If a first generation of young men making it up as they went along created great breakthroughs in nuclear weapons design, their successors in the era of underground testing took much longer to qualify themselves as experts, spent a lifetime acquiring a much more extensive body of knowledge and yet, in a very real sense, achieved much less. And now today we have a new generation of aspiring designers looking into the abyss between simulation and experiment as their mentors wonder whether it is possible for the community of designers to simply jog in place or whether, despite the expenditure of unprecedented sums of money on nuclear pedagogy, their art and science is in the process of being irretrievably eroded.

There is an obvious sense in which this involutionary momentum is the result of political developments beyond the weapons scientists' control. The end of the Cold War, followed by the international prohibition of nuclear tests, deprived the weapons laboratories of their most important experimental practice and their principal way of training new weapons designers: nuclear testing. Some weapons designers have argued that, if only nuclear testing were to resume, then a renaissance of nuclear weapons science would ensue. The reality seems to me, however, to be more complicated. After the weapons scientists' failure in the 1980s to perfect third-generation nuclear weapons, weapons scientists largely stopped talking about major new breakthroughs in their art. In the 1990s those who argued for further nuclear testing did not argue that major advances in nuclear weapons science would be possible. (Indeed it was the turn to simulations forced by the end of nuclear testing that produced the boldest innovations in the practice, and the rhetoric, of nuclear weapons science.) Instead those who opposed the end of testing argued that continued testing was necessary to ensure the reliability of old weapons as they decayed or, if they did refer to new weapons, made modest claims: they suggested that it would be possible to develop a bunker-busting mini-nuke or a sort of Maytag nuclear weapon—what designers call a "wooden bomb"—optimized to age well in the absence of testing in the future. In other words, even without the nuclear test ban, nuclear weapons science had by the end of the Cold War hit a sort of wall. It is not that weapons scientists had no ideas for improving their weapons—the reverse was true—but the refinements had become so incremental and the ratio of effort and expense to scientific advancement in the performance of nuclear testing had shifted so far that nuclear weapons science was becoming like, say, Gothic architecture at the end of the nineteenth century: an increasingly repetitious and involuted practice that was losing its energizing edge and its appeal to the best and the brightest. The Oppenheimers and Tellers had given way to smart career physicists who saw themselves largely as custodians of a settled body of knowledge. The end of nuclear testing has disrupted the established and time-tested pedagogical practices for the transmission of that settled knowledge to another generation of the nuclear priesthood but it may be that, by shaking things up, the virtual turn offers this community its last best hope of intellectual revitalization.

## Acknowledgments

My thanks go to all the participants in the two workshops at MIT, organized by David Kaiser, where this essay was first presented—especially to Mike Lynch, Kathy Olesko, and David Kaiser. My thanks also to Cyrus Mody, who took the lead in organizing a further presentation of these ideas to a joint conference of the anthropology and science studies departments at Cornell, where I received further helpful comments. Bruce Tarter, David Dearborn, and John Krige were also kind enough to give the essay close readings, despite their busy schedules. Thanks also to Allison Macfarlane for pushing me to think about the broader implications of the paper.

## Notes

1. The concept of "involution" is borrowed from Clifford Geertz, *Agricultural Involution: The Process of Ecological Change in Indonesia* (University of California Press, 1963).

2. Bruno Latour and Steve Woolgar, *Laboratory Life: The Social Construction of Scientific Facts* (Sage, 1979); Bruno Latour, *Science in Action: How to Follow Scientists and Engineers Through Society* (Harvard University Press, 1987); Bruno Latour, "Give me a laboratory and I will raise the world," in *Science Observed*, ed. Karin Knorr-Cetina and Michael Mulkay (Sage, 1983), 141–170; Thomas Kuhn, *The Structure of Scientific Revolutions* (University of Chicago Press, 1962).

3. The B53—an older model of hydrogen bomb with a very high explosive yield.

4. The B61—a newer, lower-yield hydrogen bomb designed at Los Alamos and recently modified to give it an earth-penetrating capability. The new version is the B61 Mod 11.

5. This song, named for the veteran Livermore designer Dan Patterson, is called the "St. Patterson Test Site Blu's." My thanks to the Livermore A Division Men's Chorus for providing the lyrics.

6. This story is told in Richard Rhodes, *The Making of the Atomic Bomb* (Simon and Schuster, 1988); Gregg Herken, *Brotherhood of the Bomb: The Tangled Lives and Loyalties of Robert Oppenheimer, Ernest Lawrence, and Edward Teller* (Henry Holt, 2002); Lillian Hoddeson et al., *Critical Assembly: A Technical History of Los Alamos During the Oppenheimer Years, 1943–1945* (Cambridge University Press, 1993). More eccentric accounts are by Robert Jungk, *Brighter Than a Thousand Sons: A Personal History of the Atomic Scientists* (Penguin, 1960), and Brian Easlea, *Fathering the Unthinkable: Masculinity, Scientists and the Arms Race* (Pluto, 1983). See also Jon Else's fine documentary film *The Day after Trinity* (KTEH-TV, 1980).

7. See Richard Rhodes, *Dark Sun: The Making of the Hydrogen Bomb* (Simon and Schuster, 1995), and, for an account of the political struggles around the scientific breakthrough, Herbert York, *The Advisers: Oppenheimer, Teller, and the Superbomb* (Freeman, 1976).

8. See Harvey Sapolsky, *The Polaris System Development: Bureaucratic and Programmatic Success in Government* (Harvard University Press, 1972).

9. Dan Stober and Ian Hoffman, *A Convenient Spy: Wen Ho Lee and the Politics of Nuclear Espionage* (Simon and Schuster, 2001), 36–44.

10. For works that feature several of these figures, see Herken, *Brotherhood of the Bomb*, Else, *The Day after Trinity*, and S. S. Schweber, *In the Shadow of the Bomb: Oppenheimer, Bethe and the Moral Responsibility of the Scientist* (Princeton University Press, 2000). On Oppenheimer, see J. Robert Oppenheimer et al. (eds.), *Robert Oppenheimer: Letters and Recollections* (Stanford University Press, 1995); Jeremy Bernstein, *Oppenheimer: Portrait of an Enigma* (Ivan R. Dee, 2004); Peter Goodchild, *J. Robert Oppenheimer: Shatterer of Worlds* (Houghton Mifflin, 1981). On Teller, see Teller's own autobiographies: Edward Teller, *The Legacy of Hiroshima* (Doubleday, 1962) and (with Judith Shoolery) *Memoirs: A Twentieth-Century Journey in Science and Politics* (Perseus, 2002)—as well as Stanley Blumberg and Louis Panos, *Edward Teller: Giant of the Golden Age of Physics* (Scribner, 1990); Broad, *Teller's War*; Peter Goodchild, *Edward Teller: The Real Dr. Strangelove* (Harvard University Press, 2004). Ulam's account of his own life is given in Stanislaw Ulam, *Adventures of a Mathematician* (University of California Press, 1991). Von Neumann's life is evoked by Norman McCrae, *John von Neumann* (Pantheon, 1992). For York's own account of his life and work, see Herb York, *Making Weapons, Talking Peace: A Physicist's Odyssey from Hiroshima to Geneva* (Basic Books, 1987). Ted Taylor's story is told by John McPhee, *The Curve of Binding Energy* (Ballantine, 1974) and George Dyson, *Project Orion: The True Story of the Atomic Spaceship* (Henry Holt, 2002).

11. On the x-ray laser debacle at Livermore, the definitive source is the reporting of William Broad, a science journalist for the *New York Times*. His first book on the subject has been criticized as too credulous: see William Broad, *Star Warriors: A Penetrating Look Into the Lives of the Young*

*Scientists Behind Our Space Age Weaponry* (Simon and Schuster, 1985). His second book, more than adequately compensating the blind spots of the first, is the definitive history of the rise and fall of this program at Livermore and in Washington: William Broad, *Teller's War: The Top Secret Story Behind the Star Wars Deception* (Simon and Schuster, 1992). See also William Broad, "Crown jewel of 'star wars' has lost its luster," *New York Times*, February 13, 1990; Deborah Blum, "Weird science: Livermore's x-ray laser flap," *Bulletin of the Atomic Scientists* 44 (1988), no. 6: 7–13; Robert Scheer, "The man who blew the whistle on Star Wars," *Los Angeles Times Magazine*, July 17, 1988.

12. A partial account of these safety improvements is given in Sidney Drell, "How safe is safe?" *Bulletin of the Atomic Scientists* 47 (1991), no. 3: 35–40; Jeffrey Smith, "America's arsenal of nuclear time bombs," *Washington Post National Weekly Edition*, May 28–June 3, 1990.

13. One scientist at Livermore once told me that he was never allowed to know why one of his colleagues had won the Lawrence Award, the most prestigious award within the DOE weapons complex.

14. Another anthropologist who studies Los Alamos, Joseph Masco, has written his own account of the political and emotional phenomenology of nuclear testing organized around the same tripartite chronological schema. See Joseph Masco, "Nuclear Technoaesthetics: Sensory Politics from Trinity to the Virtual Bomb in Los Alamos," *American Ethnologist* 31 (2004), no. 3: 1–25.

15. See Hugh Gusterson, *Nuclear Rites: A Weapons Laboratory at the End of the Cold War* (University of California Press, 1996); idem, *People of the Bomb: Portraits of America's Nuclear Complex* (University of Minnesota Press, 2004); idem, "Los Alamos: A Summer under Siege," *Bulletin of Atomic Scientists* 55 (1999), June: 36–41; idem, "Exploding anthropology's canon in the world of the bomb: Ethnographic writing on militarism," *Journal of Contemporary Ethnography* 22 (1993): 59–79.

16. On the experience at Los Alamos of what anthropologists would call "communitas"—a sense of community unobstructed by roles and hierarchies—see Jon Else, *The Day after Trinity*.

17. Rhodes, *The Making of the Atomic Bomb*, 454–455.

18. Ibid., 460–464.

19. The ages of the Manhattan Project scientists are taken from ibid., 118, 188, 112, and 460, and from Dyson, *Project Orion*, 29. Ulam's quote is from Ulam, *Adventures of a Mathematician*, 156 (quoted in Dyson, ibid., 22).

20. Sybil Francis, Warhead Politics: Livermore and the Competitive System of Nuclear Weapon Design (PhD dissertation, MIT, 1995), 72. By 1955 the Livermore Laboratory had only 300 physicists on its payroll out of a total staff of 1,633 (ibid., 105).

21. Herbert York, *Making Weapons*, 74. See pp. 72–73 for a discussion of the ages of his managers at the Laboratory when it first opened its doors. This initial cohort at Livermore was to play a vital role in shaping the laboratory as it evolved, since the six directors who led the laboratory through its first 35 years were all hired that first year. They are Herbert York, John Foster, Edward Teller, Harold Brown, Michael May, and Roger Batzel.

22. Francis, Warhead Politics, 63.

23. York, *Making Weapons*, 67.

24. Ibid., 74–75.

25. Until 1957 even the town of Los Alamos, not just the laboratory, was closed to the public (Jo Ann Shroyer, *Secret Mesa: Inside the Los Alamos National Laboratory* (Wiley, 1998), 7–8).

26. Interview with retired Livermore weapons designer, February 11, 1989.

27. Quoted in Dyson, *Project Orion*, 51.

28. Darol Froman to Alfred D. Starbird, May 4, 1959, quoted in Francis, Warhead Politics, 146.

29. In gun-assembly weapons two pieces of fissile material, machined so that they can mate with one another, are fired into one another at high speed within a cylinder to create a critical mass. In implosion devices a spherical (or later ovoid) piece of plutonium, roughly the size of a grapefruit, is symmetrically compressed by surrounding high explosives into a smaller, denser sphere that achieves critical mass.

30. For more detailed accounts of these advances in nuclear weapons design, see Rhodes, *The Making of the Atomic Bomb*; Rhodes, *Dark Sun*; Thomas Cochran, William Arkin and Milton Hoenig, *The Nuclear Weapons Databook: U.S. Nuclear Forces and Capabilities* (Ballinger, 1984); Sam Cohen, *Shame: Confessions of the Father of the Neutron Bomb* (Xlibris 2000); Necah Stewart Furman, *Sandia National Laboratories: The Postwar Decade* (University of New Mexico Press, 1990); Barton Hacker, *Elements of Controversy: The Atomic Energy Commission and Radiation Safety in Nuclear Weapons Testing, 1947–1974* (University of California Press, 1994); Chuck Hansen, *U.S. Nuclear Weapons: The Secret History* (Orion, 1988); Donald MacKenzie and Graham Spinardi, "Tacit knowledge, weapons design, and the uninvention of nuclear weapons," *American Journal of Sociology* 101 (1995): 44–99; McPhee, *The Curve of Binding Energy*; Howard Morland, "The H-bomb secret: To know how is to ask why," *The Progressive* 43 (1979), November: 14–45; Robert Serber, *The Los Alamos Primer: The First Lectures on How to Build an Atomic Bomb*, ed. Richard Rhodes (University of California Press, 1992); Edward Teller, *The Legacy of Hiroshima* (Doubleday, 1962). See Dyson, *Project Orion*, for a detailed history of the attempt to create a spaceship, powered by hundreds of nuclear explosions, for inter-planetary exploration and for nuclear warfighting. On the program to use nuclear explosions for civilian excavation, mining and drilling—including a plan to excavate the Panama Canal with nuclear weapons—see Dan O'Neill, *The Firecracker Boys* (St. Martin's Griffin, 1994) and the documentary film by Gary Marcuse, *Nuclear Dynamite* (Bullfrog Films, 2001).

31. Francis, Warhead Politics, 143–144.

32. Donald MacKenzie and Graham Spinardi ("Tacit knowledge," 59) make a similar point, though with a slightly different periodization, saying that "by the 1980s designing nuclear weapons had lost much of its flavor of virtuoso innovation and had become a more routine task: one, indeed, that some in the laboratories feel to have become bureaucratized, unchallenging, even 'dull.'"

33. Francis, Warhead Politics, 89.

34. Ibid., 64–65.

35. Stober and Hoffman, *A Convenient Spy*, 38.

36. Francis, Warhead Politics, 67.

37. Ibid., 115.

38. Ibid., chapter 4.

39. Ibid., chapter 7. Francis quotes marginalia scribbled by a Los Alamos scientist on the Livermore proposal: "material in newspapers like this generally carries in fine print at the bottom of the page (Advt)." (ibid., 123.)

40. Ibid., 140. The test was code-named Rainier. Los Alamos was concerned that underground testing would be too expensive and difficult to instrument, and that proof of the possibility of underground testing would make it more likely that the U.S. government would agree to a treaty banning atmospheric testing.

41. Herbert York, "The origins of the Lawrence Livermore Laboratory," *Bulletin of the Atomic Scientists* 31 (1975), July: 8–14.

42. Francis, *Warhead Politics*, 134; George Miller, Paul Brown and Carol Alonso, Report to Congress on Stockpile Reliability, Weapon Remanufacture, and the Role of Nuclear Testing (Lawrence Livermore National Laboratory, document no. UCRL-53822); Ray Kidder, Maintaining the U.S. Stockpile of Nuclear Weapons During a Low-Threshold or Comprehensive Test Ban (Lawrence Livermore National Laboratory, document no. UCRL-53820). The last round of tests before the 1958 moratorium had established that one of the Livermore Polaris designs was not inherently one-point safe: in other words, an accidental nuclear detonation of the weapon was possible if, for example, it was struck by a bullet. Rather than replace the Livermore primary in the weapon with a Los Alamos primary, Teller, Livermore's director at the time, elected to use a mechanical safing mechanism.

43. See Broad, *Star Warriors* and *Teller's War*; Scheer, "The man who blew the whistle on Star Wars."

44. James Glanz, "Panel faults laser architect for overruns," *New York Times*, January 11, 2000; idem, "Laser project is delayed and over budget," *New York Times*, August 19, 2000; "Top U.S. laser expert admits lack of a PhD and resigns," *New York Times*, August 31, 1999; David Perlman, "Test lab called $1 billion over budget," *San Francisco Chronicle*, May 9, 2001. The ostensible reason for the resignation of Mike Campbell, Associate Director for lasers, was the public revelation that he had claimed to have a doctorate from Princeton despite never having finished his PhD thesis, but most observers agreed that the timing of this revelation was tied to the increasingly evident financial and organizational problems of the National Ignition Facility, a project directed by Campbell.

45. Quoted in Dyson, *Project Orion*, 32.

46. Senate Policy Committee, Berkeley Division of the Academic Senate, University of California, The University of California, the Lawrence Livermore National Laboratory, and the Los Alamos National Laboratory (unpublished paper, 1984); Gusterson, *Nuclear Rites*, 25.

47. Interview with retired weapons designer at Lawrence Livermore National Laboratory, July 22, 1990.

48. See note 12. This is not a fanciful concern: there have been incidents in which American planes carrying nuclear weapons have caught fire on the runway, have dropped their weapons accidentally, and have collided with one another in mid-air. In at least one of these accidents, in Greenland, there was a conventional explosion that dispersed plutonium over a wide area. In another, a bomb was lost off the coast of Georgia and has never been found. For more on these accidents, the definitive source is Scott Sagan, *The Limits of Safety: Organizations, Accidents, and Nuclear Weapons* (Princeton University Press, 1993). See also Smith, "America's Arsenal"; Drell, "How safe is Safe?"; Hugh Gusterson, "Nuclear Weapons and the Other in the Western Imagination," *Cultural Anthropology* 14 (1999): 111–143.

49. Because female nuclear weapons designers are comparatively rare at Livermore and Los Alamos, describing a weapons designer as female makes it much easier to identify her and, thus, undermines the convention of anonymity in contemporary ethnography. Consequently I have over the years developed the practice of representing all weapons designers as male, regardless of their actual gender, unless it is important to the analysis to identify them as female. I continue that practice in this essay.

50. Interview with Livermore weapons designer, January 26, 1989.

51. Interview with Los Alamos weapons designer, February 11, 1992.

52. Interview with retired Livermore weapons designer, February 11, 1989.

53. At the Livermore Laboratory, B Division is responsible for the design of "primaries"—fission devices used either in tactical weapons or as triggers to ignite a larger thermonuclear explosion. A Division is responsible for the "secondaries" whose explosive yield is derived from thermonuclear fusion.

54. Interview with Livermore primary designer, January 26, 1989.

55. Interview with Livermore weapons designer, January 26, 1989.

56. Interview with Livermore manager, June 8, 1989.

57. Interview with Livermore weapons designer, January 9, 1988.

58. Interview with Livermore weapons designer, January 26, 1989.

59. Quotation from draft of Laura McNamara, "Ways of Knowing" about Weapons: The Cold War's End at Los Alamos National Laboratory (PhD dissertation, University of New Mexico, 2001), 114–115.

60. Ibid., 145–146.

61. The narrative here makes it sound as if weapons designers go straight from computer calculations and presentations to review committees to $50 million nuclear tests. As Laura McNamara points out (131–132), the reality is more complicated: "Nuclear tests were not the first stage of

empirical validation, although they were certainly the most dramatic. Rather, the validation process often began with a less expensive, less risky *hydrodynamic test*, a high explosive experiment that would provide a limited empirical benchmark for the designer's model. Often referred to as 'local shots' because they were conducted at Los Alamos proper, hydrodynamic experiments approximated full-scale nuclear tests insofar as they tested a mock-up of the design under development. Compared to a full-scale nuclear test, doing a local shot during the Cold War was a relatively simple process that moved from concept to test fairly rapidly, within a few months. . . . Local shots used no nuclear materials, just high explosives and inert components, and returned only data about implosion dynamics: e.g., how would a particular part move and change as the high explosive detonated?"

62. The person in charge of designing a device for a test was referred to as the Design Physicist (DP), and weapons designers traditionally built their reputations by being DP for devices they had designed and seeing their tests go well. In the idealized narrative above, our hypothetical designer has an exciting idea and is therefore put in charge as DP. In practice, by the 1980s, shots were designed by teams and the contributions of individual designers were harder to disentangle. After one conversation with a Los Alamos secondary designer I made the following entry in my notes: "They often make younger physicists DP now to give them experience. In the old days there were more tests, and people got more chance to truly do their own tests and be DPs of projects they'd initiated and controlled. These days, with a different design culture and less tests, the tests tend to be based more on teamwork and the decision who gets to be DP is based more on social and political factors: in part, they try to rotate the privilege and the responsibility. He [the designer] sees warhead design becoming more like plane or rocket design, where the product is so much a team effort that you're not sure who did what." On the increasing dissolution of the individual author or inventor in physics, see Peter Galison, "The Collective Author," in Mario Biagioli and Peter Galison (eds.), *Scientific Authorship: Credit and Intellectual Property in Science* (Routledge, 2003), 325–355.

63. Interview with Livermore manager, June 8, 1989.

64. MacKenzie and Spinardi, "Tacit knowledge," 62.

65. Ibid.

66. Janet Bailey, *The Good Servant: Making Peace with the Bomb at Los Alamos* (Simon and Schuster, 1995), 38, 81. The discussion here focuses on the cultivation of judgment among the elite of weapons designers at the national laboratories. Laura McNamara's work, however, makes clear that the success of nuclear tests in the Cold War also relied heavily on the judgment and tacit knowledge of experienced engineers and technicians—personnel whose contributions have received less recognition.

67. On tacit knowledge, see Philippe Baumard, *Tacit Knowledge in Organizations* (Sage, 1999); H. M. Collins, *Changing Order: Replication and Induction in Scientific Practice* (Sage, 1985); H. M. Collins, "The TEA set: Tacit knowledge and scientific networks," *Social Studies of Science* 4 (1974): 165–187; Kathryn Henderson, *On Line and On Paper: Visual Representations, Visual Culture, and Computer Graphics in Design Engineering* (MIT Press, 1999); M. Polanyi, *The Tacit Dimension* (Doubleday, 1966); Ludwig Wittgenstein, *Philosophical Investigations* (Blackwell, 1953).

68. Collins, "The TEA set," 168.

69. Baumard, *Tacit Knowledge in Organizations*, 2.

70. Henderson, *On Line*, 6.

71. MacKenzie and Spinardi, "Tacit knowledge," 45.

72. McNamara, "Ways of Knowing," 112

73. Ibid., chapter 114. See also Gusterson, *Nuclear Rites*, chapter 6.

74. Jonathan Medalia, Nuclear Dilemmas: Nonproliferation Treaty, Comprehensive Test Ban, and Stockpile Stewardship (CRS Report for Congress, Order Code 94-1007F 1994); interview with Livermore Associate Director, March 28, 1997. Donald MacKenzie and Graham Spinardi ("Tacit knowledge," 62–63) put the number of "good designers" in the U.S. at "about 50" at the end of the Cold War, while observing that one Livermore manager put the number even lower, at "about 40."

75. For an overview of this program, see Drell et al., Science-Based Stockpile Stewardship, JSR-94-345 (Mitre, 1994).

76. Randy Christensen, "Computer simulations in support of national security," *Science and Technology Review*, April 1998, available at www.llnl.gov/str (accessed May 21, 2004).

77. Interview with Los Alamos materials scientist, June 22, 1998.

78. Dan Stober, "Byting the big bang," *San Jose Mercury News*, August 11, 1998.

79. MacKenzie and Spinardi, "Tacit knowledge," 59.

80. The manager in question is Art Hudgins, quoted in MacKenzie and Spinardi, "Tacit knowledge," 60.

81. Stober and Hoffman, *A Convenient Spy*, 36.

82. McNamara, "Ways of Knowing," chapter 3, 16.

83. Michael Moore, "Wanted: Wee weaponeers," *Bulletin of the Atomic Scientists* 52 (1996), April: 8–9, quotation on 9.

84. Quoted in Matthew McKinzie, Tom Cochran, and Chris Paine, *Explosive Alliances: Nuclear Weapons Simulation Research at American Universities* (Natural Resources Defense Council, 1998), 76. McKinzie et al. quote *The Los Alamos Insider* as saying that "the students' enthusiasm is a real source of inspiration. Frequently, after class, there are a half-dozen students that remain in the classroom to ask follow-up questions, and one often hears them discussing homework problems in the hallways."

85. See McNamara, "Ways of Knowing."

86. Interview with Los Alamos computational physicist, June 23, 1998.

87. Interview with Los Alamos weapons designer, January 16, 1999.

88. Interview with Livermore manager, March 28, 1997.

89. One weapons designer, being perhaps more blasé than his colleagues would approve, told me about deterrence: ". . . in the end all you're saying is that we have a lot of really smart guys who worked really hard on these things" (conversation with Sandia weapons scientist, October 28, 1997).

90. Keith Easthouse, "Battle lines," *The New Mexican*, May 5, 1998.

91. Wen Ho Lee, commenting that "the science of nuclear weapons hasn't progressed much" since the end of the Cold War, has said of Science-Based Stockpile Stewardship that it is "like eating leftovers for dinner, [but] it's better than nothing." Quoted in Walter Pincus, "Los Alamos scientist criticizes FBI in book: Lee calls copied tapes 'crown junk,'" *Washington Post*, January 16, 2002.

92. Robert Barker, Prepared testimony before Senate Governmental Affairs Committee, Subcommittee on International Security, Proliferation, and Federal Services, October 27, 1997.

93. Interview with Los Alamos weapons designer, January 16, 1999.

# II Pedagogical Cultures in Collision

# 4 Fear, Shunning, and Valuelessness: Controversy over the Use of "Cambridge" Mathematics in Late Victorian Electro-Technology

**Graeme Gooday**

Cambridge sets the fashion in mathematics. No one but a Cambridge man is supposed to be capable of anything in mathematics. A few Dublin men have had a glimmering of knowledge, but an Oxford graduate is in outer darkness, and an unfortunate who has read by himself and never been to any University at all is doomed to eternal ignorance. In spite of this we hope that someone may write a book on practical mathematics which contains just subjects which are necessary for engineers, and nothing more. We do not mean a book of simplified explanations. They are delusive. . . .
—letter from a Leeds "Electrician" to *Industries,* April 1891[1]

The electrical engineer must not be only an overgrown wireman, a mechanical engineer with a little electrical knowledge, a mathematician, a financier, a lacquered brass and sealing-wax varnish instrument maker, a physicist, or a manager of men. He must be all of these, in different proportions in different men.
—James Swinburne, presidential address to Institution of Electrical Engineers, 1902[2]

Historians now routinely examine technoscientific controversies to map past sociocultural practices that are otherwise difficult to recover. My concern is with controversies that enable us to recover the ways in which *training* was central to articulating a new multifaceted discipline and defining the proper credentials of its practitioners. Across Europe, America, and Japan in the 1880s and the 1890s, physicists, mathematicians, electricians, telegraph engineers, mechanical engineers, and civil engineers all tried their hand in a new field designated as "electrical engineering." Practitioners of each group sought to annex this exciting and potentially lucrative new technological domain—whether in direct current (d.c.) or alternating current (a.c.) forms—as an extension of their own expertise. Yet the project of developing electrical light and power posed enormous theoretical, technical, practical, commercial, and political challenges that no single existing group had comprehensive skills to handle. Most especially there was prolonged disagreement about whose training was best suited for handling the unprecedentedly complex phenomena of a.c. technology—especially in understanding the effects of the electromagnetic parameter of "self-induction."

I focus on a sometimes heated argument about how to theorize the performance of one important but recalcitrant electromagnetic technology: the a.c. generator commonly known as an alternator. John Hopkinson's controversial 1883–84 theory of alternators was articulated using techniques he had learned as a Mathematics undergraduate at Cambridge from 1867 to 1871, and was thus intelligible only to those with a university-level training in mathematics. Nevertheless, engineers trained in the technical college and workshop could at least challenge both Hopkinson's idealistic analogical reliance on the mathematics of purely mechanical systems and the practical inferences he drew from treating self-induction as if it were as unvarying as a flywheel's "moment of inertia." This sort of treatment was second nature to a Cambridge graduate inculcated into seeking neat analytical equations within a mechanistic paradigm, but "practical" engineers more familiar with the operations of a.c. machinery than Hopkinson saw this simplifying assumption as entirely unjustified and liable to result in error. Preferring to use more trustworthy if less concise graphical methods to analyze alternator performance, these self-styled "practical men" often disparaged what they more generally saw as delusive and thus valueless equations produced by Cambridge-trained "mathematicians." As I show below, this divergence of educationally grounded practice could produce polemical results. I discuss how Hopkinson's 1894 lecture to the Institution of Civil Engineers on "the relations of mathematics to engineering" prompted criticism that led Hopkinson to suggest that too many engineers unduly "feared and shunned" the powerful techniques of Cambridge analysis.

The particularity of this dispute to the credentials of *Cambridge* mathematics and especially to training in the new field of electrical engineering is rather significant. After all, few doubted the cogency of Cambridge mathematics training for developing late Victorian theories of electromagnetism and thermodynamics.[3] Then again, Hopkinson's disagreement with the "practical men" is also clearly particular both to his own idiosyncratic mode of deploying Cambridge mathematical techniques beyond the syllabus of the university's Tripos examinations and his early reluctance to heed advice from "practical" engineers. I show that James E. H. Gordon, a contemporary of Hopkinson trained in Cambridge mathematics, was more amenable to artisanal wisdom and was thus able with his engineering workshop assistants to develop *practical* forms of a.c. technology exactly contemporaneously with Hopkinson's disputed theorizing. Examining Gordon's case enables us to see how Cambridge graduates had at least as much to learn about electrical engineering from artisan-engineers as vice versa. Indeed I argue that once constructive dialogue was established in the following decade, a new pedagogy for mathematizing electricity was synthesized from both graphical and Cambridge traditions in the discipline—one product of which featured importantly in Richard Feynman's early career.

Given the broader significance of this story, historians should see the central issue as more than just the credibility of John Hopkinson's idiosyncratic form of Cambridge theorizing. Nor need they acquiesce in contemporary characterization of opposing positions simply in terms of the *training* of individuals involved. Rather we might consider other factors that were at stake: gender (competing masculine identities), socioeconomic class (middle-class graduates versus industrial artisans), institutional power and professional credibility. In a sense education was a crucial forum for defining or redefining all these issues in late-nineteenth-century British culture. Since the early 1870s, new forms of primary, secondary, and tertiary education enabled broader competition for diversifying opportunities in an industrial society ever less constrained by traditional boundaries of social class and power.[4] Amongst all this change, one lingering traditional peak of the British educational system was the Cambridge Mathematics Tripos, which had produced many elite scientists, engineers, clergy, politicians and civil servants throughout the nineteenth century.

As Andrew Warwick has explained, the top performer in each year's Mathematics Tripos examination—the Senior Wrangler—was held in especially high public esteem, publicly celebrated in newspapers and accorded intellectual supremacy and respect enjoyed by few other contemporaries. And in this regard we should note that John Hopkinson's prima facie authority in the debate discussed below was rooted in his publicly acknowledged status as Senior Wrangler of 1871.[5] Yet as Hopkinson found to his discomfiture, especially in 1894, his educationally defined high status was questioned by an ever more confident electrical engineering community. This was defining its own new elites and prerogatives to the extent that it could challenge a Cantabrigian authority speaking on matters pertaining to their expertise in a.c. technology. Defining their identity in ways that directly opposed and limited the writ of a Cambridge Senior Wrangler was, I suggest, an effective means for electrical engineers to lessen the social power that Hopkinson and other Senior Wranglers had so freely (and hitherto uncontentiously) drawn from their Cambridge pedigree.

In two senses, then, this chapter is a corollary to Warwick's definitive historical study of the Cambridge Mathematics Tripos, *Masters of Theory: Cambridge and the Rise of Mathematical Physics*. I comment both on the demise of the now long-vanished fetishism of Senior Wrangler authority and on the previously undocumented challenges that Cambridge Wranglers faced when unreflexively deploying their university-bred practices to new contexts of enquiry.[6]

## Practice-Laden Mathematics: Cambridge Tripos vs. Engineering Graphics

Historians of electrical technology have noted that two major disputes arose in the late 1880s concerning the significance of self-induction in technological theorizing. Rather than considering broad socio-economic or educational explanations to account for the origin of these conflicts, enquiry has focused on the outcome of these encounters as the application of abstract Maxwellian "Theory" to defeat particular forms of engineering "Practice." Accordingly much attention has been paid to the ways in which a group of "Maxwellians" drew upon Maxwell's *Treatise on Electricity and Magnetism* of 1873 and adapted it to resolve problems in both physics and electrical engineering. Although John Hopkinson latterly drew selectively upon some aspects of Maxwell's work, we should by contrast see his predominant practice as epitomizing pre-Maxwellian Cambridge mathematical physics—not as propounding Maxwellian electromagnetic theory.[7]

Bruce Hunt's account of one such debate from 1888 to 1891 considers how Maxwellian theorists discredited claims by chief Post Office Electrician, William Preece, that a lightning conductor's performance was determined primarily by its electrical resistance. Maxwellians argued instead for the primacy of self-induction in governing how lightning strikes were discharged.[8] Without explicitly using the "practice" vs. "theory" dichotomy, Dominic Jordan explains similarly how, between 1886 and 1889, the same Maxwellians had appealed to the theory of self-induction to explain and resolve distortion problems in telephone lines. Both accounts take as their explanandum the defeat of Preece's view that self-induction was a "choking" effect that ought to be expunged completely from both technologies for effective working.[9] In both cases the engineering problem was solved by, among others, Oliver Heaviside, Oliver Lodge, and Silvanus Thompson using Maxwellian theory to stipulate an *optimal* rather than minimal value for self-induction.

Hunt and Jordan make few direct allusions to educational matters in explaining the anatomy of these debates. This is prima facie reasonable since none of the three key figures mentioned above acquired their Maxwellian approach through formal education: Lodge and Thompson both learned practical physics from the distinctly non-Maxwellian Frederick Guthrie in London, and Heaviside was famously an autodidact. Indeed, as Hunt and Buchwald have shown, early Maxwellian commitments were generally developed outside of formal educational experience.[10] Moreover, in cases where educational traditions did inform electrical engineers' adoption of Maxwellianism, it reflected the relative autonomy of "practical" and "Cambridge" sub-communities discussed above. For example, Ron Kline has shown that working engineers in the United

States developed effective informal "practical" theories for the alternating-current (a.c.) induction motor that diverged considerably from the Maxwellian theoretical canon. This exasperated the Cambridge-trained Serbo-American Michael Pupin who clearly expected deference from less privileged engineers to his particular Maxwellian version of electrical engineering theory.[11] From Kline's study it would seem that the historian cannot identify any single form of pedagogy to account for electrical engineers' development of efficacious theories and practices in early a.c. technologies.

Looking deeper, however, we find that some important educational issues were operative in framing the debate. Ido Yavetz has pointed out that late Victorian electrical engineers' talk about an opposition between "practice" and "theory" should be seen as a *partisan representation* of a complex socio-professional tension—not a literal fact.[12] He notes that it was Preece who rhetorically fashioned the lightning conductor debate using the "practice vs. theory" dichotomy to articulate his opposition as man of "established practice" to the Maxwellians' controversial theory of lightning conductor design.[13] Yavetz pinpoints the context-specificity of this labeling by observing that Preece invoked no such dichotomy in the telephony debate discussed by Jordan.[14] What *was* common to both these cases, however, was the molding of Preece's response by his prior training as a telegraphic electrician. His assumption that self-induction was of negligible importance (or even to be expunged) was borrowed from William Thomson's 1850s canonical modeling of telegraph signal transmission solely in terms of capacitance and resistance. Thus we might reinterpret controversies in which Preece was embroiled as focusing not so much on the virtues of Maxwellian "theory" per se but on the viability of extending the learned telegraphic exemplar to other areas of technological design. And as far as the other side in the controversies on self-induction is concerned, Silvanus Thompson and Oliver Lodge were both institutional professors—at Finsbury Technical College London, and University College Liverpool, respectively—committed to using their training programs to inculcate new approaches to the theoretical analysis of electrical technology. Pedagogy was thus a major consideration in the longer term ramifications of the "Practice vs. Theory" debate to the extent that it was the institutionalized means of extending successful strategies for theorizing machines to future generations of practitioners.

In his studies of nineteenth-century mathematical physics, Andrew Warwick has persuasively shown that such processes of theorizing were "practice-laden" in character and specific to localized regimes of training. In other words he subverts the dichotomy of "theory vs. practice" by looking in a Wittgensteinian vein at the practice *of* theory.[15] According to Warwick's education-centered account, practitioners' theorizing activities are shaped by a set of well-rehearsed problem-solving *strategies* and specific techniques learned as students to bring a disciplined and conventionalized mathematical order to bear on material situations. To that extent Warwick sympathizes with the well-known Kuhnian view that the formation of disciplinary identity hinges rather crucially on learning from common exemplars in a discipline's educational canon. Yet whereas Kuhn's account focuses on globally current paradigms, Warwick emphasizes that the characteristic strategies and techniques of theoretical practice are learned in ways *localized* to a particular educational institution, and indeed to a particular curricular scheme within that institution. He shows compellingly how difficult it was for those at other locations (German universities and British schools) who had not taken the Mathematics Tripos at the University of Cambridge to solve the characteristic problems set by its examiners—even when presented with complete specimen solutions to such problems.[16]

In further contrast to Kuhn's early account in *The Structure of Scientific Revolutions,* Warwick does not describe the long-term significance of education as the deep and deterministic inculcation of physical ontologies. After all, Kuhn's own work showed that in later stages of their careers scientists could and often did completely reject metaphysical commitments acquired in their early training.[17] By contrast, practitioners were much less inclined to abandon the higher-order strategies for problem solving they had learned in undergraduate or graduate study. Indeed Kuhn's more considered view of education emphasized its role as familiarizing students with puzzle-solving exemplars that they could use as heuristics to deal with unfamiliar problem situations. In this vein, Warwick argues that intensive mathematical training by specialist coaches for the Cambridge Mathematics Tripos equipped them with powerful techniques—algebra, trigonometry, calculus and differential equations—for extending exemplary problem-solutions to new situations.[18] He establishes this important point by showing how the later-nineteenth-century Mathematics Tripos nurtured a very characteristic kind of research publication among those of its graduates that went on to academic careers—especially those coached by Edward Routh. For example, Warwick compares the Tripos papers of John Henry Poynting and his first paper, and identifies a standard strategy to both—one that was uniquely characteristic of a Cambridge graduate.[19]

A particular feature of Cambridge-style analysis noted by Warwick was the very restricted form for what could count as a solution to a Tripos problem. A crucial desideratum for such Cambridge examinees was to achieve answers that were not only elegant and concise, but above all *analytic,* that is, closed-form algebraic expressions. Following the pattern of exemplar solutions in classical mechanics, neither an infinite series nor an overtly approximate solution would do. Tripos examiners were thus expected to set Tripos problems that had neat—if unworldly—solutions, avoiding topics that did not lend themselves to such analytic answers. As Warwick indicates, candidates were expected to make appropriate approximations so that a problem could be made to fit the pattern of learned exemplars and then solved to yield the required form of answer. Such a use of approximations to accomplish standard practice was obviously *not* dictated by the physics of the problem situations, but rather the conventions of examination setting local to the University of Cambridge.[20]

Indeed, to understand the Mathematics Tripos we should consider it as a conventionalized expression of the university's pedagogical goals to instill in its students the intellectually rigorous discipline of arriving at elegant concise solutions to abstract problems. Given the strongly self-referential nature of the Cambridge Mathematics Tripos, we cannot infer that the Tripos examiners intended or expected problem-solving strategies nurtured by it to be extended directly to extrinsic practical contexts. As we shall see, however, this was precisely what John Hopkinson did so controversially on several occasions in the 1880s and the 1890s, even going so far as to solve problems in published papers with the characteristic Tripos "rider"—specialized solutions to variants on the original problem. As will be explored further below, Hopkinson using Tripos problem-solving methods, produced equations for the performance of electromagnetic machinery precisely of the form required by Tripos examiners—over a decade after he had ceased to be a Cambridge undergraduate.

At a few points in his presentation of his Wranglerish alternator theory to fellow engineers, Hopkinson did try to adopt the techniques of graphical analysis with which the vast majority would have been more familiar. This was most obvious on the very first airing of his theory at the Institution of Civil Engineers in April 1883. Yet on that occasion his quick reversion to Wrangler tricks incomprehensible to most fellow engineers, and his condescension to them in public debates, certainly did not win their deference or enthusiasm. Few ordinary electrical engineers were likely ever to have the time or opportunity to master the simultaneous differential equations used by Hopkinson and his Tripos ilk from 1884 onwards in articulating a.c. theory.

To bridge the obvious gap between the Cambridge and graphical traditions of electrical engineering, a handful of a.c. specialists set out soon afterwards to translate Hopkinson's a.c. theory into the more familiar Euclidian idiom of graphical analysis. One was fellow Cambridge graduate Thomas Blakesley (34th Wrangler, 1869) who, after a civil engineering apprenticeship, became a freelance a.c. consultant and then Lecturer in Mathematics and Physics at the Royal Naval College in 1885. Blakesley's popular handbook *Papers on Alternating Currents of Electricity for Students and Engineers* (1889) grew out of articles written for *The Electrician,* the Physical Society of London, and the *Philosophical Magazine* since 1885. Exemplifying the use of the "geometrical method" to treat a.c. problems Blakesley devoted his opening chapter to the vexing topic of self-induction. This chapter set out to augment the elementary reader's presumed knowledge of how electromotive force was governed by resistance with an analysis of how self-induction was equally important in determining electromotive forces in alternating current circuits. For a sinusoidal form of a.c oscillation, these effects could be represented as the rotation of a straight line representing electromotive force around a point (only later referred to as a vector), the length and direction of that line being determined by the effects of self-induction and resistance. Using this geometrical technique, Blakesley painstakingly took his reader through the notion that self-induction in an a.c. circuit produces a time-lag between cyclical changes in current and potential difference. Blakesley next instructed his readers how to calculate the phase angle (as it was later known) of this retardation.[21]

The next major author seeking to bring transparency and accessibility of graphical methods to a.c. engineering was Gisbert Kapp. A mechanical engineering graduate of the Zurich polytechnic, Kapp took up British citizenship in 1875, becoming a consultant engineer and editor of *Industries* and the *Elektrotechnische Zeitschrift*. From 1887 he began to publish a series of articles for *The Electrician* titled "Induction coils graphically treated." Kapp's use of a clock diagram (see figure 4.1) drew on Blakesley's practice of using rotating vectors to represent the cyclical variation of, and inductive lag between, electro-motive force (e.m.f.) and current in an a.c. circuit.

As Ron Kline notes, even when such geometrical methods were extended to Kapp's more sophisticated clock-face diagrams, they could not give the generality of analytical techniques as adopted, for example, by John Hopkinson. Even so engineers could use trigonometric tables or a ruler and compass on a carefully constructed clock diagram to establish what they needed to know for practical work: the basic phase and magnitude parameters of a steady-state electrical circuit.[22] If engineers could get by with Blakesley's and Kapp's practical shortcuts there was no obvious benefit for busy engineers to use more technical and time-consuming mathematics just to achieve a higher level of rigor or abstraction.

As we shall see, the longer-term successful take-up of the clock-diagram technique in modeling electromagnetic machines was no mere representational cul-de-sac nor second-rate solution. Hopkinson's dogmatic insistence on the use of Wranglerish jargon, opaque derivations, and mystifying equations only reinforced the impression of ordinary engineers that Cambridge mathematics had little to offer them in advancing their understanding of a.c. technology. Overall, then, my story challenges prevailing historiographical assumptions that those educated in "advanced" mathematical techniques necessarily determine the development of techno-scientific disciplines.

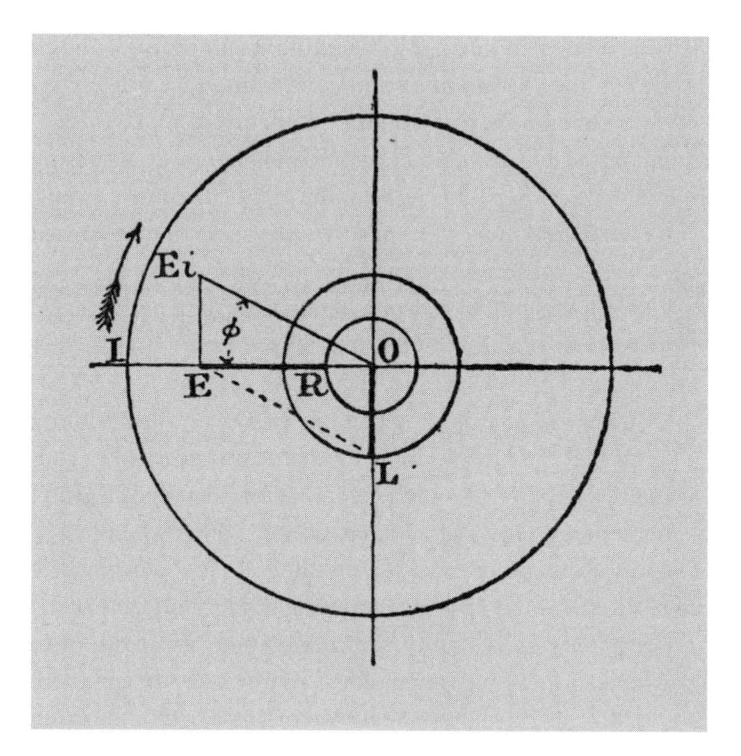


Figure 4.1
An early example of a (pre-vector) "clockface" diagram for representing the cyclical variation of—and phase displacements between—the parameters of an alternate current circuit. The maximum current is represented by the line *OI*, rotating anticlockwise so that the projection of *OI* on the vertical axis gives the actual value of current at a given moment. Similarly the line *OE* represents imposed potential difference (with similar projection onto horizontal axis), *OR* the potential difference lost through resistance, and *OL* the self-inductive element, always perpendicular to current. The parallelogram defined by the lines *OE* and *OL* gives the resultant magnitude of the induced electromotive force *OEi* and the phase angle $\phi$ between potential difference and current in the circuit. Source: G. Kapp, "Alternate current machinery," *Minutes of Proceedings of the Institution of Civil Engineers* 97 (1888–89): 1–79, figure 7 (on 10).

## Paralleling Alternators and Hopkinson's Unconvincing "Language of Equations"

From the advent of electrical lighting in the early 1880s, many problems faced those who sought to harness the behavior of electrical machinery to both financial gain and personal or communal prestige. One central dispute was whether the future lay in d.c. or a.c. technology, especially as the latter offered better business returns for larger urban scales and long-distance supply systems. Yet while d.c. technology proved relatively tractable to mechanical engineers expert in handling steam engines, and to telegraphists (such as Edison) familiar with simple electrical circuits, generating alternating current posed challenges beyond the training and experience of all. In particular while d.c. dynamos could fairly readily be connected in parallel to meet increasing consumer demand on cold winter evenings, this was perplexingly difficult to accomplish for a.c. power—at least not without huge current surges that disastrously burned out the lamps to which they were connected. Although his magisterial survey covers almost every other aspect of the "Battle of the Systems" between a.c. and d.c. from ca. 1885 to 1893, Thomas Hughes does not discuss the critical problem of paralleling alternators.[23]

John Hopkinson's 1883–84 theory of alternator operation has been hailed both by historians and by his son Bertram as definitively showing that it was possible to connect alternators in parallel and derive power from the combination, thereby (allegedly) establishing the practical and commercial viability of a.c. power.[24] Yet few contemporaries readily accepted Hopkinson's theory as proving anything at all about practical engineering; on the contrary, the seemingly false conclusion of his account gave many strong grounds for questioning the very relevance of Cambridge-style theorizing to understanding the effective operation of a.c. technology. At the time it appeared, Hopkinson's theory could not, after all, offer any explanation of one crucial and highly inconvenient empirical fact: no commercially produced alternators could actually be made to run in parallel.[25] This is especially surprising because Hopkinson, as a freelance consultant engineer in 1879, had been among the very first to promote the *empirical* study of Edison's electrical lighting and generating technology at the Institute of Mechanical Engineers.[26] Specifically, in that year Hopkinson developed what was soon known as the "characteristic curve" of a dynamo—a graph plotting the non-linear relationship between current generated and the potential difference between the terminals. This became an invaluable graphical technique in comparing the performance of dynamos for which Hopkinson was thereafter renowned—even when the rest of his theoretical work was under attack for its characteristically Cantabrigian equation-centered approach.

In "Some points in electrical lighting," a lecture given to the Institution of Civil Engineers (ICE) in April 1883, Hopkinson presented a highly a priori model of parallel alternator operation.[27] As a consultant to the English Edison Company, Hopkinson was a d.c. specialist with little experience of a.c. technologies. Without any reference to practical examples of alternator operations, Hopkinson turned immediately to the case of two linked alternators. Taking the conventional view that self-induction was the most important parameter, and drawing on his Wrangler practice of using mechanical analogies to map unfamiliar situations, he likened an alternator's self-induction to the moment of inertia of a rotating physical body. The analogy was viable to the extent that changing the speed of rotation of a spinning body was comparably difficult to changing the current through a metal conductor—in both cases an inertial "reaction" appeared to oppose any imposed change. Hopkinson used this Cambridge analogy to claim that the mathematics was of the same form for mechanical and electrical interpretations—notably that self-induction was as constant a parameter as the moment of inertia. He moved quickly, with typical Wrangler facility, to five characteristic equations, declining to "trouble" his audience with the "simple" derivations involved. He declared that these showed self-induction to be as important as resistance in determining the maximum current in an a.c. circuit and thus the work that the alternators could do—central conclusions that his audience could only accept on his Wrangler authority. Importantly, though, Hopkinson did not use such algebraic legerdemain to resolve the question of whether paralleling was possible. He instead presented phase diagrams of two alternators with voltage varying sinusoidally and examined the outcomes with staged verbal narrative. These showed that whereas a series connection would tend to produce a motor-generator pair doing no net work, a parallel-synchronized combination could in fact do useful work together.[28] Thus, he implied, it was possible for alternators to be coupled usefully together.

In November of the following year, Hopkinson presented a more developed "theory of alternating currents, particularly in reference to two alternate-current machines connected to the same circuit" to the Society of Telegraph Engineers and Electricians (STEE, which also met in the ICE building).[29] Here we see the force of Warwick's point about the pedagogical influence of Cambridge training on Wrangler research presentations (as instanced in the case of Poynting discussed above). For this highly formalized paper, Hopkinson laid out his arguments somewhat in the format of a Mathematics Tripos examination paper, "illustrating" the subject as a series of seven discrete problems. He started with the simplest cases of series and parallel connection in which the self-induction was conveniently assumed to be a constant and mutual induction effects taken to be zero. (See figure 4.2.) He then progressed to trickier variants—like the "riders" in a Tripos problem—of interesting and analytically soluble problems, albeit not closely connected to the main subject of the paper, such as the behavior of an arc lamp connected to an alternator.

Most significantly, what was common to his solutions of all seven problems was the characteristically Cantabrigian representation of the algebraic formalism as concerning "equations of motion"—as if analyzing coupled alternators with self-induction and resistance were self-evidently analogous to coupled frictional oscillators in dynamics.

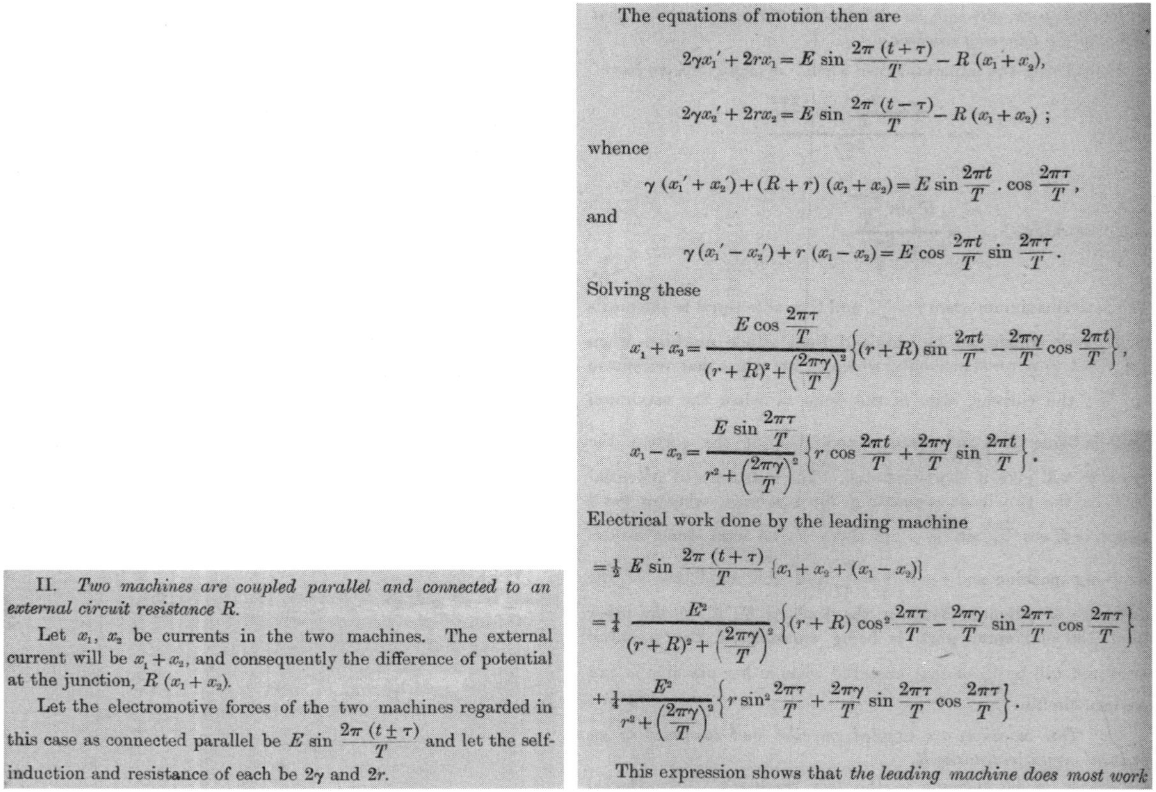

II. *Two machines are coupled parallel and connected to an external circuit resistance R.*

Let $x_1$, $x_2$ be currents in the two machines. The external current will be $x_1 + x_2$, and consequently the difference of potential at the junction, $R\,(x_1 + x_2)$.

Let the electromotive forces of the two machines regarded in this case as connected parallel be $E \sin \frac{2\pi\,(t \pm \tau)}{T}$ and let the self-induction and resistance of each be $2\gamma$ and $2r$.

The equations of motion then are

$$2\gamma x_1' + 2rx_1 = E \sin \frac{2\pi\,(t+\tau)}{T} - R\,(x_1 + x_2),$$

$$2\gamma x_2' + 2rx_2 = E \sin \frac{2\pi\,(t-\tau)}{T} - R\,(x_1 + x_2)\ ;$$

whence

$$\gamma\,(x_1' + x_2') + (R + r)\,(x_1 + x_2) = E \sin \frac{2\pi t}{T} \cdot \cos \frac{2\pi\tau}{T},$$

and

$$\gamma\,(x_1' - x_2') + r\,(x_1 - x_2) = E \cos \frac{2\pi t}{T} \sin \frac{2\pi\tau}{T}.$$

Solving these

$$x_1 + x_2 = \frac{E \cos \frac{2\pi\tau}{T}}{(r+R)^2 + \left(\frac{2\pi\gamma}{T}\right)^2} \left\{ (r+R) \sin \frac{2\pi t}{T} - \frac{2\pi\gamma}{T} \cos \frac{2\pi t}{T} \right\},$$

$$x_1 - x_2 = \frac{E \sin \frac{2\pi\tau}{T}}{r^2 + \left(\frac{2\pi\gamma}{T}\right)^2} \left\{ r \cos \frac{2\pi t}{T} + \frac{2\pi\gamma}{T} \sin \frac{2\pi t}{T} \right\}.$$

Electrical work done by the leading machine

$$= \tfrac{1}{2} E \sin \frac{2\pi\,(t+\tau)}{T} \{x_1 + x_2 + (x_1 - x_2)\}$$

$$= \tfrac{1}{4} \frac{E^2}{(r+R)^2 + \left(\frac{2\pi\gamma}{T}\right)^2} \left\{ (r+R) \cos^2 \frac{2\pi\tau}{T} - \frac{2\pi\gamma}{T} \sin \frac{2\pi\tau}{T} \cos \frac{2\pi\tau}{T} \right\}$$

$$+ \tfrac{1}{4} \frac{E^2}{r^2 + \left(\frac{2\pi\gamma}{T}\right)^2} \left\{ r \sin^2 \frac{2\pi\tau}{T} + \frac{2\pi\gamma}{T} \sin \frac{2\pi\tau}{T} \cos \frac{2\pi\tau}{T} \right\}.$$

This expression shows that *the leading machine does most work*

**Figure 4.2**

In characteristic Cambridge Mathematics Tripos fashion, John Hopkinson treated the parallel coupling of alternators as a problem in dynamics governed by "equations of motion." From this he concluded that paralleling alternators should be possible. Source: John Hopkinson, "The theory of alternating currents, particularly in reference to two alternate current machines connected to the same circuit," *Journal of the Society of Telegraph Engineers and Electricians* 13 (1884): 496–515, on 503–504.

As was typical of Wrangler practice, this strategy allowed Hopkinson at least to formulate the algebraic equations with facile dexterity. For alternators in series he equated the sum of the inductive and resistive components of circuit electro-motive force to the sum of the impressed e.m.f.; for the parallel case he argued similarly with regard to each branch of the circuit. From the latter Hopkinson showed that one alternator will always do more work than the other (in contrast to the series case), leaving readers to infer from this that it should be possible to connect alternators in parallel. This was very surprising news for the electrical engineers who, since his ICE lecture the previous year, had persistently found it impossible to do so in practice.[30]

In making these extraordinary claims, Hopkinson proposed two simplifying assumptions akin to those he had made in his previous year's presentation to the ICE. These were that the alternator's coefficient of self-induction was a constant (which he admitted to be "not exactly" the case) and that there was negligible mutual induction in the circuit (an equally convenient approximation).[31] Hopkinson acknowledged that these were questionable assumptions and that they would have been difficult to defend on the basis of Maxwellian electromagnetic theory; but—as he said some years later—without these simplifying approximations his equations would have been "unmanageable."[32] The manageability of the equations was clearly a central issue for Hopkinson to establish neat analytic solutions as embedded within the conventions of Wrangler protocol. Moreover, to acknowledge the variability of self-induction and of mutual induction as variable would have required Hopkinson to include machine-dependent expressions for these in his equations—making his solutions uninterestingly contingent on the particular construction of a particular alternator. Thus, to accomplish "universal" solutions of a generalized analytic form, Hopkinson followed a Wrangler's *theoretical practice* and not rigorous Maxwellian theory.[33]

Several expert members of his audience not trained in Cambridge mathematics—notably William Ayrton and Silvanus Thompson—found much to challenge in Hopkinson's Wranglerish simplifications concerning self-induction and mutual induction.[34] More importantly for present purposes, at least one young engineer in the audience, Llewellyn Atkinson, was skeptical about the utility of Hopkinson's "language" of equations. Atkinson later recalled hearing Hopkinson's paper as a 19-year-old trainee while studying in Ayrton's Finsbury Technical College laboratory:

> He treated [the problem] simply as the equation of motion of two moving bodies with certain forces between them—it might almost as well have been the sun and the moon. Nobody believed him, because the only alternators at that time that had any practical vogue were the Siemens alternators, and everyone knew it was practically impossible to run them in parallel. Until it had been proved in practice, this language of equations convinced nobody.[35]

Thus, it was daily familiarity with the complexity of alternators—rather than "ignorance" of theory—that made it easy for Atkinson to doubt the ability of highly self-referential Wrangler analysis to capture anything useful about the world of electromagnetic machinery. Indeed, Cambridge-trained mathematicians had much to learn from such practical men to ensure that their mathematical virtuosity engaged the material world of electrical engineering.

Later sections of this essay will cover the ways in which the opacity and disputed efficacy of Cambridge mathematical techniques became the focal point of heated discussion in the engineering community—not least in Hopkinson's own defense of Cambridge mathematical techniques as uniquely cogent in engineering work. To put that episode into perspective, however, we should not see any inevitability in the conflict between Cambridge learning and other sorts of practical engineering wisdom. In the next section I survey the work of Hopkinson's Cambridge-trained contemporary, James Gordon, who showed in 1884–1886 how university learning could be—and indeed had to be—harnessed to artisanal expertise for success in practical a.c. engineering.

## J. E. H. Gordon and the Limits of Cambridge Training in a.c. Engineering

> The public who through long winter evenings, and longer London fogs, sit reading by the cool and steady light of their electric lamps, but who are most indignant if by any chance it flickers or fails them, do not realize how intense the struggle has been for those pioneers of electric lighting who have toiled so hard and incessantly to surprise yet one more of Nature's secrets. . . . Many an engineer's wife knows how common it was four or five years ago for their husbands, who had come back late at night worn out and exhausted, to be fetched again by the message that there was "something wrong at the works."
>
> —Mrs. J. E. H. Gordon, *Decorative Electricity,* 1891[36]

Studying James Gordon's work enables us to rebut any fallacious deterministic inference that a Cambridge training in mathematics necessarily led an electrical engineer into confrontation with practical men. The importance of Gordon's case is not just that he was able to abandon the idealizing practices inculcated by the Cambridge Mathematics Tripos and learn to adapt to the recalcitrant conditions of practical engineering in ways not recognized in the formulation of Cambridge Tripos examination questions. More significantly, the contemporary engineering community also recognized, in implicit contrast to Hopkinson, that it was possible for a Cambridge-trained mathematician to make a systematically positive contribution to electrical engineering—despite (rather than because of) their high level of educational attainment. Most tellingly of all, the theories of Hopkinson were of no value to his fellow Cambridge graduate in his practical attempts to handle a.c. technology.

James Edward Henry Gordon was educated in the Cambridge Mathematics Tripos tradition, graduating with rather less distinction than Hopkinson as Junior Optime (3rd class) in 1875 before researching and publishing on electricity and magnetism for three years under James Clerk Maxwell in the Cavendish Laboratory.[37] After some years spent as a textbook writer and an independent researcher, with his wife Alice as chief collaborator, he became the Manager of the Electric Light Department of the Telegraph Construction and Maintenance Company (Telcon) in late 1882. Gordon's work for this company in managing the installation of the first British public a.c. supply for lighting at Paddington railway station (West London) between 1884 and 1886 brought him a minor place in the professional folklore of electrical engineers.[38] In this unprecedentedly grand a.c. project, Gordon found no help in Hopkinson's writing, hardly attempting to run alternators in parallel as the latter claimed ought to be possible. Moreover, Gordon found himself heavily reliant on the practical skills of mechanical engineers to supplement what he had learned in the tutorial supervisions and Cavendish Laboratory exercises of 1870s Cambridge.[39]

On June 22, 1883, Gordon read a paper to the Applied Chemistry and Physics section of the Society of Arts on his plans for a large Telcon lighting station. His main concern was with the huge practical problems to be overcome in providing customers with lighting that was "perfectly steady, perfectly safe and reliable." This was rather harder to accomplish in the contingent chaos of the world outside than in the conveniently controlled small-scale conditions of the scholar's laboratory. Tellingly, he made no appeal to the artificiality of Tripos mathematics to simplify the problem by idealizing what ought to happen if only the British public were truly deferential to his academic expertise:

> You must regard the engines, dynamos, mains etc. all as one complete system, and you must remember that if anything goes wrong, the public will not make excuses for the lamps like we may make in the laboratory. They will not say that it was not the dynamo's fault, but [rather that] it was the engine's fault or the boiler's fault. All that the public would be concerned with was that on this or that occasion the light went out.

Gordon acknowledged that perfecting such light required the coordination of a wide range of skills that were "not often united in the same person." He candidly admitted of his own status as a Cambridge-trained scholar that an "electrician" (meaning an all-round expert in electrical matters) was "very often not a first class engineer." Only by combining the requisite expertise from a team of assistants could the heterogeneous operations of an a.c. system be melded together and anything like perfection achieved in electro-technology.[40] Gordon related with striking candor his painful discovery of the insufficiency of his own skills and the necessity of supportive labor to "perfect" the a.c. technology of lighting. Over the previous three years he had tried, with the help of friends, to build his first industrial-size alternator, but encountered innumerable problems of a constructional nature. His first high-speed model was abandoned when its uncontrollable vibrations produced a "deafening roar in the dynamo room." After working well for a few hours, his second high-speed dynamo shuddered and eventually "flew to pieces with a loud explosion." Sympathetic and wealthy friends donated thousands of pounds for a third project to produce an unprecedentedly large but safely low-speed alternator. The Chief Engineer of Telcon secured for Gordon the crucial assistance of the company's mechanical engineers. After months of labor during 1882, Gordon's team completed the machine only to find that, despite a robust input of steam power, it refused to move. Only after several more days and nights of virtually constant toiling and tinkering by Telcon engineers would it produce any current at all.

With his team of Telcon assistants, Gordon conducted his first major trial with the new alternator at 9 o'clock one Monday evening in 1883 at Telcon's factory in Greenwich. After the 1,200 lights produced a "blaze of daylight" without a hitch, Gordon was appointed the Manager of Telcon's electrical light department. Gordon nevertheless remained explicitly aware of his dependence in mechanical matters on the skills of expert technicians—especially his chief assistant from 1883 to 1893, Frank Bailey. Originally trained in mechanical engineering, Bailey was employed in mineworks and locomotive construction in Northern England before picking up a knowledge of electricity in evening classes at the City of London College ca. 1882. Bailey's engineering skills in mechanical construction and human management were crucial to the successful execution of Gordon's grandiose installation plans. By the late 1880s, Bailey had acquired a reputation for overcoming "most of the difficulties inseparable from electrical lighting at that time."[41]

When site testing began at Paddington on the three alternators to be used (not in parallel) to power the station, many socio-technical problems arose. These challenged Gordon's ingenuity in ways that went far beyond any difficulty with apparatus that he might have encountered as a researcher in the Cavendish Laboratory. Frank Bailey's later reminiscences, and those of his own chief assistant, A. H. Walton, reveal that Gordon's monstrous devices failed to behave in situ as well as they had in Telcon's laboratories. Overheated armature cores had to be dismantled and redesigned with improved cooling facilities. Equally uncongenial was the irrepressible tendency of shunting locomotives to spray boiling water onto the gutta percha insulation of electrical power cables. One case of insulation breakdown was so serious that the Great West Railway's Board of Directors thought Gordon's plan was essentially unworkable; only by "very hard" pleading on Gordon's behalf did Bailey manage to avert abandonment of the whole scheme.[42] Needless to say, Bailey's case did not appeal to the authority of the 1871 Senior

Wrangler, John Hopkinson, that such a scheme *ought* to work in principle. Similarly, when local residents threatened legal action against the noise of the Paddington Station generators in November 1885, it was the practical skill of Bailey's team in soundproofing that prevented the shutdown of the new a.c. power system.[43]

After the installation began full-scale operations in April 1886, *The Electrician* lavished praise on Gordon as a university-trained scholar. Unusually he had not merely been an inventive "genius," but had also learned to be an effective manager of practical a.c. projects:

> Mr. Gordon must be congratulated upon the success attained. First known to the world as an earnest student and a successful writer, he has now shown the world that he can carry out his ideas into practice.[44]

Yet this leading electrical trade journal did not attribute Gordon's success solely to his Cambridge pedigree: Gordon had been "ably assisted" by Bailey in accomplishing success at Paddington. While the significance and nature of this assistance would probably have been obvious to most readers of *The Electrician*, the journal tactfully passed over the details. Only 35 years later, when Bailey was himself a senior figure in the field, did he gain the opportunity to offer publicly his account of the Paddington scheme's practical success. Bailey contended that it was only by hard continuous labor that he and his team had been able to convert the threatened failure into a practical success. While he paid tribute to Gordon for his "remarkable mathematical abilities, his inventive genius, and the charm of his personality," Bailey diplomatically passed no comment on Gordon's mastery of practical engineering.[45] Also contributing to retrospective discussion at the half-century celebrations of the Institution of Electrical Engineers (IEE), Bailey's assistant A. H. Walton addressed the subject of "Mr. J. E. H. Gordon's system of electric supply." Attaching no direct significance to Gordon's educational pedigree in accounting for success, Walton focused on the many "anxious times" that arose even after Paddington was operational in April 1886, recalling that his team seldom worked fewer than 12–15 hours a day.[46] Although at the time largely unreported by a sympathetic electrical press, these difficulties were manifestly obvious to Alice Gordon, as seen in the epigraph above. In the last chapter of her 1891 book *Decorative Electricity*, she observed how often her husband was called back late at night to advise on problems at Paddington. Yet, at the same time, "Mrs. J. E. H. Gordon" subtly ensured that her readers knew how much *she* had contributed to the Paddington scheme by acting as loyal assistant to her spouse—all without the university education that her husband had enjoyed.[47] Thus we see the perspectival nature of judgments of the Telcon success at Paddington, various participants claiming, with widely differing educational credentials, that their contribution was a necessary (if tacitly insufficient) component of the project.

In none of the diverse accounts of how Paddington was established as Britain's first public a.c. system in 1886 was any mention made of John Hopkinson's mathematical alternator theory, developed two years earlier. Gordon had initially tried the paralleling of alternators, but then abandoned it to operate each machine on a separate circuit (the conventional practice). In a discussion paper presented to the STEE in 1888, Gordon related his attempts to connect alternators in parallel:

> Many of us have tried . . . but they do not work together till they have run for three or four minutes; they will in that time jump, and that jumping will take months of life out of 40,000 lamps. That alone is a rather serious difficulty in coupling machines together, and I think we may take it in practice—I am not speaking about the laboratory or experiments—we do not couple machines.[48]

Even with his training in both Cambridge Tripos mathematics and Maxwellian laboratory physics, and with the constant assistance of the best mechanical engineers available and the unfailing support of Alice, Gordon and his team still could not satisfactorily instantiate Hopkinson's tendentious theoretical conclusions.[49] Indeed, so dissatisfied was Gordon with a.c. technology that in the late 1880s he experimented (as a consultant engineer) with the rather more tractable machinery of *direct current* supply—then seen by a significant lobby (including Edison) as a strong contender for winning the "battle of the systems." Importantly, though, it was not only Gordon who was puzzled by Hopkinson's Wranglerish claims in 1889 to have solved the practical problems of parallel a.c. operation. In the following section, I explain how Hopkinson's claims once again came to the fore of discussion, framed within a typically unflinching mode of Senior Wrangler condescension to fellow electrical engineers.

## Parallel Explanations: Unpacking the "Hidden Mathematical Truths" of Coupled Alternators

> The practical dynamo-builder does not care two straws whether the mystery of parallel running of alternators has or has not been packed up in somebody's mathematical equations to be extricated therefrom subsequently as from a conjuror's inexhaustible bag[;] nor does he derive much guidance from the differences of experts, some of whom consider the secret depends upon abundance of self-induction in the armature, and some of whom think it does not. The dynamo designer . . . "wants to know" exactly what are the conditions to be complied with in order to construct alternators which shall successfully perform this act.
>
> —editorial note, "Alternate current working," *Electrician* 24 (1889): 325

During the three-year period of Gordon's highly visible Paddington adventure, Hopkinson remained virtually silent on the problems of paralleling alternators, working instead on applying magnetic circuit theory to enhancing the efficiency of the d.c. dynamo for the English Edison company.[50] As Jordan has shown, Hopkinson's work on d.c. systems was much less controversial, and contemporaries subsequently awarded him joint credit with Gisbert Kapp for developing a viable and reasonably universal theory of d.c. dynamos in 1886.[51] One brief publication in the Royal Society's journal in 1887 did indicate, however, that Hopkinson now had reservations about the rigor of his alternator theory. Although his assumption of unvarying armature self-induction had been a "most useful approximation," he admitted that quantity was "not in general" constant after all—precisely the criticism made by Thompson, Ayrton, and others in 1884.[52] Hopkinson did not, however, go on to consider whether his original conclusion about alternator coupling was vitiated by the error induced in his "most useful approximation"—an error that a Wrangler might have tried to quantify.

The standing of Hopkinson's Cantabrigian theory became moot again in early 1889 when Westinghouse effectively demonstrated the paralleling of its company's alternators in its U.S. central stations. Several practitioners tried to find general accounts of the necessary and sufficient conditions that made it possible to couple alternators *in practice*. In February 1889, Gisbert Kapp presented a paper on a.c. machinery to the Institution of Civil Engineers, using a clock diagram to contend that the crucial condition for successful parallel working was a "sensible" (i.e. substantial) amount of self-induction in the armature circuit.[53] This choice of graphical analysis certainly went down well with one former Cambridge mathematics student in the audience, the freelance engineer George Forbes:

> Hitherto most persons who had taken up the subject had done so by means of analytical formulae. The geometrical method was extremely clear and simple, and the Author had so developed it as to put down all the propositions, which were usually in an analytical form, [and] in a purely geometrical form this was very instructive.[54]

Even so, Kapp was criticized by the aristocratic James Swinburne for assuming the constancy of armature self-induction. Swinburne argued instead that the best way of calculating the behavior of an alternator was to use a more painstaking and rigorous form of graphical analysis. Rather than focusing on self-induction, Swinburne suggested that the way to proceed was to find out "how many lines of force were cutting the circuit at any time. By plotting the field magnets out by this means, and working through a graphic method," he could establish the all-important "armature reaction"—his translation of what others treated as variable self- and mutual induction. This was a long and tedious iterative process, but it produced results of some accuracy.[55] Indeed, the use of armature reaction theory soon came to be a major feature in the graphical analysis techniques used in a.c. textbooks.[56]

Hopkinson soon broke his silence on a.c. matters, claiming a further instance of the commercial paralleling of alternators as a trivial instantiation of his a.c. theory. At a meeting of the Institution of Electrical Engineers in May 1889, William Mordey, manager of the Anglo-American Brush Company in London, announced the successful coupling of newly designed alternators. Mordey's lengthy paper explicitly disputed Kapp's diagnosis of the necessary and sufficient condition by stipulating that Brush-designed alternators were designed to have the *minimum* possible self-induction. For the discussion of Mordey's paper at a subsequent IEE meeting, Hopkinson sent in a haughtily written report claiming that his 1884 theory had been "quite sufficient" to explain all of Mordey's results. Hopkinson's claim amounted to the point that Mordey's design instantiated his 1884 stipulation for maximizing the power output of alternators.[57] Yet Mordey replied sardonically that he had not derived any assistance from Hopkinson's theory, and disagreed that the Senior Wrangler's theory was sufficient to predict all the results he (Mordey) had obtained.[58] *The Electrician* agreed with Mordey and, in a rather exasperated editorial in the issue dated August 16, 1889, declared that it had not "the least desire to impeach Dr. Hopkinson's mathematics," but ventured that the conditions of the problem were "not fully expressed in the formula he has given us."[59] Hopkinson's assertions that all the relevant conditions for parallel running were already locked up in his theory sounded all too much like the mathematician's conjuror's trick of which *The Electrician* had complained two weeks earlier (see epigraph above).

By 1893, when a.c. engineering seemed to have won the battle of the systems, a new view emerged that running alternators in parallel was not a question of armature self-induction after all. Rather, it hinged on mechanical matters: linking the governors of the steam turbines to get stability in rotation, and using synchronization devices to ensure that alternators were exactly in phase before linking them. In a paper on this subject presented to the IEE in February 1893, Mordey explicitly avoided theoretical speculation, addressing only the "practical" principles of machine management that he and his assistants had established for paralleling.[60] While Mordey cited Hopkinson's now-standard 1886 "back-to-back" method of comparing dynamo efficiencies, and extended it to a.c. machines, his single direct reference to Hopkinson's 1884 theory concerned the possibility of making an alternator run as a motor.[61] Indeed, in his own comments on Mordey's paper, Hopkinson himself considered only how Mordey had borrowed his comparative testing techniques. One participant who perhaps misconstrued Mordey's level of deference to Hopkinson was Hugh Erat ["Huge Rat"] Harrison, Principal of "Faraday House" training college in London. Later in the discussion, made this arch comment:

> Mr. Mordey acknowledges his indebtedness to Dr. Hopkinson's earlier papers on the subject, and states that his empirical results have there been anticipated. I am afraid that most practical engineers regard these forecasts rather in the light of that statue of great price which is said to lie in the proverbial block of marble, and I doubt if many engineers have ever chipped off the mathematics in which these great truths lay hidden.[62]

In his response, however, Mordey made clear how little benefit he had himself found in such theories. Elaborating on Harrison's metaphor somewhat further, while showing polite deference to Hopkinson (elected three years earlier to the Chair of Electrical Engineering at Kings College, London), Mordey explained:

> Mr. Harrison referred to Dr. Hopkinson's paper, and he properly points out that I give Dr. Hopkinson credit for having prophesied some of these things. No doubt I should have been able to get on a great deal better with my work if I had been fully able to benefit by Dr. Hopkinson's papers. I expect a good many of us are not able, to use Mr. Harrison's figure [of speech], to chip off the mathematics and get at the really beautiful statue inside. I have had to work round as if the statue did not exist at all.[63]

Far from being a sine qua non for practical success, Hopkinson's elaborate abstract constructions were superfluous or even chimerical as far as the ordinary engineer was concerned. More brusquely—perhaps even sarcastically—Mordey added that the responsibility for rendering Hopkinson's work more intelligible and usable lay with Hopkinson himself:

> We do not often get Dr. Hopkinson here now—I wish we saw him more frequently—but as he is here I may take the opportunity of uttering a wish that when writing the papers which always turn out long afterwards to be so very important, he would remember people like me, and condescend to our level: it would greatly increase the immediate value of his writings.[64]

Mordey's sarcasm was not far below the surface when he addressed another IEE meeting on the same subject in March 1894. For this occasion, the freelance electrical engineer George Forbes had formulated a version of Mordey's account into the algebraic language so cherished by Cambridge graduates while removing the problematic inconstant factor of self-induction from consideration. Responding to this, Mordey noted with relief that it would now be clearer to all concerned that, as he had always maintained, the practical solution to the paralleling problem was "something different" from what Professor Hopkinson had deduced from his equations: "Now that my views have received the consecration of x's and y's, I may dare say that they will be honoured with the tardy approval of those who have hitherto failed to be convinced by the logic as expressed in the vulgar tongue."[65] Mordey's arch characterization of this intra-professional tension as one opposing his own brand of "vulgar" tongue to that of analytic mathematics is very revealing. His resentment of the almost clerical elitism of Cambridge-style mathematics invokes concerns of social class only hinted at in previous discussions of whose expertise could best handle a.c. technology. To see that Mordey's position was far from idiosyncratic, let us consider broader evidence on whether contemporary electrical engineers saw Cambridge Tripos mathematics as a useful or problematic resource for handling the challenges of their subject matter.

### The Use and Abuse of Mathematics: Debating the "Cambridge Man"

In the preceding section we saw the grounds of "practical" electrical engineers' antagonism toward the apparent sterility of abstract mathematical analysis and the apparent presumptuousness of the analytical engineers who wielded it. To avoid over-generalizing from just one case study, it is important to note that the debate about Cambridge mathematicians was not confined to the topic of paralleling, or even to electrical engineering. Conversely, it is important to note that, insofar as professional disputes about theory were endemic in electrical engineering, these disputes were not only about the technological applicability of Maxwellian techniques of electromagnetic analysis. Disputes over authority and expertise in electrical engineering also hinged upon the educationally grounded credibility of the notional "Cambridge man."

From this perspective, it is very informative to consider some debates conducted in the spring of 1891 in the correspondence columns of *Industries*, a journal owned by James Swinburne and (until recently) edited by Gisbert Kapp. After a discussion (initiated by a Glaswegian student) about whether thermodynamic theory was of any use to trainee engineers,[66] a secondary debate emerged under the editorial heading "The use and abuse of mathematics." This illustrates the widespread view—hitherto undocumented by historians of electrical engineering—that the alleged ignorance of practical men was far less of a problem for the nascent profession than the hubristic assertions of mathematically trained practitioners, especially the Cambridge-trained variety.

On March 28, 1891, a Leeds "Electrician" wrote to the editor of *Industries* seeking to clarify the long-running "feud between the so-called theorist and the practical man." His concern was that theorists who wielded mathematical learning uninformed by practical knowledge had little dialogue with the practical men who "reviled" mathematics, preferring to rely instead on their inscrutable wisdom. In a normative vein, he argued that the *true* theorist and practical man should be one and the same individual. Nevertheless, he singled out the highly educated group for special criticism:

The chief real complaint against mathematicians is that they are apt to over-rate the value of mathematical work, and to forget that they are only employing deductive reasoning. Nothing is commoner than to find a simple law assumed, and then followed by pages of calculations which would be useful if the assumed law were true, but are really valueless. This vice is especially common in electrical literature.[67]

In electrical engineering the feud had indeed raged "very bitterly." For example, the study of dynamos had unhelpfully been "retarded for years" by those who wrote page after page showing hypothetically how such machines would behave if the pressure were "this function or that function" of the excitation currents—but with little reference to real engineering machinery. The target of his attack was almost certainly John Hopkinson, author of many Wranglerish treatments of idealized alternators and dynamos in the preceding decade. In ironic tones the Leeds Electrician maintained that Cambridge seemed to set the "fashion" in mathematics: none but a "Cambridge man" was supposed capable of anything in the subject, others being generally doomed to "outer darkness" or "eternal ignorance," especially if they had never attended a university.[68] Notwithstanding the alleged Cantabrigian monopoly of expertise, the correspondent hoped that somebody might yet write a book on electricity that contained the practical mathematics that was required by engineers and nothing more. Perhaps taking a particular potshot at Wrangler faith in algebraic solutions, he emphasized that he did not mean a book of "delusive" simplified explanations. Extended use of graphical methods was the important thing, notwithstanding their inelegance as viewed by the elite mathematician:

For practical work graphical methods should be much more developed than they are. The Cambridge man looks upon a graphical solution as a kind of foul play, and would as soon think of solving an equation by trial [and error], or by omission of a term, or in any other reasonable way.

The main issue was thus overcoming the Cambridge graduate's adherence to vacuous virtuosity in algebraic manipulation and the associated prejudice against recourse to graphical methods when more appropriate. Although the long-term resolution of the theory-practice feud lay in having the two parties learn from each other, the Leeds Electrician concluded on a partisan note that "a little more modesty on the part of the college man" would be of the greatest advantage to all.[69]

Subsequently published correspondence under the editorial heading "The use and abuse of mathematics" evoked a broadly sympathetic if not uncritical or consensual response. The following week, there appeared a letter credited to "Ne Sutor Supra Crepidam," an archly erudite pseudonym meaning "Let not the shoemaker judge above his shoe." While contending that the complexity of dynamo behavior could only be apprehended by similarly complex and "unpractical" mathematics, he also welcomed the new Cambridge school of engineering as a means of bringing due practicality to mathematics taught at that university. His even-handed conclusion was that the mathematician should become an engineer as much as engineers should become mathematicians.[70] Robert H. Smith of Mason College, Birmingham, wrote in to contend that the best theoreticians and practitioners in engineering had always maintained a "free, friendly and respectful" communication of ideas. Yet all reasonable observers would agree, he claimed, on the "pernicious" effects of combining knowledge of pure mathematics with ignorance of practical matters. Not only did learning unnecessarily abstract techniques waste time; it had the "evil" effect of "perverting" engineers' whole way of thinking and acting. He thus laid out a detailed agenda of the kind of mathematical learning that was appropriate to engineering education, a staged acquisition of geometry, algebra, and elementary calculus that produced no tendency to "specious mathematical showiness" and no inclination to "indulge in imaginative theory not based on fact."[71]

The debate closed with an exchange between two correspondents who differed on the significance of the appointment of a recent Wrangler—among the top twenty in Cambridge—to an electrical engineering consultancy run by Alexander Kennedy, formerly Professor of Mechanical Engineering at University College London. An "Engineering Demonstrator" from London observed this to be part of a growing trend for electrical supply companies to call in "theoretical advisors": the commercial demand for specialists with higher mathematical training was the clearest recognition of their increasing importance.[72] Writing from Bombay on June 5, "Scrutator" replied that such higher mathematical training was useful but still not necessary. Professor Kennedy himself had attained the status of chief consultant electrical engineer without any university training in mathematics, only what he had picked up in the practice of civil engineering. Scrutator added, nevertheless, that it was important for the current generation of trainee engineers to learn the equivalence between analytical, graphical, and mechanical means of calculation. Concurring with the Engineering Demonstrator, Scrutator concluded that by learning the "real" use of mathematics, such students could avoid the worst "abuse" of mathematics—which was not to use it at all.[73]

In the context of such discussions arguing the need for convergence between the two factions, I will now consider how educational plans were developed in the mid 1890s to ensure that future electrical engineers could embody a suitable unification of the "practical man" and the mathematician in one individual.

## Hopkinson and the Relation of Cambridge Mathematics to Engineering, 1894–1896

If in my lecture I have rather insisted on the uses of the calculus it was because I thought its methods were unduly feared and shunned by engineers, not because I or other Cambridge mathematicians disliked or shunned geometrical methods.
—letter from John Hopkinson to *The Electrician,* May 1894[74]

In the world of electrical engineering, Hopkinson was no diplomat; he was rarely amenable to debating his judgments on technological topics. In contrast to Mordey, Harrison, and the correspondents in *Industries*, Hopkinson was unwilling, in the early 1890s at least, to resolve the theory-practice tension by mutual accommodation; nor was he willing to translate Cambridge mathematics into more widely intelligible graphical forms, as undertaken by Blakesley and Kapp. This much is clear in his 1890 memorandum, written for the University of Cambridge, on the future of engineering education at his alma mater. Hopkinson asserted that while the majority of engineers would be "practical men" executing only a limited class of practical tasks, engineers taking a broader professional view would benefit from a mathematical education closely linked to "physical work." Indeed, he would advise a son who wished to be an engineer not to take early employment in a traditional mechanical workshop but rather to go to university first to study mathematics and laboratory physics.[75]

When the Institution of Civil Engineers (ICE) in London invited Hopkinson to deliver its prestigious James Forrest lecture in early May 1894, the supremacy of such university training was the unstated major premise of "the relation of mathematics to engineering."[76] Speaking in the same venue in which he had first pronounced on alternator theory a decade before, Hopkinson observed that mathematics had been described in that room by engineers as a good servant but a "bad master." Assuming a characteristically Cantabrigian definition of what constituted mathematics, his overt aim was to prove the first part of this proposition, and less explicitly to challenge the latter point by showing engineering's great debt to mathematics. He thus showed how the "higher" mathematics of differential equations and Bessel functions, coupled with some strategic analogies, had served engineers well in examining cases as diverse as stressed beams, compass corrections, and submarine telegraph signaling.

As one might have expected of one tutored in the rigor of the Cambridge Mathematics Tripos, Hopkinson emphasized the practice of solving all problems by first formulating differential equations to capture the material conditions, and solving these analytically to arrive at an all-encompassing algebraic expression. For the case of transmitting a musical sound down a telephone line, he cited the formula which "tells us everything" about the propagation and decay of the signal, asserting it was difficult to see how the problem could be dealt with by any means except solving a differential equation. Hopkinson was as unyieldingly committed to the necessity of Wrangler methods as he was convinced of his own authoritative originality in the complex phenomena of a.c. currents. He thus characteristically reiterated the claim that it was he who had "shown mathematically" that alternators could be run in parallel, even before—he now alleged—the matter had became one of serious practical concern in the electrical industry. Although he catalogued some of the areas in which mistakes had been caused by the "wrong" application of mathematical formulae, he insisted that such errors were only to be cured by a "more abundant supply of more powerful mathematics." It was not appropriate, by implication, to seek recourse in alternative forms of mathematics, such as approximate graphical methods, when difficulties or discrepancies arose.[77]

Some of the anticipated mixed reaction to Hopkinson's lecture can be gauged from the vote of thanks offered by the ICE's president, Alfred Giles. While he recognized this lecture would have been an intellectual treat for those already cognizant with higher mathematics, he hoped at least that younger members not so familiar with it would feel "spurred" to follow Dr. Hopkinson's "brilliant" example in applying mathematics to engineering. As Giles put it, it was not after all very often that the ICE had the opportunity of "receiving good advice from a senior wrangler."[78] Hopkinson's audience, however, had not treated his lecture with the utmost sympathy. The editors of *The Electrician*, who had long been critical of Hopkinson's views on the role of mathematics and mathematical theory in engineering, were not impressed. Reprinting the first half of his lecture in its May 11 issue, the periodical chose also to devote a substantial editorial criticizing the Cantabrigian bias of his lecture, rather pointedly highlighting the parochial nature of Hopkinson's claims with the subtitle "the relation of Cambridge mathematics to engineering."

Taking Hopkinson to be a representative of the Cambridge "school" of mathematics, *The Electrician* reproved him for assuming that Cambridge "algebraic" mathematics was the only useful form of mathematics for engineers.[79] There was of course no doubting the *general* value of such mathematics for science: it would have been "ridiculous" for Lord Rayleigh, for example, to lecture the Royal Society on the relation of algebraic mathematics to physical science—these two fields were nearly identical.[80] Yet, *The Electrician*'s staff writer contended, it was simply false to assert, as Hopkinson had done, that algebraic methods were *necessary* in engineering. Professor Hopkinson had failed to acknowledge that clock-face diagrams could readily solve a.c. problems that became highly complicated when attempts were made to solve them in the formulation of "algebraical" language. Though graphical methods were less informative and perhaps slower than analytical methods, the economic argument for using them in the engineering business was overwhelming: whatever the problem, it was generally cheaper to employ a draftsman for three hours than a wrangler for three minutes. With no little irony, and with a gesture indicating how Hopkinson might be able to redeem his position, *The Electrician* pointed to a major fertile innovation by Hopkinson in 1880—the empirical "characteristic curve" of dynamo performance—that showed the effectiveness of deploying graphical methods in electrical engineering.[81]

To the editors of *The Electrician*, the problem was particular to the educational culture of a particular prestigious British university. Although geometrical techniques were given appropriate attention at Oxford, the lamentable non-recognition of graphical methods at the higher levels of engineering would continue if Cambridge were allowed to continue to dominate the field of mathematics:

> So far is geometry behind algebra today, that it is difficult to say that the former will ever become so powerful and general a branch of mathematics. Its present condition in this country is largely due to the neglect, and we would even say the ignorance, of geometry in Cambridge.

This situation could hardly change while so many Cambridge-trained schoolmasters taught mathematics of a kind geared toward getting their pupils entrance to Cambridge. A whole political economy of mathematics teaching thus had to be reformed if any beneficial change were to be effected in engineering training.[82]

Hopkinson's response to the scathingly anti-Cambridge editorial moved him to an uncharacteristically public rebuttal. The next week, *The Electrician* published a letter from him complaining that his university had not been treated with "justice." Geometry held a specially protected place in the Cambridge curriculum, and had been used in important ways not just by Maxwell but also by such eminent graduates as Kelvin (in his method of electrical images) and P. G. Tait (in his theory of damped oscillations). Mathematicians, Hopkinson now contended (contradicting his ICE lecture), should be "ambidextrous" in both geometrical and algebraic techniques. The reason his lecture had focused on calculus was not that Cambridge mathematicians disliked geometrical methods but because calculus was "unduly feared and shunned" by engineers. Importantly, Hopkinson conceded that his own work in electrical engineering (the characteristic curve and a.c. theory) was in fact more frequently handled by *other* practitioners using geometrical methods. Nevertheless, he insisted that algebraic methods deserved paramount consideration since they had long been central to making new "discoveries" that were only later translated into geometrical form. Tellingly, Hopkinson acknowledged that this prioritization was rooted in local Cambridge history: before the analytic revolution of the 1820s, a Tripos student who solved a problem by quick algebraic methods had to translate his solution back into geometrical terms to win credit from examiners. With such a legacy, wranglers equipped with powerful analytical techniques were unlikely to capitulate ever again to the reactionary hegemony of geometry.[83]

The historian chronicling these events can hardly fail to see a considerable shift in emphasis between Hopkinson's ICE lecture and the defensive letter he wrote two weeks later in response to *The Electrician*'s review. Without overstating the significance of this embarrassing public assault on both Hopkinson and his university, it is perhaps no coincidence that his subsequent publications on a.c. theory abandoned some of their idealized Wranglerish formalism for which *The Electrician* had so strongly criticized him. One year after his ICE lecture, he produced a revised account of "alternate current dynamo electric machines" with the assistance of his Kings College demonstrators; this was published by the Royal Society in its *Philosophical Transactions*. Hopkinson may have been seeking a more sympathetic audience less dominated by skeptical engineers. Unprecedentedly, he now categorized previous work on alternators specifically as "algebraic" discussions, conceding that such research had adopted unwarranted simplifying assumptions about the invariance of self- and mutual induction. Whereas in 1884 he had strategically ignored the phenomenon of mutual induction and variable self-induction as unduly complicating the formalization of rotating electromagnetic machinery, Hopkinson now explored the effects of both these parameters in determining the perplexing performance of alternators.[84]

In characteristic Wrangler style, Hopkinson formulated the operation of an alternator as a pair of differential equations—one each for the magnetic circuit and the armature circuit. He rendered them soluble by using a standard approximation technique, requiring the assumption that the current in the magnetizing circuit did not vary greatly. These equations produced a result for a single alternator with several correction terms and coefficients, with some unanticipated higher harmonic effects—more complicated than anything in his 1884 account of *coupled* alternators. Recognizing still further the vulnerability of his assumption of constant current in the magnetizing circuit, Hopkinson then devoted the bulk of this paper to empirical studies of the characteristics of standard Siemens alternators undertaken by F. Lydall, his laboratory assistant at Kings. These Hopkinson represented in an elaborate series of *graphical* depictions. Looking at the characteristic curves that emerged, Hopkinson noted it was easily shown that the ordinary theory (implicitly his original 1884 account) "does not fully account for the facts." Hopkinson considered the discrepancies between actual machine performance and his most revisionist formalization in quantitative and qualitative depth—and found the equations unequivocally to be wanting.[85] Here was a clear illustration both of the failure of even quite sophisticated approximation procedures in theorizing alternators and of the contrasting power of graphical methods to capture the contingencies of alternator performance—contingencies alien to the orderly world of the Mathematics Tripos.

Despite this, or perhaps because of it, Hopkinson's detailed examination of working alternators greatly facilitated his initiation in the commercial production of alternators for the Mather and Platt company. In the fifth (1896) edition of his best-selling book *Dynamo-Electric Machinery,* Silvanus Thompson noted that, despite their high self-induction, Hopkinson's machines were suitable for working in parallel.[86]

## Epilogue

Hopkinson's shift toward accommodation of the analytical and practical traditions was characteristic of a wider pattern of endeavor in the later 1890s among electrical engineers from diverse educational backgrounds to communicate with one another using shared assumptions and mathematical techniques. What Gordon had accomplished in melding laboratory and workshop techniques of practical engineering a decade earlier was now undergoing an analogous process in the abstract computational practices of electrical engineering. Silvanus Thompson—ever the irenic Quaker—was, like his colleague at Finsbury Technical College, John Perry, involved in the contemporaneous project of making calculus intelligible to the next generation of student engineers in all the technical colleges. At Finsbury, since the early 1880s, Perry and William Ayrton had developed the pragmatic use of squared paper and mechanical models to teach elementary calculus to engineering students with little prior geometry or algebra. This approach was epitomized in Perry's *Calculus for Engineers* (1896), which went through several editions.[87] Thompson's most famous contribution to this project was his more broadly directed (and initially anonymous) *Calculus Made Easy* (1910). Leading his readers with great jocularity through the preliminary "terrors" of the subject, he encouraged them to adopt the motto "What one fool can do, another can." Such was the long-lasting impact of this remarkably accessible volume (continuously in print since its first publication),that its 1931 edition inspired the 13-year-old Richard Feynman to learn calculus well before his school curriculum required him to do so.[88]

With a certain poignancy, after John Hopkinson's early death in a family mountaineering accident in 1898, his surviving son Bertram—also a Cambridge mathematics graduate—completely reworked and refined his father's alternator theory in the ensuing five years. He analyzed the phenomenon of "hunting" (persistent perturbation around a harmonic oscillation) as a major machinic impediment to successful paralleling.[89] By 1903 a newly comprehensive body of a.c. theory was emerging, cultivated by Cambridge mathematics graduates working in the electrical industry, notably Hopkinson junior and also Alexander Russell in his two-volume *Treatise on the Theory of Alternating Currents* (1904–1906).[90] Despite being grounded in the Cambridge syllabus of the previous century, their new theoretical canon was free of simple Wranglerish mechanical analogies. The authors recognized too that the problems of alternator paralleling were not a matter of failing to understand Cambridge mathematics, but rather were electromagnetic and mechanical problems that needed more than a Wrangler's analytic facility to solve. Most interesting from the point of view of issues raised earlier in this chapter, they represented the analysis of a.c. machines by vector diagrams drawn directly from the "clock-face" graphical analysis long favored by "practical" engineers. The two languages of mathematics were thus amalgamated in a common pedagogy through which trainee electrical engineers learned what had been cherished by both the practical man and the theoretician of the previous generation.

In this context of pedagogical restructuring in the early twentieth century, the assessment of Hopkinson's theories of a.c. technology by his professorial contemporary W. E. Ayrton is highly revealing.[91] Reviewing Hopkinson's collected technical and scientific papers for the journal *Nature* in 1904, Ayrton suggested that, while his theory of the d.c. dynamo had helped more than "crude trial and error" in solving practical commercial problems, such was not the case for his a.c. theories. These theories had evidently not been applicable to all a.c. machines, and had been more important for what they "suggested" than what they actually proved. Indeed, Ayrton concluded that Hopkinson, in making his so-called prediction in 1883 that alternators could be worked in parallel, had been more influenced by his experimental work with commercial machinery than by his "theoretical reasoning."[92]

While revealing of how a fellow academic judged the qualified success of Hopkinson as a theoretical electrical engineer, Ayrton's account does not tell us why the controversies in which Hopkinson became ensnared happened to focus so specifically and critically on his educational credentials as a *Cambridge* graduate. Why indeed was this a controversy construed in primarily educational terms, rather than as a matter of practice versus theory, social class, or gender? The very fact that Hopkinson's much-criticized early theory of alternators persisted as a point of reference in the electrical engineering canon is indicative, I suggest, of the high social status attached to the views of a former Senior Wrangler so cogently portrayed in Warwick's recent account. Even those practical men who had been through rather different masculinity-building ordeals in the engineering workshop and found much to disagree with (in what little they understood of Hopkinson's account) felt obliged to indicate how *little* it had helped them when debating the subject in the tribunals of electrical engineering. In contrast to the other controversies over the role of self-induction, Hopkinson neither argued a case drawn from Maxwell's *Treatise* nor attracted support from academic Maxwellians committed to a more sophisticated account of self-induction in electrical engineering theory. In such an isolated position, Hopkinson's tactic was to treat challenges from mechanically trained engineers as if they had simply failed to understand the mathematics that was the prerogative of the Senior Wrangler to promote. By not working dextrously with his academic language Hopkinson afforded them no consideration when these engineers disagreed with the premises of his arguments. This reinforced the class prejudices and antipathies of those who had not had the privilege of a Cambridge mathematics education, moving them to represent Cambridge mathematics as part of the problem in a.c. technology rather than part of the solution.

Tellingly, however, the demographic predominance of practitioners who protested against the glibness of his Cambridge methods and the volubility of critical journals eventually forced Hopkinson to abandon at least some of the Wrangler techniques that he had perhaps involuntarily brought to bear on uncongenial problems of electrical engineering. Concurrently, the increasing sophistication of electrical engineering problems and the commercial pressure for solving them effectively called into the marketplace just the kind of high-level theoretical expertise that Cambridge Wranglers offered to the industry. This story, therefore, has not just been one of how economic forces determine the education needed for a "scientific industry." Nor has it been simply a story of a philosophical spat in which quasi-Platonist mathematicians argued for the underlying orderliness of machines, opposed by quasi-Aristotelian practical men arguing for the organicism of technology. Rather, I have shown that certain kinds of education—or its absence—define the very categories in which engineers of the past articulated their understanding about what was at stake in the successful prosecution of a new discipline. And the sheer heterogeneity of the skills required for electrical engineering meant that practitioners had much to learn, after leaving the academy, the college, or the workshop, from those of other educational backgrounds. Thus James Swinburne told each member of the IEE in 1900 that a practitioner had to "not be only an overgrown wireman, a mechanical engineer with a little electrical knowledge, a mathematician, a financier, a lacquered brass and sealing-wax varnish instrument maker, a physicist, or a manager of men." He had to be all these things, but in "different proportions in different men."[93]

Whatever electrical engineers did or did not say about the matter, education is not the only factor for historians to consider in understanding techno-scientific controversies of the past. But when the engineers of the past denigrate rivals specifically in terms of their deficient education, we should not merely be unsurprised; we also should seek to understand what contextual and contingent circumstances led to education's becoming a center-stage issue. After this study, historians might research how differences in practitioners' education are resolved to some extent in making collective sense of the world. After all, in no discipline of science or engineering is there ever an utterly homogeneous workforce: as David Kaiser shows in this volume, the problem of educational localism can persistently afflict even as prestigious a subject as quantum physics after World War II. And while it is hard to conceive of any practitioners claiming that their education made their accomplishments uniquely possible (are there any discoveries transcendentally reserved for the graduates of Cambridge, Harvard, Princeton, or Yale?), the very fact that techno-scientists eventually *stop* talking about the discrepant and error-inducing education of their peers unarguably merits the historian's closest attention.

## Notes

1. "The use and abuse of mathematics," letter from Leeds "Electrician" to *Industries*, April 3, 1891.

2. James Swinburne, "Presidential address," *Journal of the Institution of Electrical Engineers* 32 (1902), 13. For more information on James Swinburne and other early British electrical engineers see Graeme Gooday, *The Morals of Measurement: Accuracy, Irony and Trust in Late Victorian Electrical Practice* (Cambridge University Press, 2004).

3. Andrew Warwick, *Masters of Theory: Cambridge and the Rise of Mathematical Physics* (University of Chicago Press, 2003). See also Peter M. Harman, ed., *Wranglers and Physicists: Studies on Cambridge Physics in the Nineteenth Century* (Manchester University Press, 1985).

4. Michael Sanderson, *The Universities and British Industry, 1850–1970* (Routledge and Kegan Paul, 1972); idem, *Educational Opportunity and Social Change in England* (Faber, 1987); Graeme Gooday, "Precision measurement and the genesis of physics teaching laboratories in Victorian Britain," *British Journal for the History of Science* 23(1990): 25–51; idem, "Lies, damned lies and declinism: Lyon Playfair, the Paris 1867 Exhibition and the contested rhetorics of scientific education and industrial performance," in *The Golden Age: Essays in British Social and Economic History, 1850–70*, ed. I. Inkster (Ashgate, 2000), 105–120.

5. Warwick, *Masters of Theory*, 205–207. For discussion of Hopkinson as epitomizing some key features of Cambridge Mathematics Tripos culture, see ibid., 187, 216–217.

6. Warwick, *Masters of Theory.*

7. The key original text was James Clerk Maxwell, *Treatise on Electricity and Magnetism* (Clarendon, 1873), 2 volumes; for early readings of this in Cambridge University see Warwick, *Masters of Theory*, 286–356. For accounts of the Maxwellian community see Jed Buchwald, *From Maxwell to Microphysics: Aspects of Electromagnetic Theory in the Last Quarter of the Nineteenth Century* (University of Chicago Press, 1985); Bruce Hunt, *The Maxwellians* (Cornell University Press, 1991).

Buchwald's broad categorization includes Hopkinson as a Maxwellian, but Hunt focuses on a much smaller subgroup. Although Hopkinson's last year in Cambridge (1871–72) overlapped with Maxwell's first year as Professor of Experimental Physics, Hopkinson only became closely familiar with Maxwell's work in 1876 after Maxwell's refereeing of a paper submitted by Hopkinson to the Royal Society. See Peter Harman, ed., *The Scientific Letters and Papers of James Clerk Maxwell,* 3 volumes (Cambridge University Press, 2002), volume 3, 324–329.

8. Bruce Hunt, "'Practice vs. theory': The British electrical debate, 1888–91," *Isis* 74 (1983): 341–355.

9. D. W. Jordan, "The adoption of self-induction by telephony, 1886–89," *Annals of Science* 39 (1982): 433–461. See also idem, "D. E. Hughes, self-induction and the skin-effect," *Centaurus* 26 (1982): 123–153.

10. Hunt, *The Maxwellians;* Buchwald, *From Maxwell to Microphysics.*

11. R. Kline, "Science and engineering theory in the invention and development of the induction motor, 1880–1900," *Technology and Culture* 28 (1987): 283–313. For Pupin's 1922 reminiscences of his Cambridge education ca.1884–85, see Michael Pupin, *From Immigrant to Inventor* (Scribner, 1960), 167–210.

12. In recent historiography the various different meanings of the term "practice" cannot easily be demarcated from "theory." Whether one considers a "practice" in techno-science to be the macroscopic domain of professional activity or a localized, individuated technique, both forms of practice are obviously theory-laden in some sense. See Graeme Gooday, "Practice," in *The Reader's Guide to The History of Science,* ed. Arne Hessenbruch (FitzRoy Dearborn, 2000), 589–591.

13. I. Yavetz, "Oliver Heaviside and the significance of the British electrical debate," *Annals of Science* 50 (1993): 135–173, on 136 and 153.

14. Jordan, "The adoption of self-induction by telephony."

15. A. Warwick, "The laboratory of theory," in *The Values of Precision,* ed. M. Norton Wise (Princeton University Press, 1995), 311–351; idem, "Cambridge mathematics and Cavendish physics: Cunningham, Campbell and Einstein's relativity," *Studies in History and Philosophy of Science* 23 (1992): 625–656 (part 1), and 24 (1993): 1–25 (part 2); idem, *Masters of Theory.*

16. Warwick, *Masters of Theory,* 252–254.

17. Kuhn described normal science as a "strenuous attempt to force nature into the conceptual boxes supplied by professional education." Thomas Kuhn, *The Structure of Scientific Revolutions* (University of Chicago Press, 1962), 5. Joseph Rouse has noted the existence of two types of Kuhn: one focused on practices, and the other on global worldviews; it is the former upon which I focus here. See Joseph Rouse, "Science as practice: Two readings of Thomas Kuhn," in Rouse, *Knowledge and Power: Toward a Political Philosophy of Science* (Cornell University Press, 1987), 26–40. Many thanks to David Kaiser for bringing Rouse's piece to my attention.

18. On open-ended practice, see Andrew Pickering and Adam Stephanides, "Constructing quaternions: On the analysis of conceptual practice," in *Science as Practice and Culture,* ed. Andrew Pickering (University of Chicago Press, 1992), 139–167.

19. Warwick, *Masters of Theory*, 18–26.

20. Ibid., 273–275.

21. Thomas Blakesley, *[Papers on] Alternating Currents of Electricity*, third edition (Whittaker/ Bell, 1891), 4–6.

22. See R. Kline, *Steinmetz: Engineer and Socialist* (Johns Hopkins University Press, 1992), 38–40, 324. Graphical methods were especially congenial for engineers whose formal mathematical schooling had consisted of Euclid and elementary algebra, and whose daily working life consisted in working from scaled geometrical designs or plans. Even academic engineers who took college degrees would be trained in geometrical methods, e.g. Karl Pearson's teaching of mathematics to engineering students at University College London in the late 1880s. I thank Eileen Magnello for pointing out this aspect of Pearson's teaching.

23. Thomas Hughes, *Networks of Power* (Johns Hopkins University Press, 1983).

24. Percy Dunsheath, *A History of Electrical Engineering* (Faber and Faber, 1962); James Greig, *John Hopkinson, Electrical Engineer* (Her Majesty's Stationery Office, 1970). Brian Bowers repeated this claim in a public lecture at the IEE (October 1998).

25. A week after delivering his paper on November 13, 1884, Hopkinson learned that Henry Wilde had published on an instance of "synchronizing control" between linked alternators in the *Philosophical Magazine* for January 1869. Hitherto unaware of this article, Hopkinson apologized to the (notoriously litigious) Wilde in a footnote at the start of his 1884 publication for not having given him "the honour which was his due." J. Hopkinson, "On the theory of alternating currents, particularly in reference to two alternate current machines connected to the same circuit," *Journal of the Society of Telegraph Engineers* 13 (1884): 496–515, discussion 528–558, quotation on 496. This paper was reproduced (minus discussion) in *Original Papers by the Late John Hopkinson*, ed. B. Hopkinson (Cambridge University Press, 1901), volume 1, 133–151, quotation on 133.

26. J. Hopkinson, "On electric lighting [first paper, 1879]," in Hopkinson, *Original Papers*, volume 1, 33–46, esp. 44; idem, "On electric lighting [second paper, 1880]," in ibid., 47–56, esp. 47 , 49, and 55.

27. J. Hopkinson, "Some points in electric lighting [1883]," in Hopkinson, *Original Papers*, volume 1, 57–83, esp. 59–64.

28. Hopkinson, "Some points in electric lighting," 66–69.

29. J. Hopkinson, "On the theory of alternating currents, particularly in reference to two alternate current machines connected to the same circuit," *Journal of the Society of Telegraph Engineers* 13 (1884): 496–515, discussion 528–558.

30. Hopkinson, "The theory of alternating currents," 503–504.

31. ". . . neglecting the effect of currents other than those in the copper wire" (Hopkinson, "On the theory of alternating currents," 501).

32. See Hopkinson's discussion in the Mordey 1889 IEE paper discussed below.

33. Note that in problems V and VI Hopkinson considered results for an individual alternator with mutual induction, but importantly not the case for two connected in parallel ("The theory of alternating currents," 509–512).

34. Ibid., discussion on 531 and 538–554. See discussion of self-induction and mutual induction in Sylvanus Thompson, *Dynamo-Electric Machinery: A Manual for Students of Electrotechnics* (E. and F. Spon, 1884), esp. 387–388.

35. Llewellyn Atkinson, "Proceedings of the Commemoration Meetings," *Journal of the Institute of Electrical Engineers* 60 (1922), 441–443, on 443. For an obituary of Atkinson, see *Journal of the Institution of Electrical Engineers* 85 (1939): 769–770.

36. Mrs. J. E. H. Gordon, *Decorative Electricity* (Sampson & Low, 1891), 153–154.

37. For an obituary of Gordon, see *Electrician* 30 (1893): 117–118. On Gordon's researches, especially in the Maxwellian area of electro-optical rotation, see J. E. H. Gordon, *A Physical Treatise on Electricity and Magnetism* (S. Low, Marsten, Searle, and Rivington, 1880); the second edition (revised) was published in 1883.

38. Gordon's a.c. endeavors were overshadowed by the endeavors of the longer-lived Sebastian Ziani Ferranti (Dunsheath, *A History of Electrical Engineering*, 160–161).

39. For a discussion of the symptomatic historical erasure of technical assistants from accounts of scientific practice, see Steven Shapin, "The invisible technician," *American Scientist* 77 (1989): 554–563.

40. J. E. H. Gordon, "The development of electric lighting," *Journal of the Society of Arts* 31 (1883): 778–787, discussion 787–791, esp. 781–782.

41. Obituary of Frank Bailey, *Journal of the Institution of Electrical Engineers* 69 (1931): 1318.

42. See *The Electrician*, May 28, 1886, 51. Also see the reminiscences of A. H. Walton and of Frank Bailey in "Proceedings of the Commemoration Meetings," *Journal of the Institution of Electrical Engineers* 60 (1922): 402–406 and 416–420 (Bailey quotation on 416).

43. "A complaint," *Electrician* 16 (1885): 25. Quotations from Bailey and Walton in *Journal of the Institution of Electrical Engineers* 60 (1922): 417–418 and 405, respectively. The litigation continued later; see "The Paddington installation," *Electrician* 18 (1886–87): 500–501. Cf. Robert H. Parsons, *The Early Days of The Power Station Industry* (Cambridge University Press, 1940), 42. The Paddington installation served until 1907.

44. [Anon. editorial], *Electrician* 17 (1886), April: 51–56, quotation on 56.

45. See *Journal of the Institution of Electrical Engineers* 60 (1922): 377–500. Bailey's remarks are on 416–418.

46. A. H. Walton, "Mr. J. E. H. Gordon's system of electric supply, Great Western Railway, Paddington, 1884," in "Proceedings of the Commemoration Meetings," 405.

47. Gordon, *Decorative Electricity*, 153.

48. See Gordon's contribution to the discussion of Rookes E. B. Crompton, "Central station lighting: Transformers vs. accumulators," *Journal of the Society of Telegraph Engineers and Electricians* 17 (1888): 195–196. For a variant reporting of the discussion by *Electrician* staff, see *Electrician* 20 (1887–88): 634–637, 655–656, discussion on 656–657, 694–698, and 749–752; and *Electrician* 21 (1888): 88–93. Gordon's Paddington installation was undertaken on behalf of the Telegraph Maintenance and Construction Company. When Telcon decided to discontinue its electrical lighting projects in 1887–88, Gordon became chief engineer to the newly founded Metropolitan Supply Company: [Anon.], "Obituary: James Edward Henry Gordon," *Electrician* 30 (1893): 417–418.

49. For details of Ferranti's work at the Grovesnor Gallery in 1885–86, see J. F. Wilson, *Ferranti and the British Electrical Industry* (Manchester University Press, 1988); A. Ridding, *S. Z. de Ferranti: Pioneer of electric power* (Her Majesty's Stationery Office, 1964), 7. The heroic focus on Ferranti is sustained in Hughes, *Networks of Power*, 97–99, 238.

50. Greig, *John Hopkinson*, 17–22. See John and Edward Hopkinson, "Dynamo-electric machinery," *Philosophical Transactions of the Royal Society* 177 (1887): 331–358.

51. See D. Jordan, "The magnetic circuit model, 1850–1890: The resisted flow image in magnetostatics," *British Journal for the History of Science* 23 (1990): 131–173.

52. J. Hopkinson, "Note on the theory of the alternate current dynamo," *Proceedings of the Royal Society* 42 (1887): 167–170.

53. G. Kapp, "Alternate current machinery," *Minutes of Proceedings of the Institution of Civil Engineers* 97 (1889): 1–79, esp. 1–21. Note that Kapp's approach assumed the constancy of self-induction, but in the graphical method, this assumption could be avoided by finding a non-circular trajectory for the self-inductive "vector" indicator.

54. Kapp, "Alternate current machinery," discussion on 48.

55. Ibid., 62–63.

56. For a brief period in 1889, Swinburne acted as a supporter of Hopkinson's theory, claiming to have translated this "painfully" mathematical theory into a form intelligible to engineers. For more on Swinburne see Graeme Gooday, *The Morals of Measurement: Accuracy, Irony and Trust in Late Victorian Electrical Practice* (Cambridge University Press, 2004).

57. Several auxiliary assumptions seem to be necessary for this to be a "consequence" of the 1884 theory.

58. W. M. Mordey, "Alternate current working," *Journal of the Institution of Electrical Engineers* 18 (1889): 583–613, discussion on 613–688.

59. [Editorial], "Alternate current working," *Electrician* 24 (1889): 325, and notes in ibid., 367–368. See correspondence from a practitioner, W. B. Sayer, to whom it seemed that residual adherence to the Hopkinson theory was attributable to "parental affection" so strong that it was "too terrible" for the theorists to part with their "cherished offspring." Ibid., 387–388.

60. W. M. Mordey, "On testing and working alternators," *Journal of the Institution of Electrical Engineers*, 22 (1893): 117–134, discussion on 134–194.

61. Ibid., 117, 129.

62. Ibid., 174.

63. Ibid., discussion on 191. See Gooday, *The Morals of Measurement*, chapter 5.

64. Mordey, "On testing and working alternators," discussion on 191.

65. William Mordey, "On parallel working with special reference to long lines," *Journal of the Institution of Electrical Engineers* 23 (1894): 260–314, on 301–304, 312.

66. "Is the study of thermodynamics of use to engineers?" (letter to the editor from "Student"), January 3, 1891, *Industries* 10 (1891): 39. For further correspondence concerning this topic, see ibid., 87, 136, 160, 183, 207.

67. "The use and abuse of mathematics," letter from Leeds "Electrician" to *Industries* 10 (1891): 328.

68. In a further deconstructive vein, he noted that the introduction of women to the Cambridge examination system would beneficially prove that "many an ordinary person" might have become Senior Wrangler given a few years "free from care, work or anxiety" and tutors to help them to "do nothing but read mathematics in a regular and systematic way." Ibid.

69. Ibid.

70. [Anon.], *Industries* 10 (1891): 374.

71. Robert H. Smith, letter of April 21, 1891, *Industries* 10 (1891): 399.

72. "Engineering Demonstrator," letter of May 2, 1891, as published in *Industries* 10 (1891): 447.

73. "Scrutator," letter of June 5, 1891, as published in *Industries* 10 (1891): 616.

74. John Hopkinson, "The relations of Cambridge mathematics to engineering," May 18, 1894, as published in *Electrician* 33 (1894): 85.

75. John Hopkinson, "Memorandum on engineering education [written for a Cambridge University syndicate in 1890]," in Hopkinson, *Original Papers*, volume 1, lxiii–vi, quotation on lxivv.

76. J. Hopkinson, "The relation of mathematics to engineering," reproduced in *Electrician* 33 (1894): 41–43, 78–79, and in Hopkinson, *Original Papers*, volume 1, 269–288.

77. Ibid., 79.

78. Ibid.

79. [Editorial], "The relation of Cambridge mathematics to engineering," *Electrician* 33 (1894): 44–46.

80. Conversely, according to *The Electrician,* James Clerk Maxwell's admirable promotion of graphical methods to addressing engineering problems almost "disqualified" him from the Cambridge school. See Maxwell's use of geometrical methods to expound the theory of electrical images in his *Treatise on Electricity and Magnetism*, third edition (Clarendon, 1891), volume 1, 244–269, 280, and unpaginated appendices at the end of the volume mapping equipotential lines for various electrostatic configurations.

81. This reprises a prominent theme in the debate in *Industries* of 1891: mathematical practices in engineering theory should be adapted to the commercial conditions of engineering, not vice versa.

82. Anon., "The relations of Cambridge mathematics to engineering," *Electrician* 33 (1894): 45–46. The editorial characterized the more general problem in the following sardonic terms: "Cambridge mathematics is the result of a vicious circle to which we have alluded on a previous occasion: school-boy, scholarship, wrangler, school-master. Another smaller circle, whose viciousness is inversely as its diameter, touches this wide one, and is: examinee, wrangler, examiner." Ibid. On the influence of the Cambridge Mathematics Tripos on (private) school curricula, see Warwick, *Masters of Theory*, 254–263.

83. Hopkinson, "The relations of Cambridge mathematics to engineering," 85.

84. John Hopkinson and Ernest Wilson, "Alternate current dynamo electric machines," *Philosophical Transactions* Series A 187 (1897): 229–252, reproduced in Hopkinson, *Original Papers*, volume 1, 156–182; quotation on 157.

85. Ibid., 163.

86. Significantly, however, this was immediately qualified by the vaguely Hopkinsonian recommendation that "it is sufficient if $R$ is something of the order of $pL$," $p$ being the angular frequency at which the alternator was run. See Thompson, *Dynamo-Electric Machinery*, fifth edition (1896), 600–613.

87. As Professor of Mathematics and Mechanics from 1896 to 1913 at the Royal College of Science and School of Mines in London (part of Imperial College from 1907), Perry persistently fought the prevailing view that a mastery of Euclid was essential for all students, contending rather that for non-mathematicians, the subject should be taught primarily with a view to its "utility." See the short volume he edited for the British Association on the Advancement of Science in 1901, *Discussion on the Teaching of Mathematics*, and Graeme Gooday, "John Perry," in *Oxford Dictionary of National Biography* (Oxford University Press, 2004).

88. Jagdish Mehra, *The Beat of a Different Drum: The Life and Science of Richard Feynman* (Oxford University Press, 1994), chapter 2.

89. Bertram Hopkinson, "On the parallel working of alternators," read at Section G of the British Association for the Advancement of Science in 1903, reproduced in *Electrician* 51 (1903): 886; idem, "The 'hunting' of alternating current machines," *Proceedings of the Royal Society of London* 72 (1903): 233–252.

90. Alexander Russell, *A Treatise on the Theory of Alternating Currents,* 2 volumes (Cambridge University Press, 1904–1906). Formerly an assistant lecturer at Gonville and Caius College, Cambridge, Russell wrote as Lecturer in Applied Mathematics and Superintendent of the Testing Department at Faraday House (electrical engineering college) in London.

91. Hopkinson and three of his children died in an alpine mountaineering accident in 1898. See Greig, *John Hopkinson,* 39–40.

92. W. E. Ayrton, "The life work of a scientific engineer" [review of John Hopkinson's *Original Papers*], *Nature* 70 (1904): 169–172, quotations from 169–170.

93. James Swinburne, "Presidential address," 13. At the time of Swinburne's speech there was only one full female member of the IEE, Hertha Ayrton: educated in the Cambridge Mathematics Tripos she had been elected in 1899. See Evelyn Sharp, *Hertha Ayrton, 1854–1923: A Memoir* (Arnold, 1926); Joan Mason, "Hertha Ayrton (1854–1923) and the Admission of Women to the Royal Society of London," *Notes and Records of the Royal Society of London* 45 (1991): 201–220.

# 5 The *Geist* in the Institute: The Production of Quantum Physicists in 1930s Japan

**Kenji Ito**

## The Metaphor of Spirit

On November 22, 1901, professors from Tokyo Imperial University and bureaucrats from Japan's Ministry of Education gathered at a party celebrating the twenty-fifth year of Erwin von Baelz's service in Japan. To the guests of Japan's educators and bureaucrats, including the Cambridge wrangler and former Minister of Education Baron Kikuchi Dairoku, Baeltz, a German physician who had taught medicine in Japan since 1876, talked about the state of science in Japan at the turn of the century:

> It seems to me that in Japan erroneous conceptions about the origin and nature of Western science are widely prevalent. It is regarded as a machine which can turn out so much work every year, and therefore as a machine which can without further ado be transported from the West to any other part of the world there to continue its labours. This is a great mistake. . . . [The road of science] has been the highway of the human spirit, and the great names are written on its milestones: on one of the early milestones such names as Pythagoras, Aristotle, Hippocrates, and Archimedes; and on the recent milestones such names as Faraday, Darwin, Helmholtz, Virchow, Pasteur, Roentgen. The spirit of these is the spirit that will sustain us Europeans until the end of the world. . . . From all the lands of the West there have come to you teachers eager to implant this spirit in the Land of the Rising Sun and to enable you of Japan to make it your own. . . . But many in Japan were content to take over from these Westerners the latest acquisitions, instead of studying the spirit which made the acquisitions possible.[1]

Some historians of science quote this passage to characterize the superficiality of the introduction of Western science into Japan.[2] They claim that Japan imported science only as a form of technology, forgetting its intellectual roots and its "spirit." Others argue that Western science itself went through a radical transformation in the nineteenth century, and that Japan had simply adopted a new breed of science that was inseparably tied to technology and inevitably institutionalized by the state.[3] Rather than taking part in this dispute, here I will problematize the validity of the metaphor of the "spirit" as the conveyor of scientific knowledge and skills between different cultures by examining the case of the introduction of quantum mechanics into Japan.

The metaphor of the "spirit" that allegedly constitutes the basis of scientific developments is indeed a familiar one to historians of modern physics. Werner Heisenberg, one of the founders of quantum mechanics, wrote in the preface of his influential early textbook on quantum mechanics, *Physical Principles of Quantum Theory* (1930):

> The purpose of the book seems to me to be fulfilled if it contributes somewhat to the diffusion of that *Kopenhagener Geist der Quantentheorie,* if I may so express myself, which has directed the entire development of modern atomic physics.[4]

In Japan, the introduction of quantum mechanics, or, if we adopt Heisenberg's expression, the diffusion of the *Kopenhagener Geist,* began in the late 1920s. If we can talk about the "missionaries of the Copenhagen spirit," as John Heilbron has done,[5] then the principal "missionary" of the Copenhagen spirit to Japan was Nishina Yoshio. While working at Niels Bohr's Institute for Theoretical Physics in Copenhagen, Nishina produced one of the earliest contributions to quantum mechanics made by a Japanese. After his return to Japan in late 1928 he directed and developed a group of atomic physicists in Tokyo. Nishina's efforts resulted in a considerably stronger tradition of theoretical physics, with the rise of young and able theoretical physicists in the 1930s, including Tomonaga Sin-itiro, Yukawa Hideki, Sakata Shôichi, Kobayashi Minoru, Tamaki Hidehiko, Taketani Mituo, and Umeda Kwai.

Echoing Heisenberg and other alumni of the Copenhagen school of physics, Japanese physicists described Nishina's efforts to bring quantum mechanics to Japan as the introduction and dissemination of the "Copenhagen spirit" there (as we will see below). For Japanese physicists the "Copenhagen spirit" was a workable guiding principle that contained the methodology, knowledge, and research skills of quantum mechanics. They thought that only human mediation, not books or journals, could convey this spirit. Simply reading printed materials would not be enough to incarnate such a spirit. In this sense, Japanese physicists were aware that one could only transfer some knowledge through personal contact.

Thus, the "Copenhagen spirit" was not unlike the notion of tacit knowledge employed by Harry Collins in his discussion of the transmission of experimental skills. Collins considered experimental ability as a "skill-like knowledge, which travels best (or only) through accomplished practitioners." It cannot be "fully explicated or absolutely established." It is "invisible in its passage and in those who [possess] it."[6] If, as Japanese physicists perceived, the "Copenhagen spirit" was something like tacit knowledge and the vehicle through which physicists might transmit the "theoretical skills" of quantum mechanics, then this Spirit could provide the secret recipe to produce theoretical physicists.

Two questions then occur: (1) What was the "Copenhagen spirit" that was allegedly brought to Japan and instrumental to the production of theoretical works in quantum mechanics? (2) Did it really facilitate the dissemination of theoretical skills and the production of quantum theorists in Japan? If so, to what extent?

To answer the first question, I will analyze the accounts of how physicists at Bohr's institute conducted theoretical physics in order to try to define what the "Copenhagen spirit" was. Avoiding an essentialist approach, I will not try to determine what the "Copenhagen spirit" *really* was. Rather, I will first try to understand the way physics was conducted at Bohr's institute. By examining the memoirs of the Copenhagen physicists, I structure my description according to the values that governed physicists at this institute. What did scientists consider important and desirable, and how or why did they do so? Next, I move on to show how these values were materialized in the ways scientists generally conducted or meant to conduct their research at the institute and in various small incidents within this group of scientists. If the "Copenhagen spirit" had any effect on the way the Copenhagen physicists behaved, the characteristics of their scientific practices should reflect the "Copenhagen spirit," whatever it was. Then, I will apply the same analysis to the groups of Japanese physicists that were allegedly governed by the "Copenhagen spirit," and examine similarities and differences between Bohr's group and the Japanese groups.

These analyses will illuminate to what extent the metaphor of Spirit fails to account for the historical process in question. Instead of talking about transplantation of the Spirit, I propose to interpret the dissemination of quantum mechanics as a "resonance." Unlike spirit, resonance is a well-defined physical concept. When an oscillating system is under a periodically changing external force with a certain frequency (the eigenfrequency), the amplitude of the system becomes singularly large. An example of such phenomena is a resonance between two tuning forks. I apply this metaphor to conceptualize how similar but not identical phenomena of scientific practices followed one another. In the acoustic resonance, we do not see the vibration of one tuning fork transported to the other. A resonance occurs when there is a medium between the two tuning forks, and when the other fork satisfies necessary preconditions (for example, the same eigenfrequency). Similarly, a "resonance" of scientific practices occurs, not because the original set of practices was transported in its totality, but because certain social and cultural conditions allowed such a phenomenon to occur, and because some human or material mediation triggered a resonance of scientific practices.

## Becoming a Theoretical Physicist in Prewar Japan

Before World War II, most physicists in Japan went through a three-tier system of higher education. Potential scientists first received preliminary higher education, usually liberal arts education at a "higher school" (*kotôgakko*), after they graduated a middle school around the age of 18. Then they received education more specialized in science (or sometimes in engineering) at an "imperial university." Finally, they developed themselves as professional scientists through various forms of postgraduate training. There were several possibilities for this stage of training, such as becoming a graduate student, or a research associate at an imperial university or at the Institute of Physical and Chemical Research (hereafter Riken).[7]

The higher school was a unique prewar Japanese institution of higher education, distinct from the postwar (or American) high school.[8] These three-year colleges of liberal arts aimed to prepare students for more specialized education at a three-year imperial university. Most university students, including future physicists, went to a higher school. Entering a higher school, especially the most prestigious First Higher School in Tokyo, was extremely difficult. Students had to pass a very competitive entrance examination, or had to receive a recommendation from a middle school principal. Once accepted, students enjoyed a great degree of freedom. Going to a university was easy because the capacity of the imperial universities was about the same as the number of higher school students. Higher school students, including future physicists, received intensive liberal arts education, with strong emphasis on languages. Nitta Isamu, a chemist who entered the First Higher School in 1917, remembers that "there were ridiculously many hours of English and German."[9] If a first-year science student in the 1920s chose to take a second foreign language, his total class time per week amounted to 32 hours. Four hours were allotted for Japanese and literary Chinese, eight for the first foreign language, and four for the second foreign language. Half of all class time was devoted to language education.[10]

Except for a short period after World War II, higher schools accepted only male students. Since most students of imperial universities went through a higher school, the higher school system effectively worked as a gatekeeper for the gender segregation of Japan's higher education. As a result, the story of Japanese physicists is a male tale as much as, or even more than, the one in Sharon Traweek's work.[11] Other than higher schools, there were a few options for preliminary higher education. One was to go to a higher normal school. With the Women's Higher Normal School, this option was open even to female students.[12] Another option for a science-minded young man was to go to a higher engineering school.[13] Yet it seems that very few theoretical physicists took such an unconventional educational path.

The general arts education at higher schools allowed the students to explore their interests without future occupational concerns. They often enjoyed high-brow cultures and esoteric philosophy, of little practical use to the current problems in Japanese society. Nurturing *Bildung* and personality, and creating the elite distinguished from others with these kinds of unpractical knowledge were, if not the goals, some of the consequences of the higher school education. Theoretical physics, with its philosophical implications, was able to fit in this culture of the higher school.

The second stage of the training of physicists was the undergraduate education at an imperial university. For our purpose, it is enough to understand imperial universities as national universities with multiple colleges. In the 1920s, three imperial universities had departments of physics: Tokyo, Kyoto, and Tohoku. Tohoku University had been an important center of theoretical physics in Japan, but after it lost two theorists (Ishiwara Jun retired in 1921 and Aichi Keiichi died in 1923) its theoretical physics was considerably weakened, and its physics department produced very few important physicists. Most Japanese theoretical physicists were, therefore, trained either in Tokyo or in Kyoto.

Entering a physics department from a higher school was easier than entering another department. Very few students learned physics, for which the Japanese society did not have much use at that point. (A physics student had be prepared to be jobless after graduation.) At Kyoto University, for example, the department of physics implemented an entrance examination only after 1926.[14] Until then, the applicants for the physics department never exceeded the allotted limit, and therefore all the applications were accepted. At Tokyo University, the department of theoretical physics (later the theory major of the physics department) accommodated many students who failed to enter the college of engineering. Those students went on to learn theoretical physics, which made a good basis for engineering and was helpful to prepare for the next year's entrance examination.[15]

What I call the "culture of calculating" dominated theoretical physicists in these departments. Advanced mathematics and meticulous calculation were highly appreciated, whereas physical principles were rarely questioned.[16] For example, Tamaki Kajûrô of Kyoto University specialized in electromagnetism, relativity theory, and fluid dynamics. During the 1910s, his most active period, he published 15 papers, mostly on or related to relativity theory. Tamaki applied the principle of relativity to various problems, and derived its mathematical implications. In his 1911 paper, for example, Tamaki derived the transformation rules of electromagnetic fields between two systems, one moving with a constant velocity relative to the other. The merit of his paper was that he derived formulas for the case when the direction of velocity did not coincide with one of the coordinate axes, because "motions occurring in nature, in general, are not always of so simple a character that they take place along one of a given system of coordinate axes. When we have to deal with some problems concerning motions with component velocities, it will be found very convenient to use general equations."[17] Indeed, as Tamaki showed in his later papers, this formula could be useful.[18] Although deriving these formulas (in modern terms, applying a general Lorentz transformation) did not require much ingenuity, and the result did not come out in a beautiful form, such work was recognized as a legitimate achievement in theoretical physics.[19]

The institutional and educational settings of physics at prewar Japanese universities were in agreement with the "culture of calculating." The training provided by the physics department of Tokyo University shows that physicists there deemed mathematics singularly important. Its curriculum in the mid 1920s placed more emphasis on calculus than on any other subject. Every first-year physics student had to attend the same calculus course as a mathematics student. This one-year course consisted of five hours of lectures and two "Exercise" (problem-solving) classes per week.[20] These problem-solving classes started at 1 P.M. and usually ended at 4.[21] The first year also included four hours per week for electromagnetism, three hours of lectures and one problem-solving class for mechanics, and three hours for "Thermodynamics and Solid State Physics." In addition, two hours per week of "Application of Differential Equations" were mandatory, and students of theoretical physics had to attend a year-long course on "Theory of General Functions and Elliptic Functions," which took up three hours per week.[22]

In the third year, a student was expected to choose an advisor and conduct research. A student chose his advisor according to his interest and negotiated with the professor to become his advisor. Usually the advisor gave a topic to the student, but not always. Students were supposed to carry out research during the year and submit a thesis. The choice of the advisor at this point had a life-long consequence because it virtually determined the student's research direction. In addition, the advisor was supposed to take responsibility for the student's career immediately after graduation, no matter whether the student chose to take an academic or non-academic career. If, for example, the advisor had a strong tie with a certain line of job, the student had a better chance to get a job in that area.

The last stage of the making of physicists was the indefinite period of postgraduate training. There was no formal protocol about the training at this stage. Recent graduates of physics departments wishing to become physicists had a few possibilities. The luckiest became paid assistants or lecturers at their alma mater or at Riken. Less fortunate people went to higher schools or other institutes of higher education, such as private universities, higher normal schools, military academies, and higher technical schools. An economically less fortunate but academically more promising path was to become an unpaid assistant to a professor, or to become a graduate student. Of course this was possible only for those who could economically (and socially) afford such a position.[23] In some cases, graduate students had another appointment (such as a teaching job at a higher school, or a research fellowship [*kenkyûsei*] at Riken); this was allowed by the regulations.

Graduate education in the prewar Japanese universities had no structure. Graduate students were supposed to do research under the direction of an academic advisor. Graduates of the same university entered its graduate school without an entrance examination. The term of appointment was two years, which could be extended up to five years with permission of the dean and the faculty council. A graduate student had to submit a report about the progress of his or her research project each year. Those who wanted to receive a degree were able to submit a dissertation after a two-year stay at a graduate school. A graduate student was able to attend a course offered to undergraduates with permission of the advisor; courses specifically designed for graduate students might or might not be offered. Graduate students were required to reside near the university, but they were allowed to take other jobs if they were able to show that the job was beneficial to their research.[24] In short, graduate students were able to do almost anything they liked, with minimum support from the university and the faculty.

The best option for a recent college graduate in science was to work at Riken (the Institute of Physical and Chemical Research, established in 1917). Riken was one of a few research institutes in physics at that time, and certainly the most important one.[25] It was prestigious, even more so than Tokyo University. In contrast to those at imperial universities, Riken's scientists enjoyed an affluent research budget and freedom from teaching obligations and university bureaucracy. Many leading scientists at the institutes were also professors at imperial universities. Takamine Toshio, a senior spectroscopist, for example, originally held positions at both Tokyo Imperial University and Riken, but later retired from the former before the ordinary year of retirement. The advantage was obvious: he was then able to concentrate on his research.

Under the dynamic directorship of the second director Ôkouchi Masatoshi, Riken expanded during the first half of the 1920s. The notion of "scienticist industry" summarizes Ôkouchi's policy. Instead of the ordinary conception that science was something to be applied to industry, Ôkouchi proposed a form of industry that would serve science by marketing scientific achievements and returning the profit for further scientific research. Although initial expansion of the research budget drove Riken to the edge of bankruptcy, one of the groups in Riken enabled the mass production of vitamin

B, which allowed Riken to recover and further expand its research and manufacturing. Riken's financial basis was never secure, yet it was able to survive until the end of the war thanks to the sale of military-related products.[26]

Ôkouchi also reorganized the structure of Riken and made it more "egalitarian." Riken had consisted of a physics division and a chemistry division, each headed by an elder physicist (Nagaoka Hantarô and Sakurai Jôji, respectively). Each individual research group belonged to one of the two divisions. Ôkouchi abolished the divisions and made all the group leaders equal in status, at the same time allowing scientists to conduct interdisciplinary research. At Riken under Ôkouchi, lively and unconstrained research was possible, if not mandatory. As we shall see, this was a precondition for Nishina's research group to flourish, or for the resonance to occur.

A recent graduate of a university could be hired as a research student (*kenkyûsei*) or a research associate (*joshu*). These positions could serve as a graduate scholarship, because some graduate students could occupy them. Group leaders (*kenkyûshitsu shunin taru kenkyûin,* research scientist in charge of a research group, or *shunin kenkyûin,* chief research scientist) were able to employ research associates at their discretion within the limits of their budgets, with the approval of the administrative board. Salaries did not differ much between the two posts, being about twice as much as that of an imperial university research associate or a lecturer. A drawback of junior positions at Riken was insecurity. Besides the insecure financial foundation of Riken itself, promotion to higher positions at Riken did not always happen. Since a group leader hired more than one research associate (and sometimes many), those younger scientists could not always succeed their bosses, and most of them had to leave Riken eventually.

In contrast to Riken's junior positions, a job at an imperial university was very secure. Even a research associate position was virtually tenured. (There was no concept of a tenure process in the Japanese academic system.) Moreover, in most departments a full professor usually had only one assistant professor or lecturer and one or two research associates. A research associate had a good chance of succeeding the assistant professor and could eventually become a full professor.

Postgraduate training of physicists in prewar Japan was unstructured, and the situations varied. It is, therefore, difficult to talk about it in any meaningful way. Many of the young physicists received individual tutoring from their mentors, or trained themselves. There are, however, three instances of postgraduate training that make a relatively coherent account. The first is the school of Terazawa Kwan-iti in the 1920s and the 1930s in the College of Engineering at Tokyo University. The second is the activity of self-tutoring groups in Tokyo and Kyoto in the late 1920s. The third is Nishina Yoshio's group in Riken in the 1930s. In this section, I discuss the first two.

Terazawa Kwan-iti exemplifies how calculational approach tied theoretical physics to engineering. In a country where rapid industrialization was one of the national priorities, theoretical physicists offered their calculational prowess in engineering problems.[27] Getting a job in the College of Engineering, this graduate of the Theoretical Physics Department created a "colony" of physicists and mentored several of the most mathematically deft disciples in the department of physics, hiring them as research associates or lecturers. The Terazawa group, with Yamanouchi Takahiko, Kotani Masao, Inui Tetsuo, Husimi Kôdi, and others, formed a powerful group of theoretical physicists.

In the late 1920s, some physics students and young physicists found the situation of theoretical physics in Japan unsatisfactory and began to study on their own, in defiance of the physics establishment. Elsewhere I have written about how the poor quality of university education in Japan and various other factors at the time nourished a rebellious culture of students, pushing some social science students to Marxism and some young physicists to new sciences, such as quantum mechanics.[28] What I call the "culture of rebellion" characterizes this generation of physicists.

In Tokyo, some young physicists began studying quantum mechanics on their own in 1926, defying their old professors. The authoritarianism of the physics department disgusted them, and they despised the low quality of the official colloquium held by the physics department. They formed an independent study group, the "Physics Reading Group." Since their motivation was to revolt against the physics establishment, they chose new subjects for their readings. It happened that the foundational works on quantum mechanics appeared in 1925–26. At Tokyo University, no one was doing research on it, and there was no course on quantum mechanics until 1928. The young physicists chose the papers that interested them, made Japanese digests (usually about one-third the length of the originals), and published the digests. More than half of the digests in the first volume dealt with quantum mechanics.[29]

Similarly, four Kyoto University students under Professor Tamaki Kajûrô—Tamura Matsuhei, Nishida Sotohiko, Tomonaga Sin-itiro, and Ogawa (later Yukawa) Hideki—began studying quantum mechanics around 1926. Since Tamaki knew nothing about quantum mechanics, they had to learn it on their own. Yukawa and Tomonaga started their work on quantum mechanics with Max Born's *Probleme der Quantenmechanik,* which Yukawa read in 1926 and Tomonaga in 1928.[30] Then they read original papers on quantum mechanics.

Yamanouchi Takahiko of the Terazawa School read the foundational papers on quantum mechanics around the same time. Extremely talented and well trained in mathematics, he had mastered mathematical approaches to physics by perusing Courant and Hilbert's *Methods of Mathematical Physics.*[31] He had no problem understanding those papers in terms of mathematics. He understood Heisenberg's matrix mechanics paper as a natural extension of the correspondence principle, which he knew through Nagaoka's course on quantum theory. He could perceive that Heisenberg's calculations were those of matrices, since he knew matrices through the course at Tokyo University. Schrödinger's first papers on wave mechanics did not appear exciting to Yamanouchi, although their mathematical neatness impressed him. He had mastered the Courant-Hilbert volumes, and he found the problem of eigenvalues too familiar and too easy.[32] In short, Yamanouchi, able and loyal heir of the mathematical tradition of Japanese theoretical physics, was too talented to realize the radical significance of quantum mechanics.

The young physicists I consider here apparently opposed such a tradition. For example, when Tamura Matsuhei was a graduate student, Professor Tamaki invited Tamura to study hydrodynamics (Tamaki's specialty), but Tamura adamantly refused, saying that hydrodynamics was something that people should do in the college of engineering, not in the college of science.[33] Nor did Yukawa like calculating. Later he joked that Heisenberg's *Principles of Quantum Physics* was a good book because it contained few equations in the text, most of the equations being in the appendix.[34]

These sporadic attempts by young physicists to initiate research on quantum physics did not bear fruits at this point. Their activities, however, indicate that the condition of the Japanese physics community was ripening for quantum mechanics research. To use a resonance metaphor, young Japanese physicists in the 1930s constituted the human resource part of the tuning fork in Japan to catch sound waves from overseas. The sound wave came in the form of Nishina Yoshio.

## Nishina Yoshio in Copenhagen and in Japan

### Nishina in Copenhagen

Nishina Yoshio, born in 1890, learned electrical engineering at Tokyo University. After he graduated, he joined Riken and became a graduate student at Tokyo University. Although his official research topic was the electric furnace, Nishina shifted his interest to physics, attending Nagaoka's lectures at Tokyo University.[35]

Riken ordered Nishina to study abroad in 1921. He left Japan on April 5 and stayed at the Cavendish Laboratory for about a year, conducting experimental work on recoil electrons from x-ray scattering. In 1922 he moved to Göttingen, where he studied with David Hilbert and Max Born for half a year. Then he moved to Copenhagen. While at Cambridge, he had attended a talk given by Niels Bohr. Nishina later told the Japanese chemist Kimura Kenjirô that "Bohr's talk was murky, and it was hard to grasp what he meant. This made me think that I should definitely work with this person sometime."[36] In his letter to Bohr dated March 25, 1923, Nishina conveyed his wish to work at Bohr's institute.[37] Bohr (who according to Heisenberg never answered letters[38]) replied to Nishina rather quickly, granting permission for him to work at his institute.[39] Two weeks later, Nishina arrived in Copenhagen. In his letter to Bohr, Nishina had written that he would like to study Bohr's theory of atomic constitution, but would be willing to help others in experimentation and calculation. He did not intend to stay more than several months; his institute would not allow him to "stay for more than two terms."[40] Nishina's "several months" turned out to be seven and a half years, which led his concerned relatives to suspect he had a girlfriend there.[41]

Although Nishina started his stay in Copenhagen as an experimentalist and first worked on experimental topics related to x rays, around 1926 and 1927 he decided to shift his focus to theoretical physics.[42] In August 1927, along with I. I. Rabi, he moved to Hamburg, probably to learn quantum mechanics from Wolfgang Pauli. In Hamburg, his theoretical training bore fruit in a collaborative work with Rabi.[43] When Nishina came back to Copenhagen in March 1928, he began theoretical research in cooperation with Oskar Klein. The result was the Klein-Nishina formula for the relativistic scattering of electrons and photons, the first significant Japanese contribution to quantum mechanics.[44]

In Copenhagen, Nishina also took interest in the philosophical issues of quantum mechanics. Apparently he attended the Como conference of September 1927, at which Bohr introduced his notion of "complementarity."[45] He helped Bohr translate the Como paper from German to English in the winter of 1927–28.[46] He wrote a letter to Nagaoka about complementarity, and he proposed inviting Bohr to Japan.[47]

### The "Copenhagen Spirit" in Copenhagen

His long stay in Copenhagen must have given Nishina many opportunities to meet young and active theoretical physicists, who were supposedly imbued with the "Copenhagen spirit." Much has been written about Bohr's institute and the particular *Geist* of its physicists, yet among historians it has not been entirely clear what exactly the "Copenhagen spirit" was. John Heilbron, for example, seems to regard it as the "Copenhagen interpretation" of quantum mechanics.[48] Finn Aaserud gives a different account of the "Copenhagen spirit." He defines this term as a specific "atmosphere or style of work."[49] Perhaps the problem is that, before asking what the "Copenhagen spirit" was, we do not even know what category it belonged to. It might be a kind of mental attitude, because Léon Rosenfeld defines it as "that of a complete freedom of judgment and discussion."[50] It might be a collective delusion or the Copenhagen interpretation itself, as Heilbron suggests.[51] It might be a sort of guiding principle, since, according to Heisenberg, it led the development of physics.[52] It might be a certain set of philosophical ideas and attitudes, since Rosenfeld wrote that the "Copenhagen spirit" was coined to denote the "unity of view on the problems of science."[53] It might be a Hegelian, trans-individual *Geist*, as Tomonaga, probably inspired by Heisenberg, suggested.[54] It might be a style of doing physics, because Victor Weisskopf defined it as "the style of a very special character that he [Bohr] imposed onto physics."[55]

As I mentioned above, I analyze the practices of the Copenhagen physicists in terms of what was important and desirable among them. Two things were considered important in Bohr's group: physics and collaboration. Physics is of course central to physicists. How and why it is, however, differs from one physicist to another. I claim that physics was important for physicists in Bohr's group, not because it had practical applications, or because the knowledge that it would produce had a certain intrinsic value (such as "truth"), but because it satisfied their sense of playful curiosity. In this light, doing physics well was important in itself, just as winning a game of chess or a sporting match is important for surprisingly many people. They were the kind of scientists described by Albert Einstein as those who should be expelled from the "temple" of physicists: "Many take to science out of a joyful sense of superior intellectual power; science is their own special sport to which they look for vivid experience and the satisfaction of ambition."[56] Whereas Einstein considered himself to be one of the scientists allowed to stay inside the metaphorical temple, the physicists in Copenhagen behaved as if doing physics were playing a game, like table tennis or chess (which many of them enjoyed as well).[57] Rosenfeld, for example, wrote about the humorous and playful side of Niels Bohr:

> When he was working on the surface tension of water, he had to melt glass tubes to make jets: he took such a delight in this operation that, completely forgetting its original purpose, he spent hours passing tube after tube through the flame.[58]

Young physicists in Copenhagen practiced physics, not because it was a pursuit of the absolute truth, but because it was fun to play.

The second of the two values, high esteem of collaboration, took on a few different forms: collaboration between theorists, collaboration between theory and experiment, international collaboration, and collaboration between physics and different branches of science. Bohr's collaborative style of research and his need of a human sounding board are well known.[59] As director of the institute, Bohr promoted a collaborative working style. Weisskopf points out that there was no single paper written by Niels Bohr himself between 1920 and 1930.[60] Collaboration occurred not only between Bohr and his disciples but also between the disciples.[61] Cooperation crossed the boundary between theory and experiment. Since its inception in 1916, Bohr's institute aimed to accommodate both theory and experiment under one roof, keeping a close tie between them.[62] Interdisciplinary collaboration was not rare at Bohr's institute. The chemist George Hevesy, who used radioisotopes in biology, occupied a prominent place in the institute.[63] Internationally, Bohr considered Denmark as a prosperous meeting place of the great traditions of British and German sciences.[64] Although, as Aaserud demonstrates, the relations between theory and experiment, or theoretical physics and experimental biology, were not as idyllic as Bohr had hoped, and other socio-political factors complicated the relationships, these (attempted) collaborations serve as an example of what Bohr considered desirable.[65]

The combination of the values of collaboration and playful curiosity helped to shape an informal collaborative style of scientific research, where scientists conducted physics like a team sport, with similar enthusiasm, competition, and cooperation. Since the only rule of the game that mattered was to produce good works in physics, conventional social protocols did not regulate the activities of physicists in the Bohr group, where "a man was judged purely by his ability to think clearly and straight."[66] Rudeness and outspokenness were therefore accepted. A legendary figure of exemplary rudeness was Wolfgang Pauli, the harshness of whose criticism and sarcasm—which he directed at everyone, including Bohr—is well known.[67]

One form of collaboration was free-wheeling discussion, which was conducted in this sort of informal atmosphere. Discussion could take place on various occasions, one of which was lunch. Stefan Rozental writes that lunch was an ideal opportunity to exchange information and experiences, to hear advice at difficult times, and to discuss lines of research to pursue.[68] Obviously, discussion also took place in seminars and colloquia. Colloquia were sometimes scheduled and announced in advance, but they could also happen without advance notice when an unexpected visitor stopped by or someone found an interesting article in a new publication.[69] On these occasions, people listened to a talk or a report first and then engaged in discussion, but participants were free to interrupt the speaker with questions and critical remarks.[70] Discussion continued with no determined time limit. Intense and serious discussions were sometimes mixed with humorous remarks, eliciting bursts of laughter.[71]

The Bohr group also had fun in various activities related and unrelated to physics. Occasionally, scientists played table tennis in the library ("the readers didn't seem to mind an occasional game," recalled Otto Frisch).[72] After a symposium, they performed a satirical parody of Goethe's *Faust*.[73] Once a decade, they produced an issue of the *Journal of Jocular Physics* in honor of Bohr's birthday.[74] Nishina participated in these activities in Copenhagen. In a pamphlet about Bohr published in 1937, the year Bohr visited Japan, Nishina described Niels Bohr's institute. After emphasizing the cosmopolitan character of the institute, he wrote:

Many able people who are advancing the new physics had at some point a chance to learn from Niels Bohr, either directly or indirectly. In particular, those who gathered at Bohr's institute in Copenhagen have been bred to what Heisenberg calls *Kopenhagener Geist.* There is no doubt that this is one of the great motive forces of the current developments in physics.[75]

## Nishina at Riken

Nishina returned to Japan in December of 1928. For a short period after his return, he worked at Riken as an underling of Nagaoka's. Two things occupied Nishina's early scientific activity in Japan: the translation and publication of Heisenberg's and Dirac's lectures in Japan, and his own lectures on quantum mechanics, which he delivered at some universities. Nishina, who knew both Heisenberg and Dirac, made their trip in 1929 to Japan possible.[76] Their lectures in Tokyo concerned their current works. Nishina translated these lectures, gave meticulous annotations, and published them in a volume in 1932. Nishina distributed copies of this volume to major research centers in physics.[77]

For the dissemination of quantum mechanics in Japan, however, Nishina's own lectures on quantum mechanics had greater consequences. In particular, his intensive ten-day lecture course on quantum mechanics at Kyoto University in May 1931 played a pivotal role by encouraging young physicists in Kyoto. Moreover, it established a link between Riken and this university in Japan's old capital, making it an important source of young theoretical physicists. In the audience were Yukawa Hideki, Tomonaga Sin-itiro, Sakata Shôichi, and Kobayashi Minoru. Tomonaga and Yukawa found this lecture series based on Heisenberg's *Physical Principles of Quantum Theory* extremely clear and instructive.

In July 1931, Riken promoted Nishina to group leader (a "chief research scientist," or *shunin kenkyûin*), and Nishina finally had free reign to build his own school. Nishina wrote to Tomonaga and invited him to join his group, which Tomonaga, always insecure about his ability, accepted reluctantly.[78] Nishina's group started out on a modest scale. The theory subgroup of Nishina's group consisted only of Nishina himself and Tomonaga. The Nishina group was allotted only two rooms at the beginning: one was Nishina's office, the other, an empty laboratory (occupied by the experiment subgroup, consisting of Sagane Ryôkichi and Takeuchi Masa). Due to lack of space, Tomonaga's desk was in a storage room, which housed old issues of Riken's journals.[79]

From this humble origin, Nishina built an active group of theoretical physicists. As I mentioned above, Nishina was seen as bringing the "Copenhagen spirit" from

Copenhagen to Japan. For example, Tamaki Hidehiko, one of Nishina's earliest disciples, describes the practice at Bohr's institute, which I suspect was the ideal in Nishina's group:

> Bohr's way of working was an even more advanced form of cooperative research [than Rutherford's]. Scientists discussed scientific matters not only through writings, but also in person, discussing and working in the same place collaboratively. . . . First, they read papers from all around the world, by dividing the work of reading among them, then, they talked about and discussed what they read. Sometimes, they would continue discussing for days. Once having reached a new idea, they would again divide the work and calculate or confirm its various aspects by experimentation. Then, they would examine and compare various results. They then would repeat this whole process. . . . There was also an atmosphere, where one could say anything without hesitation. People were used to listening attentively, not blindly accepting or ignoring what others said. After listening, they would carry out scientific research with a remarkable power of execution.
>
> This new style of atmosphere and method attracted young scientists from around the world. Since Bohr's Institute of Theoretical Physics was located in Copenhagen, the style of his Institute was called the "Copenhagen spirit," and became famous internationally. Nishina Yoshio acquired this "Copenhagen spirit," and brought it back to Japan, taking the initiative of applying this spirit to practice. Hence young scientists around Japan gathered at Nishina's institute and matured as scientists.[80]

Tomonaga Sin-itiro considered what Nishina brought to Japan to be a methodology of physics: "What was brought to us by him [Nishina] was more important than scientific discoveries or cyclotrons. He brought to us awareness of the modern methodology of physics research."[81] Other physicists, including Yukawa Hideki, Sakata Shôichi, and Hiroomi Umezawa, wrote how Nishina brought the "Copenhagen sprit" to Japan.[82]

Nishina's group conducted physics in many ways parallel to the patterns established at Bohr's institute, but there were also significant differences. Just as Bohr had, Nishina valued collaboration. Nishina often said that collaboration by two was more than twice as efficient as working alone.[83] Like Bohr, Nishina promoted collaboration between theorists, collaboration between experimentalists and theorists, and collaboration between different disciplines and different nationalities.

When Tomonaga joined Riken, Nishina emulated the method he used with Oskar Klein in Copenhagen. Each made calculations independently and compared their results at the end of the day, then moved on. This way, they could not only be sure of their results and keep going but they were also able to transmit their research know-how to each other. Nishina told his students that this was the most efficient way of working, and that this method had allowed him to detect numerous errors in his work with Klein.[84] After Sakata joined the Nishina group in the spring of 1933, Nishina left most of the actual work to his disciples. First, Nishina and Tomonaga decided on the topics, and Tomonaga and another member of the group carried out the calculations.

With Tomonaga, Sakata first worked on the calculation of the pair creation of electrons from a photon. The next year Kobayashi and Tamaki Hidehiko joined the group, and Tomonaga worked with each of them.[85] Unlike when its formulation was under investigation, quantum physics at this point required extensive calculations, as much as classical mechanics and electromagnetism might have required around the turn of the century. Younger Japanese physicists were expected to carry out such calculations. At the same time, the method that Nishina introduced prevented the tedium of the task from consuming them. The cooperative approach enabled them to relax and pay attention to aspects of the problem other than straight calculations. The method allowed them to err, because they would eventually correct errors, and it prevented them from becoming mere calculating machines.

As Bohr did, Nishina had both theorists and experimentalists in his group, and they remained in close contact. Tomonaga Sin-itiro remembers that when he joined the group in 1932, Riken's colloquia for atomic physics treated both experimental and theoretical works. The experiments of that "Wonder Year" excited Tomonaga and other participants greatly. In this year, James Chadwick discovered the neutron, Harold Urey deuterium, Carl D. Anderson (and others) the positron. These discoveries set the foundation for further developments in nuclear physics.[86] Tomonaga's later writings show his great familiarity with the experimental apparatus in this area.[87]

Collaboration continued across the inner boundaries in Riken. The Nishina group kept close ties with the groups of Nagaoka, Takamine, and Nishikawa. They regularly shared seminars and colloquia. After 1935, they jointly founded the Nuclear Physics Laboratory, where the construction of cyclotrons was undertaken as a collaborative enterprise.[88] Nishina also welcomed outsiders. In particular, young physicists from the Terazawa School of Tokyo University, such as Kotani Masao, Inui Tetsurô, and Nagamiya Takeo, often visited Nishina's group, attending seminars and colloquia.[89]

Collaboration in Nishina's group crossed disciplinary boundaries, too. Nishina's first disciple, Takeuchi Masa, was in fact trained as an applied chemist at Tokyo Higher Technical School, and knew little about physics. The collaboration between the Nishina group and Kimura Kenjirô's group of chemists at Tokyo University produced eight papers between 1938 and 1942 on uranium. Kimura, another Copenhagen alumnus, had already worked with Nishina in Copenhagen on the dependence of the x-ray spectrum on chemical binding.[90] The most important result of their collaborations in Tokyo was the discovery of uranium-237, made possible by a combination of physical technique (illumination of fast neutrons generated by the lithium-deuterium reaction and the 27-inch cyclotron constructed by Riken) and chemical technique (chemical separation of non-uranium elements).[91]

Like Bohr, Nishina was interested in collaborating with biologists. Nishina often talked with Bohr over the question of life,[92] and he was also close to George Hevesy. When Nishina's group first produced a beam from the 27-inch cyclotron, on April 3, 1937, Nishina asked Murati Kôiti, a specialist of radiobiology, to see the effect of the beam on an organism. Murati and others studied the effect of neutrons on guinea pigs. Nishina delightedly wrote to Bohr about the results of the experiment, and reported that for the first time in Japan physicists and biologists were working hand in hand.[93] At the same time, the use of radioisotopic tracers in organisms started in the Nishina group. In May 1937, Nishina recruited Nakayama Hiromi, a graduate student in biology, who started replicating Hevesy's recent experiments.[94] In these projects, Nishina was obviously emulating Niels Bohr's efforts in Copenhagen to cross disciplinary boundaries in scientific collaborations.

Japan's geographical location inhibited international collaboration. Nishina, nonetheless, gave considerable effort to invite foreign (European and American) scientists to Japan. Other than Heisenberg and Dirac, Nishina helped to bring Hevesy, Bohr, and Irving Langmuir to Japan. Often, Nishina himself translated their lectures into Japanese and published them.[95]

Apparently, Nishina was too serious and too busy to appreciate playfulness. He was generally friendly when talking one on one, but at lectures Nishina was very blunt and never told any jokes.[96] Nishina's disciples were, however, able to have fun. They took particular enjoyment and excitement in competing with European and American scholars. Takeuchi writes:

> It was fun then. When we received and opened the *Physical Review,* people were doing similar things over there. We used to say, "We were just two weeks behind them, the amount of time for mail," and really had fun.[97]

In addition, the Nishina group members had fun making up nicknames for each other. Nishina was originally called "paipan" (meaning the white tile of the game of mahjong), because of his squarish white face.[98] Later he was usually called "oyakata" (meaning a boss, generally among artisans or gangsters; "oya" means "parent"). Tomonaga was called "Shako-san" (meaning squilla or mantis shrimp, which is eaten as sushi), because someone had associated this white seafood with Tomonaga's thin and pale face. Sakata was "Bonji-san" (after a famous cartoon character of the time, Tadano Bonji, whose English approximation would be Mr. Mediocre), because Sakata resembled this character. Sagane was "Gane-san" (from his real name). Takeuchi Masa was "Getaya," Tamaki Hidehiko was "Eiboko," Ymazaki Fumio was "Donchan," Sugimoto Asao was "Genji," and Kigoshi Kunihiko was "Aodaishô."[99] The scientists in Nishina's group had fun together in various activities as well. In particular, they took short excursions from time to time to scenic places near Tokyo. In town, they often went to the movies together. Tomonaga remembers that they once saw a samurai film. When Sakata left the Nishina group to join Yukawa, they all went to see the motion picture version of "Tadano Bonji," in honor of their "Bonji."[100]

To complete certain work, some of the theorists held an extended "summer camp," for which Nishina probably paid with his own money. During the summer of 1933, Nishina, Tomonaga, and Sakata spent a month at a resort near Mt. Fuji to finish their calculation of the pair creation of electrons by gamma rays. In 1935, Tomonaga, Tamaki, and Kobayashi went to a mountain resort town north of Tokyo to translate Paul Dirac's textbook on quantum mechanics.[101]

Discussion was an important feature of the Nishina group. His disciples remember that Nishina at this point was extremely fond of chatting, a characteristic that I cannot find evidence for in his earlier life. Whenever he met a member of his group he began talking and kept at it, oblivious even to the dinner hour. Once, when Nishina and Tomonaga were reading a journal in the library, they began discussion, first whispering, but soon talking loudly, unaware of where they were, until the chemist Katsurai Tominosuke reminded them that they were in the library and asked them to be quiet.[102] Taketani Mituo, when he joined the group in 1941, was amazed by Nishina's zeal in discussion. Nishina would keep asking his junior colleagues in dead earnest: "I don't understand that. Why is that so?" Discussion could start after lunch and continue, without a break or a meal, until 10 P.M.[103]

As at Bohr's institute, lunch was an important time of collaboration in Nishina's group. When established, the Nishina group was physically located in Riken's Building No. 3, which they shared with the group of Takamine. Young members of these groups (Takeuchi and Tomonaga from the Nishina group, Suga Tarô and Fujioka Yoshio after June 1932, and sometimes Tomiyama Kotarô from the Takamine group) gathered in a room of the Takamine group and talked over lunch or tea. Though the room was rather dirty, with messy shelves and half-broken chairs, they had a good time chatting over frivolous topics.[104] From time to time, Nishina and Takamine would eat lunch in the Takamine group's office with some of their disciples.[105]

The daily luncheon meeting became an important opportunity, almost a sacred ritual. It helped to maintain group integrity, especially to keep theorists and experimentalists in close contact. Everyday, all the members of Nishina's group would eat lunch together in the cafeteria of Riken and chat on various topics.[106] In the mid 1930s, when the Nishina group became larger, they no longer had lunch in Riken's cafeteria. Construction of new buildings followed the establishment of Riken's Nuclear Physics Laboratory in 1935. Beginning in the summer of 1936, they secured a new meeting room in Building 29 as a meeting place and lunchroom for group members.[107] Nakayama Hiromi, who joined Riken in 1937, writes that those who were involved in the Nuclear Physics Lab (consisting of members from the Nishina, Nagaoka, Takamine, and Nishikawa groups) ate lunch in a meeting room every day (apparently they took lunch out from the cafeteria). In particular, all the members of the Nishina group, including Nishina himself, ate lunch at the same table every day. This custom continued even during the war. When the cafeteria had to be closed down, members of the Nishina group went out to buy meals and maintained this tradition.[108] The lunchroom was the place where newcomers were initiated. Nakayama writes that the lunch enabled the group members to talk to one another in a friendly way. When he had just joined the group, Nakayama learned, by listening to what other people talked about, to recognize other members of the group (there were a few dozen of them), and to "understand the atmosphere of the group."[109] Taketani described the atmosphere of the luncheon discussion:

> All the members would meet in the dining room at noon, and eat lunch, talking boisterously. Everything was discussed. Nowhere could one find here the dull, feudalistic, and depressed atmosphere of a university. We talked about whatever we liked, freely, without considering seniority. There was no etiquette, and outspokenness was welcomed.[110]

The seminar was no different from the lunchtime discussion. New members of Riken were surprised at the free discussions. For outsiders, the heated atmosphere of the group was almost unbearable. For example, when Tomonaga moved from Kyoto University to Riken, he was astonished at the unfettered atmosphere, where the old and the young talked to each other without formality: "The seminars were lively with the contributions of young foulmouthed and quick-thinking youth, with no deference to formality and etiquette."[111]

Thus, Nishina's attempt to emulate Bohr's institute succeeded. Nishina triggered a resonance of scientific practices to occur at Riken. The significance of Nishina's return was not limited to Riken. As his students moved to other parts of Japan, some of them successfully replicated similar practices at their new workplaces.

### Dissemination of the "Spirit": Osaka University

In 1931, the Ministry of Education founded Osaka University, with the principal objective of advancing science and technology. Nagaoka Hantarô became its president. Young and talented scientists gathered at Osaka University. Yagi Hidetsugu, an able electrical engineer, famous for the Yagi antenna, headed the department of physics, which also included Kikuchi Seishi and Yukawa Hideki. Kikuchi Seishi, a son of Kikuchi Dairoku, was producing first-class experimental works. He formed a research center of experimental atomic physics at Osaka, and would later follow the Nishina group in the construction of cyclotrons.

Nagaoka also had great expectations for Yukawa. Nagaoka, who considered himself a research scientist and not an administrator, was not at all happy to "be forced to become president," but nevertheless rejoiced at having good young scientists, such as Kikuchi and Yukawa. When his son asked how bright Yukawa was, Nagaoka replied, "What is the use of those 'brightest and best,' who are just good at the subjects that the Ministry of Education decides on? . . . Yukawa has originality. That is what counts. He is not one of those 'brilliant' boys who just fit in with the rules of the old people."[112]

When Yukawa took the job of lecturer at Osaka in 1933, he must have been quite alone except for Kikuchi's experimental group. Furthermore, Yukawa was unproductive for a year or two. Yagi Hidetsugu, who had wanted Tomonaga, accepted Yukawa in the department only because Tomonaga was taken by Riken. Yagi was initially not happy with Yukawa, and Yukawa's apparent sterility made Yagi even less happy and he often scolded Yukawa for not publishing.[113]

The situation began to change in 1934. Sakata Shôichi, who had been a research student (*kenkyûsei*) under Nishina, gained a job at Osaka as a research associate to Yukawa. A research school of theoretical physics began to form around Yukawa. Aristocratic and introverted, he was the type of scientist who worked in isolation and had a completely different personality from Nishina. Uchiyama Tatsuo writes that when he, as a student, knocked on the door, Yukawa and Sakata would turn from the desks to him. While Sakata would show an amiable face with a slightly surprised smile, Yukawa always appeared ill-tempered, as if asking "What's your business here?"[114] In contrast to Yukawa, Sakata was, in Taketani's words, a "genius of organizing people into collaborative research."[115] Furthermore, Sakata, trained by Nishina and a witness to his school-building, was aware of the new cooperative method of atomic physics. He used the method of Nishina's school in Osaka. When Tanigawa Yasutaka, the earlier disciple of Yukawa and Sakata joined the school, Sakata worked with him. Tanigawa copied Sakata's calculation notes, in which Sakata meticulously wrote down necessary formulas. Then Tanigawa did his own calculations and compared the result with Sakata's.[116]

Similarly important to the theoretical physics group in Osaka University was Kikuchi Seishi and his experimental physics group in nuclear physics. Kikuchi was a young and able experimentalist who had been to Europe, spending about three years in Leipzig and Göttingen. Since his return to Japan, he had been working at Riken, until he was appointed to Osaka University. Kikuchi was working on experimental nuclear physics, and although they had a late start, this group would construct the first cyclotron in Japan, one month before Nishina's cyclotron began operation. Yukawa had a closer relation to Kikuchi's group than to his official superior, Professor Okaya Tokiharu.

The Kikuchi and Yukawa groups had lunch together in the room called "the Kikuchi Dining Hall," a simple laboratory room with a large table, located opposite Yukawa and Sakata's office.[117] It was customary for them to play board games after lunch (or for some, while eating lunch). Yukawa and Kikuchi always played the game of go. Others played Japanese chess. While the senior physicists played the game of go quietly at the end of the table, their disciples played chess with youthful clamor. The young physicists, so engrossed in the game, began to play chess longer and longer, once until tea time at 3 P.M. At this point, Kikuchi banned the game temporarily.[118]

Nothing provided a greater incentive for the development of collaborative work in theoretical physics at Osaka University than Yukawa's meson theory. When Yukawa presented his theory at a conference in 1934, the audience did not receive it very favorably. As mentioned, however, Nishina encouraged Yukawa. Yukawa, with his wife's help, wrote up the paper in English, and it was published in 1935.[119] Nishina's encouragement was important, since the significance of Yukawa's paper was not initially clear. Niels Bohr, for example, teased Yukawa by asking "Do you like new particles?" when he visited Japan in 1937.[120] In addition to Nishina, Tomonaga showed a strong interest in Yukawa's theory. Not only did he write letters to Yukawa asking for his recent works; he also affirmed the value of Yukawa's paper to Fujioka Yoshio, an experimentalist at Riken, who wondered whether or not to write a review article about Yukawa's theory for a science magazine.[121]

Thus, a resonance of scientific practices occurred in Osaka. Like Riken, Osaka University was able to accommodate a new research school because of its short history and young personnel. Sakata's role was similar to that of Nishina. He triggered what was waiting to happen.

Others who learned from Nishina further spread and created research centers in various places in Japan with varying degrees of success. When Yukawa succeeded Tamaki Kajûrô at Kyoto University, he took Sakata and Tanigawa to Kyoto with him, while Kobayashi and Taketani remained in Osaka. In 1941, Tomonaga gained a position at Tokyo Bunrika University, and later created an important research group there.[122] In the next year, Sakata founded a group in Nagoya. Two other Nishina disciples, Umeda Kwai and Ozaki Masaharu, worked at Hokkaido University and at Tohoku University respectively. The list could be expanded significantly, if the third generation were to be included.[123]

### Did the "Copenhagen Spirit" Work in Japan?

As we have seen above, some Japanese physicists perceived that the "Copenhagen spirit" brought by Nishina shaped the theoretical physics community in Japan. In many aspects, the research activity of the Nishina group in Riken was similar to that of Copenhagen, and the atmosphere in Riken seems to have been transferred to Osaka University. The importance of group lunches, informal discussion, the neglect of protocol, the lively atmosphere, and the inter- and intra-disciplinary collaboration are all characteristic of both the Japanese and the Danish labs.

There are, however, some difficulties in interpreting the process of the growth of quantum physics in Japan in terms of the dissemination of the "Copenhagen spirit." First, there were a few aspects in which the style of physics in Nishina's group was different from that of Bohr's group in Copenhagen. There was no philosophical bent among the disciples of Nishina. As I mentioned, Nishina himself was interested in complementarity and the foundational problems of quantum mechanics. His interest, however, was not infectious to his fellow Japanese physicists and disciples.[124] Nor did Nishina himself allow philosophical considerations to interfere with his physics. When Yukawa presented his seminal idea of meson theory in 1932 and 1933, Nishina was supportive. Bohr's reaction evidently indicated that Bohr thought solving a problem by creating a new particle was a naive realist attitude and was therefore unacceptable. Such a philosophical nicety did not concern Nishina.

Another aspect in which Nishina differed from Bohr was in Nishina's way of directing his group. Nishina was much more dictatorial than Bohr. He could fire group members who disagreed with him and could impose his ideas on the group. The relation between Nishina and his disciples was much more formal than the relation between Bohr and other scientists in his institute.[125] In response to this situation, Nishina's disciples had to devise a way to "control" Nishina, without his knowing. For example, occasionally Nishina would demand that a group member carry out an impossible task. A smart disciple, such as Tomonaga, would not immediately object that his demand was unreasonable. Rather, he would initially accept the demand. Then, he would go back to Nishina in a few days, tell him that the requested task would take several hundred sheets of papers to calculate and several hundred hours of time, and ask Nishina whether he should continue the work. This way, Nishina's disciples managed to "control" Nishina, but the very necessity of such a tactic indicates the difference between Nishina and Bohr.[126] We can also say the same thing about Yukawa's group in Osaka, as the post-lunch games scene in the "Kikuchi Dining Hall" would suggest. The group leaders played a different game, in a different manner, at the end of the table, without joining their disciples. It was a clear indication of the fact that they were not equal to the other group members.

Additionally, the relation between the Osaka group and Riken, or rather Yukawa, cannot be understood if the skills necessary to quantum theorists dubbed as the "Copenhagen spirit" could only be transmitted through personal contacts. We have already seen that Yukawa was not a person like Nishina or Bohr, who were both managers of organized research groups. It was Sakata who created the Osaka group of theoretical physicists. On the one hand, the "Copenhagen spirit" explains the rise of the Osaka group well. It fails, however, to account for Yukawa, who had only little contact with Nishina. Yukawa attended Nishina's intensive lecture course on quantum mechanics in 1929, asked him a few questions after the lectures, and dined with him. Yukawa presented his theory at a conference, which Nishina commented on encouragingly. These occasions were important for Yukawa, who considered Nishina his mentor. Yet it is hard to believe that Yukawa could know the "Copenhagen spirit" through these occasions. More definitive evidence is the conversation that took place in 1967 (15 years after he wrote about the "Copenhagen spirit" in his autobiography, *Tabibito,* as quoted above), between Yukawa, Tomonaga, and their college friend Tamura Matsuhei. Yukawa asked Tomonaga what the "Copenhagen spirit" was: "I have been wanting to ask you, Tomonaga-san. Until today, I have never understood what the 'Copenhagen spirit' is."[127] This conversation suggests that Yukawa was outside the tradition of the Copenhagen school, from Bohr to Nishina to Tomonaga. Yet Yukawa was able to produce his theory of mesons, the first important quantum-mechanical work completed in Japan. This implies that there were already conditions in Japan that allowed someone to become a successful theoretical physicist without any contact with the "Copenhagen spirit."

Also, Nishina might not have been the only person responsible for the atmosphere of his group. There were other physicists, especially young physicists who came back from Europe in the early 1930s, who greatly contributed to the atmosphere of the group. In particular, Fujioka Yoshio studied in Leipzig from 1929 to 1932 with Heisenberg, and Kikuchi Seishi studied with Max Born and Heisenberg from 1929 to 1931. Tomonaga testifies that these two young physicists were crucial in shaping the atmosphere of the colloquia in Riken.[128] In particular, Fujioka, a member of Heisenberg's group, was impressed with the way collaboration was carried out there. Around 1930, Fujioka wrote to Nagaoka about the theoretical physics seminar at the University of Leipzig:

> Discussions at the seminar are often inspiring. The most important thing about this gathering of these first rate scientists is its scholarly atmosphere. . . . Discussion and research are carried out very smoothly and naturally, which creates a certain atmosphere. It is very enviable, and something that we should learn from them.[129]

When Fujioka came back to Japan, he organized a symposium format presentation and discussion at the semi-annual conference at Riken, in an attempt to emulate the Leipziger Vorträge under Heisenberg. Tomonaga remembers that Kikuchi and Fujioka were sometimes too forthright with older professors. Fujioka once criticized Kimura Masamichi, a professor at Kyoto University and an authority in spectroscopy, and Kikuchi also made a harsh comment after a talk by Honda Kôtarô, a revered senior experimental physicist famous for his metallurgical studies.[130] Kikuchi played an important role in creating a research environment when he moved to Osaka, as we have seen.

Not all the Japanese physicists were so sympathetic to Bohr and the Copenhagen school of physics. Physicists around Yukawa, especially Taketani Mituo, developed a methodology of physics that was alien to Bohr's philosophical tendencies. Taketani, a Marxist, regarded Bohr's philosophy as a kind of "bourgeois Machian idealism," while attributing the success of Yukawa's theory and the Yukawa group to its realistic attitude (such as in the prediction of a new particle).[131]

As a resonance in physics does not always imply a reproduction of the same occurrence, a resonance of scientific practices does not mean their precise replication. What happened was not a transportation of something (be it a spirit or tacit knowledge), but recreation of an event under reasonably similar conditions, with some mediation.

## Conclusion: Revisiting the Metaphor of "Spirit"

While physicists in Niels Bohr's group and Japanese physicists under Nishina's leadership considered themselves imbued with the "Copenhagen spirit," their versions of the "Copenhagen spirit" had both similarities and differences. As for the collective working style, they both shared the ideal of collaboration between the members of their group, between theorists and experimentalists, and between scientists in different disciplines. Nishina's group, however, did not realize much international collaboration, while at Bohr's institute in Copenhagen, internationalism was an essential part of the "Copenhagen spirit" both in theory and in practice.

Both groups operated under a single male leader. Leadership played a crucial role in both groups. Although never imperious, Bohr was the boss of his institute, and scientists were under his leadership.[132] Both Bohr and Nishina can be seen as paternal figures to the younger physicists of their groups. While Bohr's management of the group was subtler, Nishina was much more patriarchal, probably reflecting Japan's more autocratic society.

Playfulness was an integral part of those groups, but in a somewhat different way. Physicists in both groups enjoyed solving physics problems together, and doing other related activities as a group. Nevertheless, the leaders were different from each other. Niels Bohr participated eagerly in youthful whimsy, sometimes surpassing his younger colleagues in this respect. Nishina was much more sober and restrained, while his disciples matched the physicists in Copenhagen.

The style, the method, and the values that Japanese physicists perceived in the "Copenhagen spirit" were certainly instrumental in creating a successful research school of theoretical physics in Japan and in inspiring Japanese theoretical physicists to explore new problems and new fields. Yet the "Copenhagen spirit" was not the only factor that enabled the creation of a productive research tradition of theoretical physics in Japan. It was one of a few major factors. Nor was the "Copenhagen spirit" that was brought by Nishina exactly the same as the original one.

The metaphor of spirit, therefore, does not work in this case. Our tendency to reify intellectual activities seems to be hiding behind this metaphor. Yet intellectual activity is not something that one can carry and move around. It is an event, or a phenomenon, rather than an entity. Instead of talking about the transplantation of "spirit," therefore, I see the dissemination of quantum mechanics as a "resonance," which occurred through various kinds of mediation. This model implies three things. First, it involves mediators, some of them human (such as Nishina) and others material (such as physics journals). Second, the process of translation across cultures transforms the practice, incorporating the new into the old. Quantum mechanics, like any element of scientific knowledge, was not a stable entity one could trivially transport across continents and simply insert into pre-existing educational institutions and cultural settings. Rather, the transmission of quantum mechanics from Europe to Japan was a resonance of two events, mediated on multiple levels, including formal mathematical theories, cultural values, skills, techniques, and meanings. Third, such mediation does not have to take place on all the levels, and the mediation does not imply a global or a total relocation of contexts. Tuning forks do not need to be identical for them to resonate. To resonate to the progress of theoretical physics, Yukawa did not have to know the "Copenhagen spirit."

If Erwin von Baelz felt that many Japanese in the early twentieth century regarded science as if it were a "machine which can turn out so much work every year, and therefore as a machine which can without further ado be transported from the West to any other part of the world there to continue its labours,"[133] he probably had good reasons to think so. Yet it was a mistake on his part to think that there was such a thing as a "spirit" of science that could simply be "implanted" into Japan by Western practitioners. Even in this relatively small scale of knowledge-transfer, the situation was more than simply transplanting the "Copenhagen spirit." The worldly, concrete institutions of training—with their architectures, values, and mores—molded quantum mechanics and its Japanese practitioners far more extensively than any elusive and ethereal "spirit."

## Notes

The following abbreviations are used in the notes: *NY*: *Nihsina Yoshio: Nihon no genshi kagaku no akebono*, ed. Tamaki Hidehiko and Ezawa Hiroshi (Misuzu Shobô, 1991); *TSC*: *Tomonaga Sin-itiro chosakushû* (Misuzu Shobô, 1981–1985). Japanese personal names appear here in the traditional order except as the authors of European language works. Transliteration of Japanese names followed Koh Masuda, ed., *Kenkyusha's New Japanese-English Dictionary* (Kenkyusha, 1974) or the preference of the name holder when it is known.

1. Erwin Baelz, *Awakening Japan: The Diary of a German Doctor*, ed. Toku Baelz, tr. Eden Paul and Ceder Paul (Viking, 1932), 149–150; emphasis added.

2. See Masao Watanabe, "Science across the Pacific: American-Japanese scientific and cultural contacts in the late nineteenth century," *Japanese Studies in the History of Science* 9 (1970): 115–136.

3. For example: Hirosige Tetu, *Kagaku no shakaishi: Kindai Nihon no kagaku taisei* (Chûôkôronsha, 1973).

4. Werner Heisenberg, *Physical Principles of Quantum Theory* (University of Chicago Press, 1930), preface.

5. John Heilbron, "Earliest missionaries of the Copenhagen spirit," in *Science in Reflection*, ed. Edna Ullmann-Margalit (Kluwer, 1988), 201–233.

6. Harry M. Collins, *Changing Order: Replication and Induction in Scientific Practice* (Sage, 1985), 103–104.

7. *Riken* is short for *Rikagaku Kenkyûjo*.

8. In Japanese, *kôtôgakkô*. "Higher school" is now the standard translation, and this word, though it sounds unfamiliar, is convenient to distinguish it from the American or postwar Japanese "high school."

9. Nitta Isamu, *Nagareno nakani: kagakusha no kaisô* (Tokyo Kagaku Dôjin, 1973), 103–104.

10. *Daisan Kôtô Gakkô ichiran Taishô 12-nen yori 13-nen ni itaru* (Kyoto: Daisan Kôtô Gakkô, 1924), 63–66.

11. Sharon Traweek, *Beamtimes and Lifetimes: The World of High Energy Physicists* (Harvard University Press, 1988).

12. Yuasa Toshiko, probably the only important Japanese female physicist during this period, took this path. She graduated from the Women's Higher Normal School, and made herself an experi-

mental physicist in Paris. See Kenji Ito, "Gender and physics in 20th century Japan: Yuasa Toshiko's case," *Historia Scientiarum,* 2005.

13. Takeuchi Masa, for example, the first disciple of Nishina, was a graduate of Tokyo Higher Technical School. He managed to enter Riken as a non-paid research fellow (*kenkyûsei*). Takeuchi Masa, "Nishina kenkyûshitsu monogatari," in *NY,* 209.

14. Tomonaga Sin-itiro, "Taidan kagaku no imi," in *Butsurigaku to watashi,* volume 2 of *TSC* (Misuzu Shobô, 1982), 343.

15. Katsuki Atsushi, *Ryôshirikigaku no shokkô no nakade* (Seirinsha, 1991), 40.

16. Kenji Ito, Making Sense of Ryôshiron: Introduction of Quantum Physics Into Japan, 1920–1945 (Ph. D. dissertation, Harvard University, 2002), chapter 2.

17. Kajûrô Tamaki, "Note on general equations for electromagnetic fields in a moving system," *Memoirs of the College of Science and Engineering, Kyoto Imperial University* 3 (1911): 111.

18. Kajûrô Tamaki, "Reflexion and refraction phenomena relating to a moving medium," *Memoirs of the College of Science, Kyoto Imperial University* 2 (1916–17): 59–104.

19. For other examples, including Aichi Keiichi, Sano Shizuo, and Terazawa Kwan-iti, see, Ito, Making Sense of Ryôshiron, chapter 2.

20. Dong-Won Kim, "Emergence of theoretical physics in Japan," *Annals of Science* 52 (1995): 948–949.

21. Katsuki, *Ryôshirikigaku no,* 45.

22. Kim, "Emergence," 948–949.

23. For example, both Yukawa and Tomonaga had a professor at Kyoto University as their father. Their families were naturally supportive of their choice to pursue academic careers, and their decision to be an unpaid assistant in science caused no apparent conflict. This would not be the case for a person from a poor or non-academic family.

24. *Tokyo Teikoku Daigaku ichiran, Taisô 8 yori Taisho 9 ni itaru* (Tokyo Teikoku Daigaku, 1920), 98–101.

25. For the establishment of Riken, see Kiminobu Itakura and Eri Yagi, "The Japanese research system and the establishment of the Institute of Physical and Chemical Research," in *Science and Society in Modern Japan: Selected Historical Sources,* ed. Shigeru Nakayama, David L. Swaine, and Eri Yagi (University of Tokyo Press, 1974), 158–201.

26. Itakura and Yagi, "Japanese research system," 195–196.

27. According to Karl Hall, Soviet theoretical physicists took similar roles in the midst of Stalin's five-year plans. See Karl Hall, Purely Practical Revolutionaries: A History of Stalinist Theoretical Physics (Ph. D. dissertation, Harvard University, 1999).

28. Ito, Making Sense of Ryôshiron, chapter 3.

29. Butsurigaku Rinkôkai Dôjin, *Butsurigaku bunkenshô 1* (Iwamami Shoten, 1927).

30. Max Born, *Probleme der Atomdynamik* (Verlag von Julius Springer, 1926). The Japanese translators of this book erroneously write the title as "Mechanics of the Atom." See Hideki Yukawa, *Tabibito (The Traveler)*, tr. L. Brown and R. Yoshida (World Scientific, 1982), 220. See also Tamura Matsuhei, Tomonaga Sin-itiro, and Yukawa Hideki, "Zadankai: Nihon ni okeru soryûshiron no reimei," *Kagaku* 38 (1967): 390.

31. Richard Courant and David Hilbert, *Methoden der Mathematischen Physik* (Springer, 1924).

32. Nishijima Kazuhiko, Yamaguchi Yoshio, Yamanouchi Takahiko, and Tomonaga Sin-itiro, "Zadankai ryôshirikigaku no shôgeki to taiken," in *Ryôshirikigaku: Tanjô kara 60-nen* (Saiensusha, 1984), 11–12.

33. Tamura Matsuhei, Tomonaga Sin-itiro, and Yukawa Hideki, "Nihon ni okeru soryûshiron no reimei," 389.

34. Tamura Matsuhei, Tomonaga Sin-itiro, and Yukawa Hideki, "Nihon ni okeru soryûshiron no reimei," 394. As a perusal of Heisenberg's book reveals, Yukawa was exaggerating. There are still many equations in the text. The fact that this book gave such a strong impression on him supports my point even more eloquently.

35. See Ito, Making Sense of Ryôshiron, chapters 5 and 3.

36. Kimura Kenjirô, "Kopenhâgen no Nishina hakase," in *NY*, 35.

37. Yoshio Nishina, letter to Niels Bohr, March 25, 1923, in *Y. Nishina's Letters to N. Bohr, G. Hevesy, and Others*, volume 21 (Nishina Kinen Zaidan, 1985), 1.

38. Stefan Rozental, *Niels Bohr: Memoirs of a Working Relationship* (Christian Eilers, 1998), 12.

39. Bohr, letter to Nishina, March 29, 1923; Koizumi Kenkichirô, "Yôroppa ryûgaku jidai no Nishina Yoshio: Rironbutsurigaku no sendatsu no kiseki," *Shizen*, November 1976, 67.

40. Nishina, letter to Niels Bohr, March 25, 1923.

41. Nishina Kôjirô, "Chichi Yoshio no ryûgaku seikatsu," in *NY*, 270.

42. Kimura Kenjirô, "Kopenhâgen no Nishina hakase," 40.

43. Yoshio Nishina and I. I. Rabi, "Absorptionskoeffizient der Röntgenstrahlen nach der Quantentheorie," *Verhandlungen der Deutschen Physikalischen Gesellschaft* 9 (1928): 6–9.

44. Oskar Klein and Yoshio Nishina, "Über die Streuung von Strahlung durch freie Elektronen nach der neuen relativistischen Quantendynamik von Dirac," *Zeitschrift für Physik* 52 (1928): 853–868.

45. The participants of this conference left their signatures on a sheet of paper, which is reproduced in Nishina Yoshio hakase wo shinobu kai, ed., *Zenryokushissô no jinsei* (Kagaku Shinkô Nishina Zaidan, 2001), 26. Nishina's name is there among the participants.

46. Riken's archive holds Nishina's manuscripts for this translation.

47. The letter is reproduced in Tamaki Hidehiko, "Butsurigakushi shiryô iinkai dayori, No. 7," *Butsuri* 39 (1984): 160.

48. Heilbron, "Earliest missionaries," 200.

49. Finn Aaserud, *Redirecting Science: Niels Bohr, Philanthropy, and the Rise of Nuclear Physics* (Cambridge University Press, 1990), prologue.

50. Léon Rosenfeld, *Niels Bohr: An Essay Dedicated to Him on the Occasion of His Sixtieth Birthday, October 7, 1945*, second edition (North-Holland, 1961), 3.

51. Heilbron, "Earliest missionaries."

52. Heisenberg, *Physical Principles*.

53. Rosenfeld, *Niels Bohr: An Essay*, 3.

54. Tomonaga Sin-itiro and others, "Zadan kisobutsurigaku kenkyûjo wo megutte," in *Hiakareta kenkyûjo to shidôsha tachi*, volume 6 of *TSC* (Misuzu Shobô, 1982), 246–247.

55. Victor Weisskopf, "Niels Bohr, the quantum and the world," in *Physics in the Twentieth Century: Selected Essays* (MIT Press, 1972), 55.

56. Address delivered at a celebration of Max Planck's sixtieth birthday (1918) before the Physical Society in Berlin; Albert Einstein, "Principles of research," in *Ideas and Opinions*, tr. Sonja Bargmann (Bonanza, 1954), 224.

57. In regards to the contrast between Einstein and Bohr in terms of personal tastes and values in physics, see David Kaiser, "Bringing the human actors back on stage: The personal context of the Einstein-Bohr debate," *British Journal for the History of Science* 27 (1994): 129–152; Ito Kenji, Bôa Ainshutain Ronsô (B.A. thesis, University of Tokyo, 1990, in Japanese).

58. Rosenfeld, *Niels Bohr: An Essay*, 19.

59. Rozental, *Niels Bohr*, 34, 37.

60. Weisskopf, "Niels Bohr, the quantum and the world," 55.

61. Peter Robertson, *The Early Years: The Niels Bohr Institute 1921–1930* (Akademisk Forlag, 1979), 134.

62. Rozental, *Niels Bohr*, 15.

63. On Hevesy, see Hilde Levi, *George de Hevesy: Life and Work: A Biography* (A. Hilger, 1985).

64. Rosenfeld, *Niels Bohr: An Essay*, 6.

65. Aaserud shows how Bohr had to change his approach "from one based on a unity between theory and experiment to one based on theoretical discussion" in the 1920s and began a new effort to unite theory and experiment in the 1930s. See Aaserud, *Redirecting Science*, chapters 4–6; quotation on 102.

66. Otto R. Frisch, *What Little I Remember* (Cambridge University Press, 1979), 101.

67. See, for example, Gamow, *Thirty Years That Shook Physics: The Story of Quantum Theory* (Doubleday, 1966), 62.

68. Rozental, *Niels Bohr*, 21.

69. Robertson, *The Early Years*, 135.

70. Rozental, *Niels Bohr*, 21.

71. Robertson, *The Early Years*, 135.

72. Frisch, *What Little I Remember*, 90.

73. Robertson, *The Early Years*, 136–137. For the transcript of this play, see Gamow, *Thirty Years*, 167–218. See also Mara Beller, "Jocular commemorations: The Copenhagen spirit," *Osiris* 14 (1999): 252–273.

74. For the analysis of this "journal," see Beller, "Jocular commemorations." See also Hans Bohr, "My father," in *Niels Bohr: His Life and Work as Seen by His Friends and Colleagues*, ed. Stefan Rozental (North-Holland, 1967), 325–399.

75. Nishina Yoshio, *Niels Bohr*, Iwanami kôza butsurigaku, gakusha denki (Iwanami Shoten, 1938), 2.

76. Tamaki Hidehiko, "Sekai no gakkai ni mado wo hiraku," in *NY*, 84.

77. Ibid., 84–85.

78. Tomonaga Sin-itiro, "Wagashi wagatomo," in *Chôjûgiga*, volume 1 of *TSC* (Misuzu Shobô, 2001), 200.

79. Tomonaga Sin-itiro, "Omoide banashi," in *Chôjûgiga*, volume 1 of *TSC* (Misuzu Shobô, 2001), 204.

80. Tamaki Hidehiko and Iwasaki Masao, *Nishina Yoshio* (Kokudosha, 1976), 90–91.

81. Tomonaga Sin-itiro, "Nishina sensei," *Kagaku* 21 (1951): 212.

82. Yukawa, *Tabibito*, 177; Sakata Shôichi, "Chûkanshi riron kenkyû no kaiko," 91; Hanada Keisuke, Umezawa Hiroomi, and Shizuma Yoshitsugu, "Nihon no kagaku to shisô," in *Kindaikato dentô*, volume 8 of *Kindai Nihon shisô kôza* (Chikuma shobô, 1959), 334. As for Umezawa, see *Selected Papers of Hiroomi Umezawa*, ed. Hiroshi Ezawa et al. (Syokabo, 2001).

83. Tomonaga Sin-itiro and others, "Zadan Nishina sensei wo shinonde," in *Hirakareta kenkyûjo to shidôsha tachi*, volume 6 of *TSC* (Misuzu Shobô, 1982), 80.

84. Kobayashi Minoru, "Riron kenkyû," in *NY*, 99.

85. Yoshio Nishina, Sin-itiro Tomonaga, and Shôichi Sakata, "On the photo-electric creation of positive and negative electrons," *Supplement to Scientific Papers of the Institute of Physical and Chemical Research* 17 (1934): 1–5; Yoshio Nishina, Sin-itiro Tomonaga, and Minoru Kobayashi, "On the Creation of Positive and Negative Electrons by Heavy Charged Particles," *Supplement to Scientific Papers of the Institute of Physical and Chemical Research* 27 (1935): 137–177; Yoshio Nishina,

Sin-itiro Tomanaga, and Hidehiko Tamaki, "On the Annihilation of Electrons and Positrons," *Supplement to Scientific Papers of the Institute of Physical and Chemical Research* 18 (1934): 7–12; Yoshio Nishina, Sin-itiro Tomonaga, and Hidehiko Tamaki, "A note on the interaction of the neutron and the proton," *Supplement to Scientific Papers of the Institute of Physical and Chemical Research* 30 (1936): 61–69.

86. Helge Kragh, "The rise of nuclear physics," in Kragh, *Quantum Generations: A History of Physics in the Twentieth Century* (Princeton University Press, 1999), 174–189; Emilio Segrè, "The wonder year 1932: Neutron, positron, deuterium, and other discoveries," in Segrè, *From X-Rays to Quarks: Modern Physics and Their Discoveries* (Freeman, 1980), 19–67.

87. Tomonaga Sin-itiro, "Kenkyû seikatsu no omoide," in *Butsurigaku to watashi*, volume 2 of *TSC* (Misuzu Shobô, 1982), 302–303.

88. Takeuchi Masa, "Nishina kenkyûshitsu monogatari," 218.

89. Tomonaga Sin-itiro and others, "Kizobutsurigaku kenkyûjo wo megutte," 248.

90. Shin'ichi Aoyama, Kenjirô Kimura, and Yoshio Nishina, "Die Abhängigkeit der Röntgenabsorptionsspektren von der chemischen Bindung," *Zeitschrift für Physik* 44 (1927): 810–833.

91. Saitô Nobufusa, "Nishina Yoshio to aisotôpu," in *NY*, 128–129; Saitô Nobufusa, "Nishina Kimura no mei konbi," *Nihon Genshiryoku Gakkaishi* 32, no. 7 (1990): 697–698.

92. Takeuchi Masa, "Nishina sensei to hôshasen seibutsugaku," in *NY*, 154.

93. Ibid., 155.

94. Nakayama Hiromi, "Torêsâ to shokubutsu seiri no kenkyû," in *NY*, 135–137.

95. Werner Heisenberg and P. A. M Dirac, *Ryôshiron sho mondai*, tr. Nishina Yoshio (Keimeikai, 1931).

96. Tomonaga Sin-itiro, "Kagaku no imi," 344.

97. This comment by Takeuchi refers to the subgroup of experimentalists, but Tamaki Hidehiko confirms that it was also the case in the theory subgroup: Tomonaga Sin-itiro and others, "Nishina sensei wo shinonde," 65.

98. Tomonaga Sin-itiro and others, "Nishina sensei wo shinonde," 84.

99. Partially based on the reminiscence by Yokoyama Sumi (Nishina's secretary): *Shôwa shi no tennô*, ed. Yomiuri Shinbunsha, volume 4 (Yomiuri Shinbunsha, 1968), 89.

100. Tomonaga Sin-itiro, "Sakata-kun no koto," in *Butsurigaku to watashi*, volume 2 of *TSC* (Misuzu Shobô, 1982), 244.

101. Tamaki Hidehiko, "Nishina Kenkyûshitsu rironbu no omoide," Matsui Makinosuke, ed., *Kaisô no Tomonaga Sin-itiro* (Misuzu Shobo, 1980), 167–173. Also see Nogami Yaeko, "Tomonaga-san to Kitakaru," ibid., 186–194.

102. Tomonaga Sin-itiro and others, "Nishina sensei wo shinonde," 68–69.

103. Yomiuri Shinbunsha, *Shôwa shi no tennô,* 164.

104. Tomonaga Sin-itiro, "Tomiyama san no omoide," in *Butsurigaku to watashi,* volume 2 of *TSC* (Misuzu Shobô, 1982), 253–254.

105. Takeuchi Masa, "Nishina kenkyûshitsu monogatari," 217.

106. Riken provided either a bowl of rice or a roll free of charge. See Takeuchi Masa, "Nishina kenkyûshitsu monogatari," 216.

107. Takeuchi Masa, "Nishina kenkyûshitsu monogatari," 218.

108. Nakayama Hiromi, "Torêsâ to shokubutsu seiri no kenkyû," 138.

109. Ibid.

110. Taketani Mituo, "Soryûshiron gurûpu no keisei: watashi no me de mita," in *Shinri no ba ni tachite,* ed. Yukawa Hideki, Sakata Shôichi, and Taketani Mituo (Mainichishimbunsha, 1951), 187.

111. Tomonaga Sin-itiro, "Wagashi wagatomo," 201.

112. Nagaoka Haruo, "On'yônaru bishô," in *Tsukiai,* ed. Yasutaka Tanigawa (Kôdansha, 1968), 211–216.

113. Ibid., 168–169.

114. Ibid., 167.

115. Taketani Mituo, "Soryûshiron gurûpu no keisei," 165.

116. Taniakawa Yasutaka, "Aru jidai," in *Tsukiai,* ed. Tanikawa Yasutaka (Kôdansha, 1968), 157.

117. Uchiyama Tatsuo, "Handai kyôju Yukawa hakase," 168.

118. Taniakawa Yasutaka, "Aru jidai," 159–160.

119. Hideki Yukawa, "On the interaction of elementary particles, I," *Proceedings of the Physico-Mathematical Society in Japan, Series III* 17 (1935): 48–57.

120. Tamura Matsuhei, Tomonaga Sin-itiro, and Yukawa Hideki, "Nihon ni okeru soryûshiron no reimei," 395.

121. Fujioka Yoshio, "Yukawa-san ni hajimete atta koro no hanashi," in *Tsukiai: Yukawa hakase kanreki kinen bunshû,* edited by Yasutaka Tanigawa (Kodansha, 1968), 7.

122. Silvan S. Schweber, *QED and the Men Who Made It: Dyson, Schwinger, Feynman, and Tomonaga* (Princeton University Press, 1994), chapter 6.

123. Yoshinori Kaneseki, "The elementary particle theory group," in *Science and Society in Modern Japan: Selected Historical Sources,* ed. Shigeru Nakayama, David L. Swain, and Yagi Eri (University of Tokyo Press, 1974), 221–235.

124. See Ito, Making Sense of Ryôshiron, chapter 7.

125. If a disciple disagreed with Nishina on an important issue, such as how to construct a cyclotron, he had to leave the group. See Ishii Chihiro's remark in Tomonaga Sin-itiro et al., "Nishina sensei to kakubutsuri no hatten," in *Hirakareta kenkyûjo to shidôsha tachi*, volume 6 of *TSC* (Misuzu Shobô, 1982), 115.

126. Tomonaga Sin-itiro and others, "Nishina sensei wo shinonde," 69–70.

127. Tamura Matsuhei, Tomonaga Sin-itiro, and Yukawa Hideki, "Nihon ni okeru soryûshiron no reimei," 394.

128. Tomonaga Sin-itiro, "Keikaku onchi: Fujioka-san no omoide," in *Butsurigaku to watashi*, volume 2 of *TSC* (Misuzu Shobô, 1982), 278–279.

129. Tomonaga Sin-itiro, "Ryôshirikigaku to watashi," in *Nihon no butsurigakushi*, Nihon Butsurigakkai (Tôkai Daigaku Shuppankai, 1978), 649.

130. Tomonaga Sin-itiro, "Fujioka-san no omoide," 278–279.

131. See Ito, Making Sense of Ryôshiron, chapter 7.

132. For an analysis of Bohr's leadership, see Beller, "Jocular commemorations," 254–257.

133. Baelz, *Awakening Japan*, 149.

# 6 Instruments in Training: The Growth of American Probe Microscopy in the 1980s

Cyrus C. M. Mody

The replication and spread of laboratory techniques is an old concern in science and technology studies.[1] Similarly, the field has learned much about processes of teaching and generational shift in science.[2] Few researchers, however, have worked the intersection of these themes.[3] This essay proceeds from this intersection, taking seriously the interweaving of replication issues with pedagogical environment. I will try to illustrate this through a story about the development of scanning probe microscopy—especially scanning tunneling microscopy and atomic force microscopy—in corporate research labs and academic groups in the mid 1980s.[4]

In the early 1980s, no one knew how to build a probe microscope or how to interpret a scanning microscope image. These questions became tied to problems in figuring out who was qualified to build and operate the microscopes, how probe microscopists should behave, what kind of training they should have, and what kind of career might be appropriate for them. The notion of a moral economy of experimentation (as laid out by Robert Kohler, drawing on E. P. Thompson) is a fitting way to describe the settings in which these questions were answered.[5] Analytically, we can use "moral economy" as a way of talking about participants' expectations of how tools and tool users should behave, how various actors take up different social positions and give meaning to those roles, and how social differentiation and inequality in the laboratory—underwritten and/or challenged by an evolving moral order—affect (and effect) the production of knowledge. Pedagogically oriented laboratories are paradigmatic examples of moral economies—they abound with social inequalities and unstated expectations and they subsist on the ongoing exchange of information, techniques, and materials between different kinds of participants. Laboratories are places where—simultaneously—new experimentalists are trained and new techniques are forged, in ways that are framed by (but also rewrite) moral order.

With scanned probes, we have a comparative study to clarify the roles of pedagogy and moral economy in crafting new techniques. The first "STMers" in North America quickly segregated into two distinct experimental regimes. Though these groups maintained a common core set for several years, they held very different views on the relationship between training and experimental practice, the primacy of tacit or formal knowledge, and the legitimacy of looking beyond institutional and disciplinary boundaries for help. Some of these differences pre-dated scanning tunneling microscopy, others were co-constructed along with the new instrument. In the end, these groups differed radically in the knowledge they produced, and in what knowledge *could be produced* with the instruments they built. This difference originated, in part, in their methods for creating locally adequate skills and training subjects to possess them.

## The First STMs

Scanning tunneling microscopy began at the IBM research lab in Zürich in the late 1970s. A team building a computer based on Josephson junctions asked a colleague, Heini Rohrer, to help them learn how to grow thin, defect-free oxide films.[6] Rohrer hired a newly minted PhD, Gerd Binnig, and together they planned an instrument that would use electron tunneling to a sharp tip as a probe of the electrical properties of the films. By scanning the tip, the instrument could map the conductivity of the film and locate defects.[7]

The Josephson project soon died, but Binnig and Rohrer continued to develop the scanning tunneling microscope. It was conventionally assumed that the STM's resolution would be limited by the radius of curvature of the tip, but Binnig believed that on rough tips virtually all the tunneling would be to a single atom, providing atomic resolution both vertically and horizontally. So they proceeded to demonstrate that the STM could measure a tunneling current, and that it could image single atomic steps on gold, work that appeared in several journal articles along with a short note in *Physics Today*.[8] These results, however, drew little response.

Thus, Binnig and Rohrer were confident in their microscope, but had little idea what samples to train it on (in both senses of "train") that would get it noticed. So they approached various colleagues in the Zürich lab for suggestions and samples. In the end, their most fruitful exchanges were with IBM's surface scientists. Neither Binnig nor Rohrer were trained in surface science, but the STM seemed well suited to investigating the metal and semiconductor samples favored by that field. The Zürich surface scientists, though initially skeptical, proposed several materials about which their discipline already knew a great deal, but which the STM might elucidate even more.

Surface science arose as a diverse postwar constellation of research interests—friction, adhesion, electron physics, thin films, interfaces, and other phenomena were all under its umbrella. In the late 1950s, this nascent discipline found an institutional home in the American Vacuum Society—thus cementing a long-standing tie between surface science and vacuum engineering. In the late 1960s, this connection strengthened, as the field underwent a reorganization that some participants describe as a Kuhnian paradigm shift.[9] IBM and Bell Labs held the lion's share of the discipline both before and after this shift, and so their researchers were able to guide the field to a new focus on the structures of metals and semiconductors (and their interfaces) in the flat geometries typical of the new integrated circuit technologies that were becoming vital to both companies. Just as IBM and Bell's dominance of their respective markets prompted inward-looking technological culs-de-sac, so too did their shared dominance of surface science mean that they looked largely at one another to establish the field's standards and values.

The new surface-science paradigm (especially as practiced at IBM, Bell, and a few other corporate labs) was a tightly knit, self-vindicating complex of materials, preparation techniques, theoretical machinery, computing power, and an ever-growing range of diffraction and spectroscopic instruments.[10] All of these were understood as requiring "well-defined" surfaces and very clean environments, making ultra-high vacuum (UHV) technology a sine qua non. With the new paradigm came new problematics, above all the quest to understand surface reconstructions (the reordering of surface atoms from the bulk arrangement). A few of these reconstructions became canonical for the new surface science. One, the silicon (111) 7 × 7, was especially powerful—it had been known from diffraction studies since the 1950s, yet its structure remained a mystery.[11] By 1980, a few models existed, with scant or contradictory experimental evidence to discriminate between them. Still, the 7 × 7 was a standard test of an individual surface scientist's ability to prepare a specimen, employ a theoretical tool, or construct a new instrument.[12]

Thus, in order to make their microscope interesting to surface scientists, Binnig and Rohrer had to enter into this weave of tools, techniques, and samples—and so, they accepted the necessity of UHV operation, learned the rudiments of specimen preparation, and took up the 7 × 7 to test the microscope's abilities. When Binnig finally produced a "golden image" that purported to show the atoms of two unit cells of the 7 × 7, and offered a plausible model of the reconstruction based on this image, surface scientists took notice.[13] The STM had not solved the 7 × 7 question, but it had shown its ability to fit into a surface-science form of life. The STM was now part of two distinct moral economies. For Binnig and Rohrer, it was a site for somewhat untutored modification and tinkering, as well as a letter of introduction to more disciplined communities of experimentalists. For surface scientists, it was now a recognized, if immature, tool, ready to be made to contribute to the positive accumulation of surface scientific knowledge.

### STM and Surface Science in America

The remainder of this essay is about the groups that took up the STM in North America. My analysis centers on two different styles of research. One worked out of surface science, was centered primarily in corporate and national research laboratories, and focused on scanning tunneling microscopy. The other, located in West Coast academic physics and electrical engineering departments, engaged in building general-purpose microscopes and quickly moved beyond scanning tunneling microscopy to encompass atomic force microscopy and other probe microscopy techniques. For both traditions, methods for training researchers were crucial in guiding experimental choices.

After the 7 × 7, IBM management accelerated efforts to replicate the instrument and grow an STM community. Big Blue began sending scientists to Zürich to learn to build an STM and return to begin programs at labs in Almaden, California and in Yorktown Heights, New York. It is unclear what was gained from this "pilgrimage to Zürich."[14] Binnig and Rohrer's machines were notoriously unreliable in this period, and one could stay at the lab for weeks and never see a working STM.[15] At first, IBM sent Oliver Wells, an electron microscopist; but it is notable that his six-month stay in Zürich yielded an unsuccessful replication attempt at Yorktown.[16] Later, IBM management decided that surface scientists, not microscopists, were the people to replicate the STM.

IBM had a ready store of people who could be encouraged to become tunneling microscopists. These were postdocs and young staff scientists, part of a steady stream of new PhDs flowing through Yorktown and Almaden. As graduate students, many of these people had learned the basics of surface science—sample preparation, UHV technology, operation of characterization tools. Many had also built characterization instruments from scratch.[17] They were acquainted with control electronics, instrumentation software, and machine tooling, and could unite these ingredients with some sample to generate new surface-science knowledge. There were a few exceptions—Randy Feenstra, for example—who did not come to IBM trained in surface science, but even these people had backgrounds in related instrumentation, and they soon became orthodox surface scientists when they turned to scanning tunneling microscopy.[18]

The surface-science tradition in which these postdocs and young scientists were immersed in graduate school and at IBM and Bell Labs was in many ways a harshly critical one valuing dogged, even tedious, work.[19] Once at IBM and Bell, they began by bouncing between projects, looking for one that would garner attention from their managers and provide a roadmap to building a career. This meant projects that (with many long nights) could produce two or three publications in prestige journals in the 12 to 24 months before their positions were reviewed. If successful, they continued on, making careers out of the instruments or techniques they forged at IBM or Bell.[20] From their postdoctoral outing at Bell/IBM they usually went to a staff scientist position in the same small world (many IBM postdocs became IBM staff scientists; about as many moved to Bell—and vice versa) and/or at various national laboratories around the world.[21] Academic posts, usually in physics or chemistry departments, were a final option, exercised more as the corporate labs declined in the 1990s.

This surface-science ecology, then, framed graduate education as the place to acquire the formal knowledge and tacit skills of experimentation; in the postdoc or early staff scientist position these skills were attached to new projects, remaking graduate students into productive surface scientists and corporate researchers. This newly re-formed subject would be less concerned with acquiring new tacit skills than in creating new surface-science knowledge that would fit with the disciplinary body of knowledge in a generative way, allowing for a planned and methodical mode of experimentation (through the step-by-step variation of materials and parameters and instruments) for years to come.

One feature of surface science (and other) research at IBM and Bell Labs in the early 1980s that influenced the shape of STM work was an air of intense and institutionalized competition at every scale.[22] At the level of the individual researcher, being a postdoc or a new staff scientist was an anxiety-ridden and liminal (though also exciting) existence. As Sharon Traweek points out, time works against these people, and they must use this stage of their career to aggressively carve out an individual experimental identity.[23] This meant eagerly locating potentially hot projects like the STM; coming up with new technical solutions to problems that had already been tackled by other junior researchers; frantically appropriating equipment, assistants, and even physical space; and, most importantly, knowing when to seek help from other groups and when to withhold or spurn such advice.[24] Similar jockeying occurred at the level of the research group. This was particularly institutionalized at Yorktown, where official policy was to assign multiple groups to similar projects and give official evaluations of groups based on their achievements relative to other teams in the same lab.[25] Group competition was also the rule, however, at Bell and (somewhat less so) Almaden. Groups raced each other to achieve various milestones (e.g., the first atomic resolution of different reconstructions); group leaders recruited each other's personnel; they lobbied newcomers to align with their style of STM use rather than that of other groups at the same institution; and, again, they (re)crafted and appropriated the physical sites of experiment in order to insinuate their groups more advantageously into the moral, bureaucratic, and technoscientific orders of the labs.[26]

At the level of the labs themselves, Bell and IBM were locked in a long-standing competitive but also constitutive and mutually validating dyad. Bell Labs' contributions to semiconductor technology in the 1950s and the 1960s made it the superpower of surface science; the Bell Labs facilities in Murray Hill, New Jersey had achieved so much that IBM managers had, by the early 1980s, become somewhat obsessed with besting Bell at its game (a fact that clearly aided the use of STMs at both sites). Bell surface scientists cultivated the tradition of harsh skepticism to a fine degree, and this, combined with the lab's continued successes, meant that techniques and findings from outside Murray Hill often aroused initial suspicion.[27] Indeed, early forays into STM use by Bell scientists, sparked by seeing Binnig or Rohrer give early presentations, were discouraged by their peers and management. Here, again, the 7 × 7 made the difference—when a theorist, Don Hamann, refereed Binnig and Rohrer's first 7 × 7 paper, he became interested in giving a theoretical justification for atomic resolution. He went back to their earlier paper on gold monosteps, which surface scientists had universally ignored, and convinced his postdoc, Jerry Tersoff, to treat the problem of how a sharp metal tip could image a flat metal surface via tunneling. This was the first theoretical treatment of an STM outside Zürich, and it gave early STMers a handy, durable justification for their images.[28]

Likewise, experimentally, a young x-ray physicist, Jene Golovchenko, was establishing himself in a surface-science group and latched onto scanning tunneling microscopy as a promising technique. His managers blocked the work, however, until Golovchenko brought Rohrer to Murray Hill to present the 7 × 7 data to an overflowing crowd of enthusiastic Bell researchers.[29] Replicating the STM and competing with IBM in this new field became a priority, and Bell Labs' management fostered their own nascent tunneling microscopy community. As at IBM, the work of building and operating STMs fell to postdocs, technicians, and junior researchers. And again (as at IBM), these people worked against the clock and each other (though sometimes also cooperatively) to generate interesting surface scientific results and create distinctive experimental personae.

So what did "replication" of the Zürich STM mean at IBM and Bell? In the first North American instruments, much of the design traveled intact from Switzerland; but the Zürich group had aimed at simply proving the STM's worth, rather than producing working surface scientific knowledge. Thus, operation of their microscope was laborious and not geared to the rapid characterization of materials and production of journal articles. Its electronics, for example, were entirely analog, with a proliferation of controls that an operator could use, after much practice, to tune the instrument and coax it to higher resolution.[30] At Yorktown and Almaden and Murray Hill, on the other hand, researchers needed more user-friendly electronics, with some automation and digital control, in order to achieve a rapid throughput of samples.

By drawing on skills learned in graduate school, the American groups built electronics and wrote software that not only made it easier to characterize samples swiftly, but also increased the amount of information generated from each sample. They wrote line-filling software, for instance, to turn a set of scans into a gray-scale, top-down, topographical, digital image that could be generated more quickly, interpreted more easily (particularly in conjunction with simulated images provided by theorists), and processed more artfully than those used in Zürich.[31] They then reinterpreted the tunneling probe as a channel for multiple kinds of information, and so produced not just one image of a surface, but multiple, juxtaposed layers of images showing both topographical and spectroscopic characteristics. Indeed, devising ways of using the STM for spectroscopy became a locus for intense, sometimes bitter, competition among the Yorktown and Murray Hill groups. Constructing a new spectroscopy technique was one of the best ways for a new researcher to stand out from the crowd, and the variety of scanning tunneling spectroscopies quickly bloomed into a confusing mass.[32]

Tunneling spectroscopy of any sort, however, was merely a projected outcome when these groups first began merging the STM into surface science. Replication took a long time, and there was an awkward period (1982–1984) when no one could reproduce the original results.[33] All these groups started by chasing the 7 × 7, for several reasons—the Zürich group had shown atomic resolution with it; it was a standard material for testing surface-science instrumentation; several of the groups had particular expertise in preparing it; and Binnig and Rohrer's cursory study had left a great deal more information that might be derived from this particular surface using a tunneling microscope. Still, atomic resolution was elusive; all that most of these researchers (and their managers) knew was that they and others were working on the problem for months, even years, and getting nowhere.

In late 1984, in this uncertain state, Calvin Quate organized a workshop so that a handful of researchers might gather and pick each other's brains, and Rohrer's, and overcome the replication hurdle. At the time, Golovchenko was furthest along (he had seen the corner holes of the 7 × 7 but still no atomic resolution). At issue at Quate's workshop (held in Cancun) were the now-recognized usual suspects of STM difficulties: tips, piezos, surface preparation, coarse and fine approach.[34] Whatever was learned in Mexico, it seems to have helped. By the time of the March 1985 American Physical Society meeting, the panic was over. Several groups were routinely getting atomic resolution, and a standing-room-only panel was organized to show off images of the 7 × 7 of such clarity and expanse that Golovchenko described them as "pornographic"—the atomic structure of this puzzling reconstruction, long visible but disguised in inverse-space diffraction images, was now plain to see.[35]

With replication achieved, the STMers began a routine course of experimentation guided by the moral economy of corporate surface science. This meant a quick succession of samples passing through the new microscopes, in a manner that could be easily extended through planned, methodical variation framed within surface science's formal body of knowledge. The road ahead of the STM had been traveled before by other instruments, and so could be advanced along quickly. Once the silicon 7 × 7 had been seen, and operation of the microscope was adequately understood, the new STMers could chart a variety of courses. At Bell, Golovchenko's group stayed with the silicon (111), moving through a set of reconstructions (5 × 5, 9 × 9, 11 × 11) that resembled the original 7 × 7. At Yorktown, Joe Demuth's group pushed to other unit cells of silicon: the (100) 2 × 1, the (111) 2 × 1, the (001) 2 × 8, etc. And Randy Feenstra's team moved to other semiconductors: germanium, gallium arsenide, and eventually gallium nitride.

Notably, the track laid for this research emerged from traditional surface-science theory and experiment: basic crystallography told STMers to vary the crystal index of their surfaces—(111), (100), (011), etc.; and low energy electron diffraction (LEED) and other techniques showed the possible range of reconstructions—7 × 7, 9 × 9, 2 × 1, $\sqrt{3} \times \sqrt{3}$, and so on. Techniques of sample preparation that had co-emerged with LEED (i.e. preparation recipes counted as different if they generated different LEED patterns) were thus ready to hand. The STMers were disciplined by the knowledge, skills, and technologies of surface science into experimenting with materials in a well-defined, generative way. Indeed, this disciplining bore fruit, producing large numbers of papers, a heap of new knowledge about semiconductor and metal surfaces, and a promising start to the STMers' careers.

To use Ian Hacking's terminology, these STMers were making the new microscope part of the "instrumentarium" of surface science.[36] Binnig and Rohrer had started down this road, but only to entice more credentialed surface scientists to finish the job. The moral economies of experimentation at IBM and Bell gave rise to expectations about what work a young scientist should do, how many papers they should produce and in what quality of journals, and how they should build and operate instrumentation. Tying the STM to an established body of disciplinary knowledge seemed the easiest way to fulfill these expectations. Thus, in the hands of these junior researchers, the STM began its transformation into a standard surface-science instrument.

Importantly, various *design* changes accompanied this transformation. For instance, it became possible to move samples and probes in and out of the instrument while still in a fully pumped-down UHV chamber. A reliably working instrument could remain in an evacuated chamber for months, even years; samples could be dropped in, characterized, and exchanged at an ever-increasing pace. What's more, samples (and even probes) could be moved around the chamber to be prepared, cleaned, and characterized by *other* surface scientific tools before merging with the STM. It became standard practice to clean and prepare samples in the chamber (since most surface scientists—unlike Binnig and Rohrer—believed exposure to air, tweezers, fingers, etc. would cause disastrous levels of contamination), do a quick LEED image to check the reconstruction, and move to the STM, all in UHV. Some groups, particularly ones doing spectroscopy (where tip shape is most important) also cleaned the tip and characterized it with a field ion or electron microscope before and after STM runs.

STM designs also changed to vary parameters that had meaning in surface-science discourse, such as temperature and adsorbate composition. Bell Labs groups built first low-temperature instruments, then variable-temperature ones that could deposit adsorbates on standard surfaces such as silicon and gold.[37] At IBM, researchers used in situ preparation to passivate surfaces and grow nanoclusters and thin films, again on standard substrates that had already been imaged with the STM.[38] These were routine surface-science experiments that had been done before with LEED and spectroscopic instruments. Repeating them with an STM quickly gave results that spoke to the earlier data. Indeed, some STMers Fourier-transformed images into inverse space equivalents, so as to simulate LEED patterns—STM images could (almost literally) be *translated* into LEED results, facilitating the instrument's integration with the established discourse of surface science.[39]

One sign of the STM's entrenchment was its replication by surface scientists outside Bell and IBM. By 1985, groups at Ford and Philips research labs, at the National Institute of Standards and Technology, at Lawrence Berkeley National Laboratory, at Lawrence Livermore National Laboratory, and at the National Research Council in Canada were all building their own, as were academic surface-science groups at Berkeley, Illinois, Wisconsin, Penn State, Penn, Harvard, Cornell, and elsewhere. Some of these instruments derived from Binnig and Rohrer's designs; as at Bell and IBM, researchers often constructed a first instrument inspired by Zürich, then built a second to work more reliably in a surface-science idiom. Almost all these machines were UHV-compatible, and were usually combined with various traditional surface preparation and characterization techniques.

Meanwhile, at Yorktown, Joe Demuth organized an effort to "mass"-produce his group's microscope and sell it internally to IBM researchers. The Central Scientific Services workshop at Yorktown churned out these instruments, which were taken up for experiments with adsorbates, surface reactions, passivating and depassivating silicon, etc. As usual, postdocs performed these experiments; often, they took the STM with them (literally and figuratively) after they left IBM to form their own research groups.[40] When they moved out of the research lab, these people founded or seeded further centers of scanning tunneling microscopy.[41]

In the process, an STM community took shape. The first informal STM workshop at Cancun in 1984 was followed the next year by a meeting organized by IBM Europe at Oberlech, Austria. IBM-connected researchers dominated the meeting, but outsiders included both those doing surface-science work (John Pethica at Cambridge, Golovchenko at Bell, and Bill Kaiser and Bob Jaklevic at Ford) and representatives from academic instrument-building groups in California (Cal Quate at Stanford, Paul Hansma at UC Santa Barbara, John Baldeschwieler at Caltech, John Clarke at Berkeley).[42] The next year, at Santiago de Compostela in Spain, STMers met in the first annual "STM Conference."[43]

For many surface-science STMers, though, such conferences were a temporary solution to temporary problems—a way to advertise the new microscopy, to bring people together to aid replication and map out what experiments could be (and were being) done with the new instrument. Once STM became an established, black-boxed technique, like electron microscopy or LEED, the conferences could dissolve and researchers could present results at their own disciplinary conferences—AVS, APS, SemiCon, ACS, etc. Indeed, by the 1987 STM Conference at Oxnard, California, the surface-science contingent was clearly bowing out. Scanning tunneling microscopy was now an accredited part of surface science, rather than an erratic and ambiguous technique on the margins of the discipline; and within the probe microscopy community the corporate lab researchers were becoming respected (and somewhat aloof) elders, rather than the avant-garde of SPM development.

Crucially, the moral economy of corporate surface science, of which the STM was now a part, pre-dated tunneling microscopy and easily took the STM in stride. "STM" took its place as just one acronym among the many used for various surface-science instruments. The young researchers who effected that integration had risked much early on in becoming STMers. In the 1980s, they simultaneously showed that they were competent corporate researchers and that the instrument could reliably answer outstanding surface-science questions. (As a testament, the early 1990s saw many older, more entrenched surface scientists move from LEED to STM). In taking these early risks, the early STMers were creatively yet dutifully answering the give and take of life at Bell Labs and IBM; in doing so, they rose quickly through the ranks of the corporate and national laboratories.

## The STM Moves West

Although a few North American academic research groups started building STMs alongside Bell and IBM, most lagged behind the corporate labs.[44] The ones that joined the STM core set most quickly and most influenced the development of the technology all happened to be based in California: Calvin Quate at Stanford (in electrical engineering and applied physics), with a distinguished background in microwave electronics and acoustic microscopy; Paul Hansma (in physics) at the University of California at Santa Barbara, with extensive experience in inelastic electron tunneling; and John Baldeschwieler (in chemistry) at Caltech, with an interest in catalytic and biological materials.[45] Quate habitually changed directions every decade, and a 1982 *Physics Today* article on the STM prompted his transition from acoustic to tunneling microscopy.[46] Hansma became involved through his electron tunneling work and acquaintance with Rohrer.[47] Baldeschwieler, always the most marginal, decided to build his own STM after hearing a report of an early Binnig talk.[48]

Early on, then, these groups had little in common. This was a crucial difference from the corporate labs, where the STM folded into long-standing and widely shared traditions of training; the West Coast groups had to construct commonalities in their instrument-building traditions almost from scratch. Of the three, Quate had a head start and a large group of graduate students to assign to the task. Moreover, his involvement with scanning acoustic microscopy meant he could draw on in-house skills and equipment (piezoelectric scanners, feedback circuitry, and imaging technology) that were relevant to using an STM. Still, replicating the STM proved no easier for Quate than for the IBM and Bell groups. When atomic resolution of the 7 × 7 was not forthcoming, Quate suggested the Cancun workshop to discuss the obstacles to replication. Interestingly, Yorktown researchers interpreted Quate's problems as stemming from his lack of surface-science knowledge about the meticulous preparation of clean samples and the operation of instruments in UHV.[49] Crucially, though, Quate's ties to other wings of IBM were strong—he had several former students at the nearby Almaden lab, and his personal contact with the Zürich group expanded throughout this period.[50] Binnig and Rohrer found Quate's experimental style more comfortable than that of the surface scientists, and in 1985–86 Binnig and his technician, Christoph Gerber, took sabbaticals at Almaden to work with Quate's group and help them achieve routine use of the STM.[51]

Hansma, meanwhile, concentrated on a much simpler problem—the construction of controlled-gap tunnel junctions. This was an old idea in the tunneling community, though only rarely and unsuccessfully attempted.[52] But Hansma drew on the STM example and was able to adapt traditional metal-oxide-metal (MOM) tunneling experiments to the new controlled-gap setup. Through continued contact with Quate and Rohrer, Hansma parlayed this work into membership in the STM community. Yet Hansma's squeezable tunnel junctions were only tangentially related to STMs. They lacked the complex mechanical and electronic design of a tunneling microscope—no scanning, image formation, sharp tips, control software, etc. Instead, Hansma's squeezable junctions were simply two (or four) crossed wires that could be mechanically brought close enough for tunneling to occur.

Hansma's most important deviation from the STM was to operate his tunnel junctions in air, rather than vacuum.[53] Traditional metal-oxide-metal tunnel junctions were made in UHV, but operated in air. While testing the squeezable junctions, Hansma concluded that they too could work in an ambient environment—the current-voltage curves that can be obtained from *any* tunnel junction are erratic, and the curves from the new devices looked no more so than those from traditional MOM junctions. For Hansma, this suddenly made tunneling *microscopy* attractive, since using an STM in air was a simpler prospect than using one in UHV—no vibrations from vacuum equipment, no need to "harden" the microscope's materials for vacuum, full bodily access to the instrument when tinkering (as opposed to the restricted access through manipulators in UHV chambers), and no lengthy pump-down and bake-out routines when transferring instruments or materials into and out of UHV.

But what to look at with this new STM? The 7 × 7 and most paradigmatic surface-science specimens were out, since in air they grew an oxide layer and became non-conducting. Instead, Hansma used graphite, a conducting, non-oxidizing material readily available from laboratory supply companies. The traditional customers for graphite, though, were chemists and molecular biologists, not surface scientists. Thus, there were interpretive problems in imaging graphite in air—most STMers at the time were unfamiliar with graphite and distrustful of any work done in air. When his images of graphite showed atomic heights 10 to 100 times larger than expected, Hansma withheld them for a year rather than risk the skepticism of the corporate STMers.[54]

Still, an air STM had many attractions, especially for those—such as Binnig, Quate, and Hansma—who approached the STM from outside surface science. Within surface science, an explanation for the graphite anomaly came from John Pethica, an early British STMer interested in friction and adhesion.[55] Pethica conjectured that an STM image contained not only the topographic and electronic information of interest at Yorktown and Murray Hill, but also information about attractive and repulsive *forces* acting between probe and surface. STMers knew the probe could interact mechanically with the surface by crashing into it, but with hard semiconductor surfaces such contact was relatively obvious. Pethica informed the STM community that, with graphite (particularly when coated by a contaminant layer in air), a microscopist might unknowingly contact the surface, causing the probe to compress the graphite sheets and exaggerate atomic heights.

The air STM was a crucial turning point, in that it offered an alternative to surface scientific instrument design and the "epistemic materials" associated with surface science.[56] With validation from Pethica, graphite images began to seem real and the air STM viable for certain applications. This rallied Hansma, Quate, Binnig, and Baldeschwieler to construct a shared experimental culture, particularly between Santa Barbara and Palo Alto, oriented to explorations of further alternative microscope designs and uses. Binnig, Hansma and Quate began exchanging information, materials, tools, and even students, thereby forging a new style of probe microscopy that soon swamped Yorktown and Murray Hill.

By moving away from surface science, though, Hansma, Quate, and their collaborators lost the disciplinary structure that guided the corporate STMers. Instead, they embarked on blind variation-and-selection, relying on their tacit or informal knowledge about building instruments to tweak designs in unexpected and unplanned ways to examine samples about which they had little prior knowledge. Indeed, it became common practice (particularly after the invention of the atomic force microscope, which could examine insulating materials) to choose samples from ordinary experience: ice, table salt, hand cream, soft drinks, bone from rib-eye steak, blood from lab workers, leaves from house plants, Polaroids from cubicle walls.[57] Hansma and Quate's students knew little about how to prepare these "found objects," nor how to make their images meaningful in any disciplinary discourse.

For the corporate STMers, this chaotic style of experiment lacked both rigor and standards. But questionable work became a rhetorical advantage, as summed up in two oft-repeated Hansma proverbs: "Do everything as poorly as you can" and "Make as many mistakes as you can as fast as you can"—meaning, roughly, do the experiment blindly and quickly, knowing there will be false turns but that you can learn skills and hints along the way. This quick pace didn't mean the academic groups *worked* any faster than the corporate labs researchers; indeed, especially in Santa Barbara, in the summer the pace became slower and more relaxed in a way that is hard to imagine at Yorktown or Murray Hill. Rather, the Stanford and UCSB style rewarded quick dashes into experiments with little prior knowledge or preparation; research was done in the moment, cheaply and with whatever materials were at hand.

Thus, Hansma and Quate's groups cultivated a style of playful experimentation that expanded beyond the boundaries of the lab. Group meetings, for instance, were often held at the Hansmas' home; group members took up juggling, ultimate frisbee, baseball, and surfing, and at Stanford they were quickly enrolled into a group volleyball team; and (especially at Santa Barbara), lab culture emphasized self-improvement through photography, meditation, and travel.[58] Experimental work, too, became defined by teamwork, embodied skill, and hobby-like or self-cultivating behavior. Hansma's instruments, for instance, often started as wooden models crafted in his woodworking shed at home, which were then handed to his technician, Barney Drake, to translate into metal prototypes.[59] Baldeschwieler's students half-jokingly published images they obtained using pencil lead for the probe.[60] Quate group veterans fondly remember scrabbling together microscopes in a few hours from parts borrowed from other microscopes in progress in the lab, and speak admiringly of Gerd Binnig's proficiency at spewing out brilliant (if mostly unworkable) ideas and turning them into reality with glue-gun and blow-dryer.[61] Othmar Marti, a former Binnig protégé, played a similar bricoleur role as a Hansma postdoc at a time when the Santa Barbara group was making atomic force microscopes (AFMs) by crushing fifty-cent pawn shop diamonds and using eyebrow brushes to glue them to tungsten probes.[62]

Thus, these groups valued tacit, embodied knowledge as a source of the skills and intuition needed to make quick, educational mistakes. There were, however, limits to work done without a disciplinary framework—in particular, it was difficult to publish images of a material and have them be meaningful to researchers for whom that material was an epistemic thing. This was a big difference from the corporate labs—at Yorktown and Murray Hill, IBM and Bell brought together a substantial fraction of the surface-science community under one roof. Thus, surface-science STMers had a ready-to-hand community to learn from, argue with, and have results validated by. It might be necessary to publish in the external literature, but disciplinary approval could be measured long before results were submitted to journals. The California academic groups, on the other hand, were disciplinarily more isolated. Their only "colleagues" (in any strong sense) were each other, scattered along the West Coast. They could hammer out local techniques, and generate results they themselves found interesting, but acceptance by any wider community (or even knowing what community to talk to) was difficult.

In answer, Quate, Hansma, and Binnig found they could borrow a disciplinary cover by forming ad hoc collaborations. In a sense, Binnig had already done this by initially enrolling surface scientists into the early STM work on the 7 × 7. Quate's acoustic microscopy work in the 1970s, too, had involved sending students to labs in other departments to learn new sample preparation techniques, new ways of interpreting images, and some formal knowledge about their samples.[63] For Hansma, enrolling other disciplines started through personal connections and only later became routine. Even in his tunnel junction days he had searched for experiments that would draw on his wife Helen's training in molecular biology, and, indeed, the Hansma group eventually became the center for biophysical use of the STM and the AFM.[64] Hansma first attempted to build a probe microscope specifically for a practitioner outside the STM core set, though, when he and Drake took a sabbatical at the University of Virginia with Bob Coleman, a friend from graduate school. There, they learned a modest amount about Coleman's superconducting layered compounds and constructed a low-temperature air STM to study them, then left the microscope with Coleman and returned to Santa Barbara.[65]

This set the pattern for future collaborations—as news of the STM (and later the AFM) spread, (usually junior) professors or postdocs from various disciplines (molecular biology, biophysics, materials science, mineralogy, etc.) contacted Hansma to inquire about spending time in the lab. Usually, they brought or sent some material or system (some "epistemic thing") about which they were knowledgeable to Santa Barbara to try out in the new microscopes. When such visitors came to UCSB, they learned how to operate the microscopes from Hansma's students, while in return teaching them how to prepare and understand the samples peculiar to the visitor's discipline. Students were continually building new generations of microscopes or new pieces of hardware and software to adapt the microscopes to particular applications; visitors were enrolled in this process, with student and visitor implementing the new design, testing it, and generating images of the samples the visitor had brought or made. Often, the visitors would take that instrument, or a copy (or, later, a commercial version), back to their home lab and enter the field of probe microscopy on their own, while the Hansma group wound down their involvement with that discipline and took up new ones.[66] In Quate's group a similar style evolved, with a variety of collaborators (usually professors and postdocs drawn from within Stanford) and numerous microscopes in various stages of design, construction, or operation at once. Students and materials, and occasionally postdocs, circulated quickly from project to project, both within Stanford and between Palo Alto and Santa Barbara.[67]

This style of experimentation—with its emphasis on the circulation of goods and people; its prizing of the production and consumption of experience, knowledge, and microscopes; and its reliance on trading with other groups—encouraged the commercialization and commodification of the instruments. At first, like many of the other activities in these groups, commercialization was done semi-playfully. Doug Smith, a Quate student who had worked closely with Binnig, began building air STMs after hours at home to sell to groups trying to enter the STM game. Materially, the "Scanning

Tunneling Microscope Company" produced only a few instruments and some bright red ball caps with the company logo that Smith gave away with them.[68] These were unpolished microscopes, intended for people with the expertise but not the time to build one themselves. Reportedly, Smith earned just enough to buy a sports car before moving to Munich for postdoctoral work with Binnig.[69]

Meanwhile, in Santa Barbara, a more serious venture was beginning. For several years, Hansma's department administered a Master's in Scientific Instrumentation degree program run by a fellow professor, Virgil Elings. Given disagreements with his colleagues over his prickly personality and their low regard for the degree program, Elings was already prepared to leave the university, and he had previously commercialized two of the instruments invented by his master's students.[70] Now, through contact with Rohrer, Hansma, and Niko Garcia (a Spanish STMer with Zürich ties), he came to believe that a commercial STM would be successful. He approached Hansma about joining the company, but what emerged instead was a symbiotic relationship where the Hansma group and Elings's enterprise shared people, resources, and knowledge while maintaining semi-formalized ties. As a first step, Hansma showed Elings many of the tricks and shortcuts of building an STM. From there, Elings and his son adapted and reworked Hansma's design and built an STM in their garage that doubled as the prototype for a commercial instrument and as a junior high science fair project.

Soon, Elings co-founded a company, Digital Instruments (DI), with Gus Gurley (a former student), gathered together a small staff, and began churning out STMs. These were, of course, air instruments, designed as stand-alone, general-purpose microscopes rather than surface scientific tools. Thanks to graphite, air STM now had its 7 × 7, i.e. a material for which atomic resolution could be a relatively easy test of both instrument and user. By the Oxnard conference in 1987, every other talk was on graphite—the ease with which air STMs could be made or bought, and the ease with which graphite could be purchased and prepared, opened the doors of the STM core set.

One reason for the air STM gold rush was that Quate, Hansma, and their collaborators had reinterpreted the STM as a stand-alone microscope, akin to optical or electron microscopes, that could investigate single objects placed on some substrate (usually graphite). Anyone with an air STM and some graphite could put whatever microscopic object they wanted onto the substrate and image away. The problem was, it became increasingly difficult to put materials (particularly biomaterials) onto graphite, image them in air with the STM, and understand what the images meant. A gap existed between understanding the materials and understanding the instrumentation, and in the quickly circulating world of California STM groups it was difficult to bridge the two. For a time, bridge-building was not attempted—virtually any image of biomaterials on graphite could be published with or without a full, credible interpretation. Collaboration with biophysics groups existed, but extended mostly to learning how to get biomolecules to stick on a substrate without being moved around by the probe. STM images, with their convolution of information about topography, electronic structure, and force interactions, were particularly difficult to interpret for the soft, insulating materials of interest to biologists. Still, as probe microscopists like to say, an SPM "always gives you an image"—the community was swamped with images of biomaterials with no well-understood way to interpret them.[71]

## Controversy and Closure

The grounds were set, therefore, for the first real STM controversy. The corporate STMers had always thought that much of the academic groups' work flew in the face of established facts—tunneling doesn't work in air, surfaces need to be kept clean, etc. For the West Coast groups, their methods, though chaotic, sufficed for generating a high volume of "good" results. In 1987–88 and again in 1990–91, two such results, though, came under scrutiny. The first was a Quate/Smith investigation with low-temperature scanning tunneling microscopy of the vibrational spectroscopy of single molecules.[72] This was a holy grail for surface scientists—several STMs at Bell, IBM, and other corporate and academic labs were built for this purpose. In all cases, though, corporate STMers discarded the spectra before they faced the critical eye of other surface scientists. Quate, however, forged ahead, controversially publishing vibrational spectra for sorbic acid. After all, surface scientists had accepted Binnig and Rohrer's 7 × 7 work because it adopted the "essential" methods of surface-science experiments, and they acknowledged Hansma's contributions in air because he had not attempted major experimental tasks within surface science—but Quate's vibrational spectra attempted to solve a major problem without following surface scientific methods and values.

The response was quick, and after a round of replies and letters to the editor Quate retracted the result.[73] In Pasadena, meanwhile, a similar story unfolded around Baldeschwieler's claims of clear, atomic-resolution images of DNA on graphite, as published on the cover of *Nature*.[74] Many STMers wanted to believe Baldeschwieler's images. Indeed, stranger things had already happened with STMs, and, if true, the rewards were great (above all, Hansma and others hoped to sequence DNA with an STM).[75] This contradicted all surface scientific knowledge, however, about how an STM should work and what it could image. The corporate STMers occasionally tried putting graphite into their UHV instruments, but what they found bore no resemblance to the graphite images circulated by air STMers. The results didn't really disagree; rather, they were incommensurable. The standards of instrument design, specimen preparation, and image interpretation were so divergent that there was no way to begin comparison.

Unfortunately for air STMers, they couldn't find a discipline with which they *could* begin a dialogue about graphite. Atomic resolution of graphite simply wasn't relevant to the practice of any field; and the fields that might be interested in the materials (such as DNA) that could be deposited on graphite were skeptical about the utility of an STM. Image interpretation was complicated and counter-intuitive, and required such heavy theoretical machinery that most biologists and biophysicists shied away. Moreover, even among experienced air STMers, doubts about DNA on graphite began to emerge; as they gained the tacit knowledge needed to prepare and characterize samples, they increasingly saw that on graphite (unlike other materials) the connection between preparation techniques and the images they yielded was tenuous and vexed. Eventually, critiques emerged from Binnig and others showing that graphite, the wonder substrate for an air STM, was an extraordinarily problematic material.[76] In particular, it was prone to crystalline defects that, in air, could mimic the adsorbates researchers were trying to image. Thus, Baldeschwieler's "DNA" came to be seen as a defect masquerading as a biological molecule.

How closure was achieved in these controversies says much about the relevant moral economies of experimentation (the Bell and IBMers, the West Coast academic groups, and a nascent, diffuse "user" economy populated by customers of the startups associated with the West Coast groups); in particular, closure illuminates how experimental culture and pedagogical orientation were linked, and how these groups created microscopists at the same time as microscopes. In some ways, these controversies set back the West Coast groups and their allies; Baldeschwieler, in particular, never withdrew his claims and so faded into the background. Quate and Hansma, on the other hand, acceded to their critics' point—it was a given of their experimental style that some claims would be challenged. *Debating* such challenges required learning their detractors' language and methods and answering in kind; but an easier and more natural course for these groups was to *abandon* controversial techniques and move on, making as many mistakes as possible as quickly as possible.

Thus, they reinterpreted the STM again, as a technique with severe limitations—where surface scientists saw vacuum STM as enabling an almost endlessly generative series of experiments, Hansma, Quate, and Elings now saw the air STM as a dead end. As Hansma puts it, looking around his office, the only thing in the room that can be imaged with an (air) STM is a gold wedding ring (whereas vacuum STMers say that any metal or semiconductor, with proper sample preparation, can be imaged).[77] Eventually, corporate and academic STMers even came to the mutual understanding that the air

STM was not relying on tunneling—that instead the probe was measuring some other interaction with the surface. Thus, the West Coast groups opted to deviate from scanning tunneling microscopy just enough that they could retain the skills and informal knowledge they had learned about the technique, while finding new ways to employ undisciplined variation in the materials they imaged. This meant a probe microscope that could image insulating materials—not merely isolated non-conducting adsorbates but whole samples.

This desire led, first of all, to the atomic force microscope—Binnig, Quate, and Gerber's attempt to adapt the STM to image insulating materials. The AFM evolved from Pethica's explication of the effect of surface forces on the probe—Binnig imagined a probe with a spring constant smaller than the force between atoms on a surface; such a probe could scrape along a surface, causing minimal damage to hard materials, and bending with the topography of the surface. A sensitive detection scheme could measure this bending and build up an image, just as an STM did. So, Binnig put a small, flexible tungsten wire tip under an STM probe and imaged away.[78] But the first AFMs were, by all accounts, fantastically difficult to operate.[79] Both the tunneling probe and the AFM tip were difficult to control and subject to strange changes in shape and quality, so that stable images were rare.

Despite this, Binnig and Quate declared the AFM the next logical step in probe microscopy. The logic and inevitability of AFMs, though, had to be interpreted in particular experimental contexts. With its lower resolution than the STM, and its intended application to insulating materials, the AFM offered little to surface scientists interested in atomic-scale phenomena on metals and semiconductors; atomic force microscopy was a relatively uninteresting curiosity. And for users of air STMs from Digital Instruments or its rivals (including one startup from the Quate group—Park Scientific Instruments—and another from Baldeschwieler's—Quanscan), the STM, however problematic, could be operated easily and gave mountains of images very quickly. Groups *buying* STMs were unlikely to *build* an AFM, given how tedious and troublesome the new technology appeared. For Quate, Hansma, and some of their associates at IBM Almaden, though, the AFM was the answer they were looking for.

Their preference for the AFM emerged from the evolving nexus between commerce and pedagogy in Santa Barbara and Palo Alto. AFMs seemed to lift the constraints on blind variation that made STMs problematic in two interwoven ways. On the one hand, the wider range of materials that could be imaged meant a wider array of disciplines to borrow from and appropriate. On the other hand, where before instrumental variation seemed limited to changing the environments the STM operated in (air, water, oil, electrochemical solutions), with the AFM Binnig and Quate suggested modifying the instrument to measure a range of *properties*—friction, hardness, magnetism, attractive and repulsive forces, etc. Each variation represented (for Quate and Hansma's students) a potential dissertation topic, as well as (for Digital and Park) a potential new product and market.

This became an educational template at Stanford and UCSB that continues to today, and that (through commercialization) dominates probe microscopy and defines the knowledge produced with scanned probes. Graduate students coming through Quate and Hansma's groups learn the basics of SPM instrumentation (usually from other students), and pick up the informal, tacit techniques of building microscopes. For their dissertations, they find some aspect of SPMs to tinker with or improve; at the same time, they learn a little about the entities that will be measured or imaged with their new instrument, how to prepare them, what they look like, how they will react with the probe. When done in conjunction with some visitor from another discipline (a postdoc or junior professor from electrochemistry, biophysics, molecular biology, materials science, geology, electrical engineering, etc.), then two new microscopists are created, and SPMs can join the toolkit of another discipline (with the academic credit and commercial gain that implies). After graduating, Hansma and Quate students (at least until the mid 1990s) often joined Digital or Park and continued creating new instruments, new microscopists, and new markets.

This reveals an important difference between the corporate and academic moral economies surrounding early probe microscopy in how to view the proper relationship between the local lab group and wider epistemic or disciplinary communities. The corporate STMers saw their roles as tied strongly to their home institutions; they worked in parallel, but not necessarily in concert, with other similar groups at their own and similar institutions, working toward similar goals using a similarly handed-down toolkit of techniques and knowledge. The academic groups saw themselves as starting from a more rootless and atomistic point, with no natural epistemic community to call home. Yet they also began to see themselves as the natural centers of a new instrumental community, one which would encompass a variety of disciplines. Where postdocs at the corporate labs were trained to reference their work to institutions of surface science, graduate students at UCSB and Stanford were taught to take an almost touristic attitude to the disciplines. The exchange of knowledge and materials central to a moral economy were conducted in house in the corporate labs, whereas the academic groups had no "house" and so cobbled together shared understandings and practices by appropriating cultural materiel from the communities they encountered.

Thus, one hallmark of the California groups, particularly Santa Barbara, was the variety of actors within this moral economy. Because of their outward-looking orientation, and because tacit skills were thought to eclipse formal knowledge in microscope-building (formal knowledge being only a token of exchange in exporting the microscopes to other disciplines), formal training mattered little to someone who joined DI or the Hansma group. Elings's philosophy in his scientific instrumentation program, for instance, had been "you can't know what you're doing"—i.e., people embarking on building an instrument should just try to build one, rather than formally learning the technique of instrument design, since they probably already possessed most of the knowledge they needed (and what they lacked could only be learned by doing, not by reading books or listening in class).[80] Thus, DI's early employees were former engineering undergraduates, master's of instrumentation students, and a hodgepodge of history PhDs, Eastern-religion majors, and skater rats who just wanted to hang out in Santa Barbara.[81] Early on, these people were assigned tasks for which they had little training, let loose to learn as they went. Elings institutionalized weekly "inventing sessions" where employees gathered to brainstorm inventions having nothing to do with probe microscopy.[82] The chaotic experimentation driving the academic groups reigned here also, and many variants of probe microscopy started as Elings's half-serious suggestions, or as late-night soldering sessions.

Similarly, Barney Drake, Paul Hansma's "hands," profoundly shaped probe microscopy through the enormous tacit, extra-disciplinary knowledge at his disposal. Drake not only realized most of Hansma's designs but was also responsible for fine-tuning their operation so graduate students could use them, as well as for liaising with DI to transfer designs to be repackaged commercially.[83] Drake's story illustrates the Santa Barbara epistemic culture—after building microscopes during the school year, he usually disappeared over the summer to be a river guide in the West. Eventually, wanting to cultivate business skills and see if he could run his own start-up, he founded a surface analysis company, first in a back room at DI, then later in the bedroom of his ski chalet in Truckee, California (where the "specimen preparation equipment" was the kitchen sink).[84] There, he continued the Hansma group's work, taking samples from researchers and companies, figuring out how to image them in a probe microscope (without knowing much about the nature of the samples—sometimes even, for proprietary reasons, knowing nothing at all), then walking customers through the images and results. Again, this work extended the AFM's colonization of disciplines begun in the Quate and Hansma groups—Drake's customers were companies or researchers who might want to buy an AFM, but had little understanding of what such an instrument could do or how it would fit with their current equipment. When Drake's business grew enough that he needed an assistant, he didn't seek out another AFMer; he recruited a fellow river guide and trained him from scratch (this person now designs instruments for a DI spinoff).[85]

Drake's story exemplifies the Quate and Hansma groups' orientation to tacit knowledge and, by extension, the heterogeneity (in terms of training) of the people whom they enrolled. By comparison, I don't wish to say that the corporate STMers did not value tacit knowledge or building instruments as an integral part of learning to create scientific knowledge. For surface scientists, though, acquiring experimental "hands" was the first step in tailoring an instrument to generate disciplined surface-science knowledge. On the West Coast, meanwhile, building instruments became a pedagogical *end*, not a *means*—a semi-disciplined end centered on recipes and tacit skills, but also reliant on disciplined knowledge borrowed or skimmed. Thus, the Quate and Hansma groups routinized a division of knowledge and labor between instrument "builders" and instrument "runners." In turn, the startups associated with these groups envisioned their customers as possessing *disciplined knowledge*, but not the *tacit skills* needed to build an instrument. These companies crafted microscopes for their imagined users by blackboxing many of the embodied skills needed in home-built instruments.[86] SPM manufacturers, especially DI, disciplined their users by guiding them into particular ways of using the instrument and facilitating some ways of seeing rather than others.

At the same time, the manufacturers saw a need to seem disciplined themselves if they wanted images to make sense to potential customers. This meant extending the methods of the academic groups—for instance, bringing postdocs and post-postdocs through the companies to work on instruments or modes aimed at some disciplinary niche.[87] At DI, these researchers built a vast applications lab, housing biologists, physicists, and chemists to forge new applications and publicize them to their colleagues.[88] The applications department also housed skilled operators with some of the most subtle and tacit knowledge anywhere about using AFMs.[89] Together, these people further disciplined users through a kind of distributed cognition—users learning to operate the instrument, still unclear about its parameters or what their materials should look like under its gaze, would send DI's applications people samples to be imaged, then talk with them about how to "see."[90] Ironically, DI often published images of such samples as ads in journals like *Nature* even before the original researcher could publish their own disciplined results in the same journals.

## The End of the 1980s

So what happened to the epistemic cultures that brought probe microscopy to North America? At Almaden, Yorktown, and Murray Hill, the close of the 1980s ended the first phase of surface-science STM experiments. With the recession of the early 1990s and the consequent downturn at IBM and Bell, modes of competitive basic research prevalent since World War II shifted radically. Many researchers left for academic positions or government labs, where they continued on more or less as before. Those who stayed became more preoccupied with IBM's actual product lines than the dictates of surface science. Since most working microelectronic components operate in air and with insulating oxide layers, and since the critical dimensions of integrated circuits are still considerably larger than atomic scale, AFMs have largely replaced STMs even at IBM and Bell.

For DI, the 1990s brought a considerable shift as well, more from economic success than hardship. By pushing probe microscopy into more variants, for more disciplines, with more operating skills black-boxed, Digital became the behemoth of scanning probe microscopy, thereby making the AFM (particularly using DI's patented technique, tapping mode) the mainstream of American probe microscopy.[91] As sales boomed for each successive generation of SPMs (with more modes and variants, and more disciplines providing customers), the company grew, and the swift, playful, circulating experimental style that generated early designs became slower and more managed.[92] Turning embodied, tacit skills into hardware and software became a more consuming task, with larger teams of engineers and code-writers involved; this was especially so as DI turned to markets seen as "conservative" (i.e., as wanting push-button microscopes requiring no "hands" or in-depth understanding of the instrumentation to operate), such as biology, the pharmaceutical industry, and, most notably, semiconductor manufacturing, where the culture of production emphasized disciplining cheaply paid and cheaply trained workers by giving them fully automated, single-button instruments. To meet this need, DI spent the 1990s building process-line AFMs with robots to exchange tips and samples, do the approach and calibration, and run preset inspection routines.

This enabled DI's merger with Veeco, an East Coast, Manhattan Project–vintage semiconductor processing equipment company with the full complement of middle management and MBAs and the concomitant prizing of the formal, disciplined knowledge of business and management lacking at DI. (Elings prized tacit knowledge—particularly his own—not only in instrument design but in all corporate operations, scorning competitors' use of modern business methods.[93]) It highlights where DI came from, and how it has changed, that spinoffs (once a rarity) are now cropping up. Most notable is Asylum Research, formed over lunch breaks by veterans of the Hansma group and DI to restore the small-scale, undisciplined, playful moral economy of the 1980s. "Asylum," of course, is an overt double entendre—the company is an asylum for those escaping from Veeco's corporate atmosphere, but also an institution run by self-described "lunatics."[94]

## Conclusion

This essay has addressed the intertwining of pedagogical concerns with concerns about the roles of tacit and formal knowledge in experimental settings, particularly when a new technique—a new way of generating knowledge—is at stake. When tacit knowledge entered science studies through Harry Collins and the sociology of scientific knowledge (borrowing, of course, from Michael Polanyi), the term was used to question a formalized, abstract, disembodied version of scientific knowledge usually associated with analytic philosophies of science.[95] At the time, Collins and others left unclear the relation between tacit and formal knowledge—formal knowledge was apparently found in books, expressible purely in symbolic forms, could be transmitted unproblematically, survived apart from the contexts of its origin (indeed, apart from any context at all). Tacit knowledge, on the other hand, couldn't (entirely) be written down, could only be transmitted through embodied, social interaction, and only had meaning in its context of use. Most importantly, not everybody could learn it, and, what was worse, you couldn't always tell if you had learned it or not. To ascertain if you possessed the proper tacit knowledge, you had to be accountable to some community, some set of practitioners who negotiated what counted as good practice. Controversy studies attempted to watch that negotiation in action, since disagreements within core sets are places where the boundaries of tacit and formal knowledge are set.

Probe microscopy presents both a technological and a scientific controversy, albeit not a very heated one.[96] Here we see (at least) two quite different orientations to tacit and formal knowledge. For the surface scientists, knowledge that could be published in papers and found in books was both the means and the ends of experiment. The formal knowledge of solid-state theory and surface scientific practice reliably mapped out what experiments satisfied managers' expectations of postdocs and young staff scientists within the moral economies of national and corporate research laboratories. Tacit knowledge was still important—building a new instrument, after all, required all kinds of embodied skills—but subordinate. Formal knowledge bounded embodied skill, limiting surface scientific habitus—if, for instance, formal knowledge said tunneling wouldn't work in air, then one might as well not try.[97]

The California groups reversed this paradigm. For them, tacit knowledge preceded formal knowledge, both temporally and in importance. In their view, people approach tasks with informal understandings and practical skills, including some they may be unaware of—the way to liberate those abilities, and create more, was to cultivate a variety of experiences, wandering unrigorously through an array of activities, disciplines, tasks, etc. Learning meant attaining goal-oriented skills as well as more general, tacit knowledge about how to attain further skills in the future. Ostensibly, anyone could perform this kind of extra-disciplinary learning; thus, the shape and boundaries of this laboratory culture were as fluid and various as the activities that composed it. Still, these groups found tacit knowledge alone *too* tied to context and *too* undisciplined to make sense to any external community; thus, the disciplines were pulled in as resources, repositories of knowledge, tools, and people with which to construct new meanings around instruments and images. Thus, this epistemic culture worked through enrollment, circulation, and commodification. Like the trading zones described by Peter Galison, this experimental style drew together different kinds of actors to exchange knowledge and tools and hammer out local meanings.[98]

This story's lessons for better understanding replication and knowledge transmission lie in seeing how pedagogical imperatives shaped the boundaries between tacit and formal knowledge. What became known with vacuum STMs, for instance, was really what was already known (via LEED and other techniques), only more so (i.e., an STM could fill knowledge gaps already known to exist). In crafting STMs to merge with an established body of practice, the corporate microscopists initiated *themselves* into an established body of practitioners. What became known with the instruments built in the California academic groups, on the other hand, was sometimes unexpected, and in some cases it later became un-known (or even disowned). The first task of the academic STMers and AFMers was to build an autonomous practice (and corps of practitioners) for probe microscopy; for that, they felt, tacit knowledge was the essential ground. Formal results were the price of making probe microscopy an indispensable tool, in that they generated the credibility, material resources, collaborators, epistemic materials, and tasks necessary to train new generations of probe microscopists. In both the corporate labs and the academic groups, though, turning "awkward students" into masters was coterminous with turning awkward instruments into reliable tools. What could be seen, what could be known, what could be said—these were *outcomes* of the need to create new knowing subjects and new citizens of the laboratory.

## Notes

1. I focus most here on the work of Harry Collins. See Collins, *Changing Order: Replication and Induction in Scientific Practice*, second edition (University of Chicago Press, 1992); idem, "The seven sexes: A study in the sociology of a phenomenon, or the replication of experiments in physics," *Sociology* 9 (1975): 205–224. Collins adverts to pedagogical issues in his "awkward student" example, but neglects educational contexts in describing the social contingencies of knowledge transmission. This essay tries to fill this gap, but also makes a stronger claim—that *replication* is often

carried out in training contexts, and that the logic of pedagogy can be a vital guide to exiting the experimenter's regress.

2. See Stuart Leslie, *The Cold War and American Science: The Military-Industrial-Academic Complex at MIT and Stanford* (Columbia University Press, 1993); Sharon Traweek, *Beamtimes and Lifetimes: The World of High Energy Physicists* (Harvard University Press, 1988).

3. Most relevant to this essay are David Kaiser, "Making tools travel: Pedagogy and the transfer of skills in postwar theoretical physics" (in this volume); Robert Kohler, *Lords of the Fly: Drosophila Genetics and the Experimental Life* (University of Chicago Press, 1994).

4. What little science studies work there is on STM and AFM is fairly cursory. For a good introduction, see Arne Hessenbruch, "History of recent science and technology: Materials research" (2001), available at http://hrst.mit.edu.

5. Kohler, *Lords of the Fly*; see also E. P. Thomson, "The moral economy of the English crowd in the eighteenth century," *Past and Present* 50 (1971): 76–136.

6. R. K. Bassett, *To the Digital Age: Research Labs, Start-Up Companies, and the Rise of MOS Technology* (Johns Hopkins University Press, 2002) describes IBM corporate lab culture generally, with some discussion of the Josephson project. Bassett claims that Big Blue's dominance of business computing promoted an inward-looking, idiosyncratic research style often disconnected from the exigencies of commerce (as exemplified by the massive and disastrous Josephson project). Bill Leslie makes a similar point about AT&T and Bell Labs in S. G. Knowles and S. W. Leslie, "'Industrial Versailles': Eero Saarinen's corporate campuses for GM, IBM, and AT&T," *Isis* 92 (2001): 1–33; S. W. Leslie, "Blue collar science: Bringing the transistor to life in the Lehigh Valley," *Historical Studies in the Physical and Biological Sciences* 32 (2001): 71–113. I find this picture compelling, and it underlies much of my discussion of corporate surface science.

7. G. Binnig and H. Rohrer, "The scanning tunneling microscope," *Scientific American* 253 (1985): 50–56.

8. G. Binnig and H. Rohrer, "Scanning tunneling microscopy on crystal-surfaces," *Journal of Crystal Growth* 65 (1983): 679–680; G. Binnig, H. Rohrer, C. Gerber, and E. Weibel, "Tunneling through a controllable vacuum gap," *Applied Physics Letters* 40 (1982): 178–180; B. M. Schwarzschild, "Microscopy by vacuum tunneling," *Physics Today* 35 (1982): 21–22.

9. C. B. Duke, "Atoms and electrons at surfaces: A modern scientific revolution," *Journal of Vacuum Science and Technology* A2 (1984): 139–143.

10. I draw on Ian Hacking, "The self-vindication of the laboratory sciences," in *Science as Practice and Culture*, ed. Andrew Pickering (University of Chicago Press, 1992), 29–64, in talking about the self-vindicating instrumentarium (which I take to mean the dense, complex thicket of mutually reinforcing instruments, techniques, recipes, etc. that bring into being and lay out for study the particular experimental entities of a laboratory discipline). Hacking clearly has the STM in mind at several points in his essay.

11. The shorthand for describing a reconstruction like the silicon (111) 7 × 7 has three parts: the material (e.g. silicon); the bulk crystalline orientation (e.g. 111), a notation developed by crystallographers and used extensively in both materials science and solid-state physics; and the configuration of the unit cell of the crystal at the surface (e.g. 7 × 7). This last number is easily available from low energy electron diffraction (LEED) and other diffraction techniques. LEED can tell roughly how large the unit cell is, and how many atoms it contains, but gives only elusive indications of how the atoms are arranged in those unit cells.

12. LEED manufacturers, for instance, would ship a 7 × 7 sample with the instrument to aid in verifying that the diffractometer was working. Interview with academic surface scientist, May 9, 2001.

13. "Golden image" is borrowed from Peter Galison, *Image and Logic: A Material Culture of Microphysics* (University of Chicago Press, 1997). The "image" and "logic" dichotomy seems never to have been as severe in surface science as in high-energy physics, but certainly the STM amplified the value of golden images in this field.

14. Interview with government scientist, July 6, 2000.

15. Interview with former IBM scientist, May 2, 2001; interview with IBM scientist, March 15, 2001.

16. Interview with IBM scientist, February 22, 2001; interview with former IBM scientist, November 13, 2001.

17. Interview with former IBM scientist, June 28, 2000; interview with former IBM and Bell scientist, May 22, 2001.

18. Interview with former IBM scientist, May 2, 2001.

19. Interview with former IBM and Bell scientist, May 22, 2001; interview with academic surface physicist, May 3, 2001.

20. Interview with former IBM scientist, June 28, 2000.

21. Yorktown's funding structure encouraged importing postdocs from IBM's international subsidiaries. Many of these brought STM back with them to national or corporate labs in their home countries.

22. For a famous episode of how such competition shaped the production of technical knowledge, see Michael Riordan and Lillian Hoddeson, *Crystal Fire: The Birth of the Information Age* (Norton, 1997).

23. Traweek, *Beamtimes and Lifetimes*.

24. Interview with former IBM and Bell scientist, May 22, 2001.

25. Ibid.

26. Interview with IBM and Bell scientist, October 11, 2001; interview with IBM scientist, February 23, 2001.

27. Interview with IBM and Bell scientist, October 11, 2001; interview with former IBM and Bell scientist, May 22, 2001.

28. J. Tersoff and D. R. Hamann, "Theory and application for the scanning tunneling microscope," *Physical Review Letters* 50 (1983): 1998–2001. After his postdoctoral tenure at Bell, Tersoff went to Yorktown and collaborated with STMers there.

29. Interview with former Bell scientist, February 20, 2001.

30. Interview with former IBM scientist and Hansma group member, November 16, 2001; interview with IBM scientist, September 26, 2000.

31. Famously, the Zurich team refused to use computers for either control or imaging. In real time, they would watch an oscilloscope and try to assemble its traces into an image in their heads. Later, they would cut out a set of oscilloscope traces on paper, paste them onto cardboard, and glue the cardboard cutouts together to form a three-dimensional representation of the surface (an acutely labor-intensive process).

32. Interview with former IBM and Bell scientist, May 22, 2001.

33. Interview with former Bell scientist, February 29, 2001.

34. Interview with former IBM scientist, November 13, 2001; interview with former Bell scientist, February 20, 2001.

35. Interview with IBM scientist, November 12, 2001.

36. See Hacking, "Self-vindication of the laboratory sciences."

37. R. A. Wolkow, "A variable temperature scanning tunneling microscope for use in ultrahigh-vacuum," *Review of Scientific Instruments* 63 (1992): 4049–4052.

38. E. J. Vanloenen, J. E. Demuth, R. M. Tromp, and R. J. Hamers, "Local electron-states and surface geometry of Si(111)-($\sqrt{3} \times \sqrt{3}$)Ag," *Physical Review Letters* 58 (1987): 373–376; R. J. Wilson and S. Chiang, "Structure of the Ag/Si(111) surface by scanneling tunneling microscopy," *Physical Review Letters* 58 (1987): 369–372. In fact, these two groups both (unknowingly) imaged the same classic substrate-adsorbate system (silver on silicon), and published similar images but came to contradictory interpretations in these back-to-back articles.

39. J. E. Demuch, U. Koehler, and R. J. Hamers, "The STM learning-curve and where it may take us," *Journal of Microscopy* 152 (1988): 299–316. LEED and other surface-science instruments generally survey much larger areas of a surface than an STM. Thus, throughout this period there was constant pressure to increase the scan range of tunneling microscopes, to make STM more compatible with the surface-science instrumentarium.

40. Interview with former IBM scientist, February 26, 2001.

41. This is an instance of the kind of "postdoc cascade" that David Kaiser has described in the case of Feynman diagrams. See Kaiser, "Making tools travel."

42. For the proceedings of this conference, see G. Binnig and H. Rohrer, "Scanning tunneling microscopy," *IBM Journal of Research and Development* 30 (1986): 355–369.

43. For the proceedings of this conference, see N. Garcia, "STM 86: Proceedings of the 1st International Conference on Scanning Tunneling Microscopy, Santiago-De-Compostela, Spain, 14–18 July 1986. Preface," *Surface Science* 181 (1987): R9 and following articles.

44. For the purposes of this essay, I leave aside European work, which began early but never showed the same communal dichotomy as the American groups.

45. A more detailed analysis might include other researchers in this subculture: John Clarke at Berkeley, former Quate students John Foster (and his collaborator, Jane Frommer) and Dan Rugar (and Clarke's student Jon Mamin) at IBM Almaden, and Hansma collaborator Stuart Lindsay at Arizona State, among others.

46. C. F. Quate, "Vacuum tunneling: A new technique for microscopy," *Physics Today* 39 (1986): 26–33.

47. Interview with Hansma group member, March 19, 2001; interview with former IBM scientist, November 13, 2001.

48. Interview with Baldeschwieler group member, March 28, 2001; interview with former Baldeschwieler group member, March 30, 2001.

49. Interview with IBM scientist, February 22, 2001.

50. Interview with IBM scientist, September 26, 2000; interview with former IBM scientist, November 13, 2001.

51. Interview with IBM scientist, September 26, 2000; interview with former IBM scientist, November 13, 2001; interview with IBM scientist, November 12, 2001.

52. Interview with former corporate lab scientist, June 27, 2001.

53. J. Moreland, J. Drucker, P. K. Hansma, J. P. Kotthaus, A. Adams, and R. Kvaas, "Air as an adjustable insulator for C-V and G-V analysis of semiconductor surfaces," *Applied Physics Letters* 45 (1984): 104–106; interview with Hansma group member, October 18, 2001.

54. Interview with Hansma group member, October 18, 2001.

55. J. B. Pethica, "Interatomic forces in scanning tunneling microscopy: Giant corrugations of the graphite surface, Comment," *Physical Review Letters* 57 (1986): 3235.

56. I draw on Hans-Jörg Rheinberger's discussion of how disciplines construct "epistemic things" around which to weave an elaborated discourse: Rheinberger, *Toward a History of Epistemic Things: Synthesizing Proteins in the Test Tube* (Stanford University Press, 1997).

57. Interview with former Quate group member, June 28, 2001; interview with former Quate group member and Park employee, October 12, 2001; interview with Hansma group member and Digital employee, March 19, 2001. For analyses of variation-and-selection, see Walter Vincenti,

*What Engineers Know and How They Know It: Analytical Studies from Aeronautical History* (Johns Hopkins University Press, 1990); Karen Knorr-Cetina, *Epistemic Cultures: How the Sciences Make Knowledge* (Harvard University Press, 1999).

58. Interview with former Hansma group member and DI employee, March 27, 2001; interview with former Quate group member and Park employee, October 12, 2001. My argument in describing these aspects of the West Coast groups draws inspiration from Andrew Warwick, "Exercising the student body: Mathematics and athleticism in Victorian Cambridge," in *Science Incarnate: Historical Embodiments of Natural Knowledge*, ed. Christopher Lawrence and Steven Shapin (University of Chicago Press, 1998), 288–326.

59. Interview with Hansma group member and Digital employee, March 19, 2001.

60. Interview with Baldeschwieler group member, March 28, 2001.

61. Interview with former Quate group member and Park employee, October 12, 2001.

62. Interview with former Hansma group member and DI employee, March 27 2001; interview with Hansma group member, March 19, 2001.

63. Interview with former Quate group member and IBM scientist, October 19, 2001.

64. Interview with Hansma group member, March 19, 2001; interview with Hansma group member, March 19, 2001.

65. C. G. Slough, W. W. McNairy, R. V. Coleman, B. Drake, and P. K. Hansma, "Charge-density waves studied with the use of a scanning tunneling microscope," *Physical Review B* 34 (1986): 994–1005.

66. Interview with Hansma collaborator, November 14, 2001; interview with former IBM scientist and Hansma group member, November 16, 2001; interview with Hansma group member and Digital employee, March 19, 2001.

67. Interview with former Quate group member and Park employee, October 12, 2001; interview with former Quate collaborator and IBM employee, April 3, 2001.

68. Interview with former Quate group member and Park employee, October 12, 2001; interview with IBM scientist, March 14, 2001.

69. Interview with former Quate group member, June 28, 2001.

70. Interview with Digital employee, February 26, 2001; interview with former Digital employee, October 18, 2001.

71. Interview with former government scientist and Digital employee, October 12, 2001; interview with IBM scientist, November 12, 2001.

72. D. P. E. Smith, M. D. Kirk, and C. F. Quate, "Molecular images and vibrational spectroscopy of sorbic acid with the scanneling tunneling microscope," *Journal of Chemical Physics* 86 (1987): 6034–6038.

73. Interview with former Quate group member, June 28, 2001; interview with former Quate group member and Park employee, October 12, 2001; interview with IBM and Bell scientist, October 11, 2001.

74. R. J. Driscoll, M. G. Youngquist, and J. D. Baldeschwieler, "Atomic-scale imaging of DNA using scanning tunneling microscopy," *Nature* 346 (1990): 294–296.

75. Interview with Hansma group member, March 19, 2001.

76. W. M. Heckl and G. Binnig, "Domain-walls on graphite mimic DNA," *Ultramicroscopy* 42 (1992): 1073–1078; C. R. Clemmer and T. P. Beebe, "Graphite: A mimic for DNA and other biomolecules in scanning tunneling microscope studies," *Science* 251 (1991): 640–642.

77. Interview with Hansma group member and Digital employee, March 19, 2001; interview with IBM scientist, February 22, 2001.

78. G. Binnig, C. F. Quate, and C. Gerber, "Atomic force microscope," *Physical Review Letters* 56 (1986): 930–933.

79. Interview with former Hansma group member and DI employee, March 27, 2001; interview with Hansma group member, October 18, 2001.

80. Interview with former Digital employee, March 20, 2001.

81. Interview with Digital employee, February 26, 2001; interview with former Digital employee, October 18, 2001.

82. Interview with former Digital employee, March 20, 2001; interview with Digital employee, October 18, 2001.

83. Interview with Hansma group member, October 18, 2001.

84. Ibid.

85. Ibid.

86. The automation of how the microscope's tip approaches is a good example of such black-boxing. Handling the approach of probe to surface, for instance, had started out as an embodied task similar to tuning a guitar. Indeed, many groups converted signals from the feedback circuitry into an audible output so that they could listen to the probe as it approached, and coordinated ear, eye, and hand to avoid crashing the tip. (The approach worked by turning a knob to increase the voltage to the z-piezo until the probe came close enough to the surface.) With commercial instruments, however, this task became an automated, push-button job handled by software that repackaged embodied skills into code. For the black-box literature, see Bruno Latour, *Science in Action: How to Follow Scientists and Engineers through Society* (Harvard University Press, 1987); Kathleen Jordan and Michael Lynch, "The sociology of a genetic engineering technique: Ritual and rationality in the performance of a 'plasmid prep,'" in *The Right Tools for the Job: At Work in the Twentieth-Century Life Sciences*, ed. Adele Clarke and Joan Fujimura (Princeton University Press, 1992), 77–114.

87. Interview with European surface scientist and former Park employee, November 16, 2001; interview with former Hansma group member and DI employee, March 27, 2001.

88. Interview with Digital employee, March 23, 2001; interview with former government scientist and Digital employee, October 12, 2001.

89. Interview with Hansma group member and Digital employee, March 19, 2001; interview with Digital employee, February 26, 2001.

90. On "distributed cognition" see Edwin Hutchins, *Cognition in the Wild* (MIT Press, 1995).

91. With Digital's emphasis on air and ambient instruments, and its abandonment of STMs and non-contact AFMs, these techniques have become almost completely associated with vacuum. Thus, the largest manufacturer of STMs (still quite small compared to Digital) is a German company, Omicron, that specifically targets surface scientists by selling complete vacuum systems combining STM, AFM, LEED, Auger electron spectroscopy (AES), and scanning electron microscopy (SEM).

92. Interview with Digital employee, March 23, 2001.

93. Interview with former Digital employee, March 20, 2001.

94. Interview with Digital employee, October 18, 2001; interview with former Digital employee, March 23, 2001; interview with former Hansma group member, March 20, 2001.

95. H. M. Collins, "Tacit knowledge and scientific networks," in *Science in Context: Readings in the Sociology of Science*, ed. Barry Barnes and David Edge (MIT Press, 1982), 44–64; Michael Polanyi, *The Tacit Dimension* (Doubleday Anchor, 1967).

96. In talking about the interpretive flexibility surrounding probe microscopy, I have largely been relying on Harry Collins's sociology of scientific knowledge framework. It should be clear, however, that controversies over STMs and AFMs concerned not simply what one could *know* through these instruments, but also their material form, the practices surrounding them, and what one could *do* with them—i.e., that these were *technological* controversies. For an extension of the sociology of scientific knowledge to technology, see W. E. Bijker and T. Pinch, "The social construction of facts and artifacts: Or how the sociology of science and the sociology of technology might benefit each other," in *The Social Construction of Technological Systems: New Directions in the Sociology and History of Technology*, ed. W. E. Bijker, T. P. Hughes, and T. Pinch (MIT Press, 1987).

97. Pierre Bourdieu, *The Logic of Practice* (Stanford University Press, 1990).

98. Peter Galison, "Computer simulations and the trading zone," in *The Disunity of Science: Boundaries, Contexts, and Power*, ed. Peter Galison and David Stump (Stanford University Press, 1996).

# III The Action of Textbooks

# 7 The Power of Didactic Writings: French Chemistry Textbooks of the Nineteenth Century

**Antonio García-Belmar, José Ramón Bertomeu-Sánchez, and Bernadette Bensaude-Vincent**

Textbooks have a bad reputation in science studies because they are considered the variety of scientific literature most remote from the creative source of knowledge. They are usually placed at the bottom of a long chain of writings stemming from laboratory notebooks, moving to oral presentations in larger and larger circles to journal articles, then to popular magazines, and finally to textbooks. If "the writing of textbooks is the last existential act in science," as John Brooke put it, it is clear that textbooks have nothing to tell us about the enthusiasm of creation, about the tangled labyrinth of the construction of scientific facts, or about the struggles with instruments and colleagues—all interesting facets of "science in action."[1] Historians of science consequently pay little attention to textbooks and use them only insofar as they provide a window onto "normal science."

Only a limited group of textbooks are mentioned by historians of science. In the case of chemistry, Lavoisier's *Elements of Chemistry* and Mendeleev's *Principles of Chemistry* are described as sources of knowledge. Lavoisier's decision to address beginners by proceeding from the simple to the complex and from the known to the unknown encouraged the foundation of a modern chemistry based on analysis. Mendeleev discovered the periodic law while he was striving to order the chapters of a general chemistry textbook intended for his students at St. Petersburg University.[2] These brilliant exceptions do not question the established general view of textbooks as repetitive, uninspired, and standardized expositions of pre-established knowledge.

In stark contrast with this long-standing tradition of despise for textbooks, Owen Hannaway argued in 1975 that the science of chemistry was shaped by textbooks.[3] The break between alchemy and chemistry was less a change of paradigm, such as the end of the belief in the possibility of transmutations or the rejection of alchemical and mystical symbolism, than a change in the exposition of chemistry. Chemistry became public knowledge as opposed to secrets transmitted from master to disciple. More specifically, chemistry became a teachable subject organized in a rational way. This was a real challenge because chemical knowledge mainly consisted in a tangled collection of recipes for preparing mixtures, and descriptions of their most striking properties or medical virtues. Hannaway located the didactic origins of chemistry as a science in Andreas Libavius's *Alchemia* (1597), although that book advocated a return to Aristotle. Starting with a set of definitions, Libavius attempted a classification of the recipes and processes in order to organize the chapters of his textbook. Thus chemistry gradually became a scientific discourse based on a set of general principles rather than a chaotic collection of empirical data. Hannaway did more than simply point out another famous textbook. He emphasized the heuristic power of didactic writing: as soon as a first classification was recognized it presented a new problem that others—readers and colleagues—would attempt to solve through new hypotheses and experiments. Libavius initiated a process of confrontation between organizing hypotheses and traditional empirical procedures. In thus revealing the creative power of writing a textbook, and the specificity of chemical knowledge, Hannaway acted as a pioneer. He opened up a Pandora's box of questions about the relations between didactic and scientific discourses. How could writing a textbook bring about scientific change? In which conditions could a textbook be creative? To what extent does the audience help shape the profile of a discipline? How did textbooks become established as a genre? How did they differ from other forms of chemical literature?[4]

This essay summarizes some results of our investigation of a corpus of about 500 textbooks published in France between 1789 and 1860. First we present the philosophical roots of the current image of textbook science in a brief preliminary section in order to contrast this image with the conclusions of our analysis of nineteenth-century textbooks. Then we describe the complex process of consolidation of French chemistry textbooks as a particular genre of scientific literature, paying attention to four actors: textbook writers, publishers, readers, and educational institutions. Finally, we discuss the issue of the creativity of textbooks through two typical subjects dealt with in nineteenth-century chemistry textbooks: classification and atomic theory.

## Toward an Archeology of the Standard View

The distinction between creative science and expository science seems intuitive and quite natural nowadays, yet it is a rather recent view whose historical origin can be dated to the early nineteenth century. In France, it was clearly formulated by the founder of positivism, Auguste Comte. In the second lesson of his popular *Cours de philosophie positive*, Comte explicitly stated that the most advanced sciences could no longer be taught according to the chronological order of discovery. Teaching required a consistent and organic reconstruction of knowledge independent from the actual process of production of knowledge. Comte named it "the dogmatic order" by contrast to the "historical order" of exposition.[5] Although Comte stated that a science cannot be fully understood without its history, he considered the dogmatic order necessary for educative purposes. It is impossible, Comte argued, to raise an individual intellect, most often a mediocre one, to the level of knowledge acquired by many generations of geniuses. No individual, in a lifetime, can go through all the steps made by mankind. Hence the need for a shortcut—a logical reconstruction of the present state of knowledge. More than a simple gap, the dichotomy between the dogmatic and the historical orders required making a choice: either the historical presentation or the dogmatic presentation. Comte admitted that most textbooks were in fact a mixture of the two orders, but he insisted that such compromises were either inconsistent because the dogmatic order was incompatible with the historical exposition of science, or they relied upon biased accounts of the past.

Given this positivistic origin one might expect that anti-positivistic philosophers of science would question the dichotomy between the generation of knowledge and its didactic exposition. On the contrary, a century of textbook tradition seems to have reinforced it. Gaston Bachelard, for instance, insisted that the divorce was a characteristic feature of the scientific age, the unavoidable consequence of the break (the rupture) between the pre-scientific and the scientific spirits. While the former was unmethodical and wandering, the scientific way of thinking was "trained in official laboratories and codified in school textbooks." Bachelard emphasized that physics textbooks were repetitive and under a strict control. They supplied a "socialized and fixed science that could pass for natural only because of unchanging school syllabuses."[6] They were not descriptive but prescriptive, not really meant for transmitting science but rather commandments.

Thomas Kuhn added conservatism to the features of textbook science. Textbooks are meant for the perpetuation of the paradigm, for training students in solving the puzzles raised within the paradigm rather than inventing new problems.[7] Kuhn argued that they assume their conservative function through various ways. They present only established and incontrovertible knowledge, the stable results of past revolutions. They regularly occult revolutions either by eliminating history or by presenting the present state of knowledge as the end product of a linear accumulation of data. They consequently disguise the actual procedures both of discovery and justification. As training tools and rituals of introduction in a community, they are powerful precisely because they stabilize the discipline in denying scientific changes.

Despite doctrinal differences between them, Comte, Bachelard, and Kuhn agreed on various points. They assumed that the existence of textbook science (or of didactic exposition of science) is a necessity. They emphasized that textbooks did not mirror science in action.

Textbooks, they all agreed, deliver a biased image of science, distorting the real nature of scientific activity for didactic purposes. Although Comte developed a gradualist, continuist view of scientific change while Kuhn and Bachelard advocated a discontinuist view with radical breaks and revolutions, they all agreed that textbooks belong to a regime of accumulation and reproduction of knowledge rather than to a regime of innovation and creation. While they characterized textbooks as mere transmission tools, none of them envisioned a possible impact of the audience on the contents of a discipline.

From this quick survey of the standard view of textbooks we derived a number of methodological decisions. First, we would not take for granted that the distinction between "textbook science" and "science in action" was necessary or natural. Rather we wanted to examine in which contexts the dichotomy between the creation and transmission of science was generated. We consequently focused on the circumstances and constraints in which textbooks emerged as a genre of scientific publication. The middle of the nineteenth century appeared as a turning point, at least in France. Second, we would not discuss textbook science from the unique standpoint of science studies. As tools intended for the transmission of a set of knowledge, textbooks are written and manufactured for the specific audiences created by educational measures and reforms. We had to identify the heterogeneous agents that interact in the production of textbooks: the contents of chemistry, the students attending courses in various educational situations, and the authors with their backgrounds and professional activities, which shaped their personal views on chemistry and teaching. Insofar as chemistry textbooks are located at the intersection between the advancement of science and pedagogical views, they are under strong social, economic, and political pressures. In teaching spaces, as Kathryn Olesko emphasized, economic, social, and political forces rush into the structure and function of scientific knowledge.[8] This means that we had to revise the present image of scientific teaching as a second-rate scientific activity. Moreover, textbooks are material and commercial products subjected to the technical, financial, and political constraints of the publishing market. As such they belong to the history of books and reading. Therefore a historical study of science textbooks should intertwine the history of scientific disciplines with the history of science education as well as the history of books and publishing.

## The Consolidation of an Editorial Genre

How did textbooks turn into an independent and characteristic genre of scientific publication? The answer to this question raises a methodological difficulty: how to define the object under study. In the first half of the nineteenth century, various types of scientific literature acquired identities of their own as editorial products and as instruments of scientific communication. Science journalism, popular books, technical handbooks, encyclopedias, dictionaries, and various other types of publications settled in the publishing market while scientific communication was expanding.[9] These scientific genres involved sets of conventions which were shared by authors, publishers, and readers. As T. H. Broman remarked, "any writer who sets out to write a textbook, or a dissertation or scientific research article for that matter, must take into account those conventions and practices that permit a writing to be recognized and accepted by readers as an exemplar of a particular genre."[10] However, scientific genres were never stable realms with sharp and airtight boundaries. On the contrary, they were continuously negotiated and transgressed by readers, authors, and publishers, who might produce literary experiments in order to broaden their audiences.[11] Textbooks interacted and sometimes overlapped with other scientific publications because they shared publics, uses, and objectives. How to distinguish a textbook for primary education from a popular text, a general course of chemistry from a treatise, or even a handbook of experimental chemistry from a catalogue of instruments?

Two possible definitions are available. Either textbooks are defined by their uses—a textbook is every text practically used as a didactic tool in teaching institutions—or by their purposes—a textbook is every text especially and explicitly designed to be used as a didactic instrument in teaching institutions. Each definition suggests different sources and questions.[12] The former would lead us to focus on teaching practices and on the various uses of the text as a didactic tool. The latter, on the contrary, takes us to a history of the concept of textbook as it was designed and materialized by the various actors converging on the production of this type of text. The latter definition will be adopted in this essay. This means that our main subject is a group of books that, according to the indications that authors and editors included in the titles, covers, or forewords, were expressly written to be used in chemistry teaching in specified institutions. The election of a definition based on purposes and not on uses is determined by the sort of questions that lead our analysis. We aim to determine when chemistry texts conceived for teaching purposes started to have characteristics of their own, and we attempt to clarify a periodization of the emergence of chemistry textbooks as an autonomous editorial genre. First we try to assess the impact of the changing audiences for chemistry. Then we discuss the shifting biographical profiles of textbook authors and publishers as well as the main features of the French nineteenth-century textbook market. Whereas historians are usually more concerned with the description of the social aspects of this process of institutionalization, we will emphasize the consequences of institutional changes and consider how the contents and practices of a scientific discipline were adapted to new audiences, changing spaces, and teaching practices. As traces of teaching practices, textbooks are a privileged source. In less than six decades they were transformed into didactic tools with distinctive features.

The educational reforms introduced during the French Revolution transformed science teaching. In the eighteenth century chemistry had been taught in public courses with no regulations. Chemistry lectures were attended by medical and pharmacy students, artisans and craftsmen interested in practical applications, as well as by the gens du monde attracted by the spectacle of experimental demonstrations. After the French Revolution, by contrast, chemistry became an integral part of the official syllabus in a number of teaching institutions. It was taught at the secondary level in the écoles centrales and lycées, as well as in the institutions of higher education—écoles normales supérieures and facultés des sciences—dedicated to the training of teachers.[13] It was also an integral part of medical and pharmaceutical studies, which reinforced its importance in secondary teaching.[14] Simultaneously, chemistry was fully integrated in engineering schools, from the Ecole polytechnique and its écoles d'application to the military academies.[15]

The institutionalization of chemistry teaching had a tremendous impact. First, it increased the number of chemistry students and teachers.[16] However, the most important changes were less quantitative than qualitative. The heterogeneous audiences of public courses became a captive public, with specific capacities and training necessities that varied according to the different levels and institutions in which chemistry had been integrated. Classrooms reserved for regular students replaced the auditoriums and private laboratories open to everyone interested in acquiring chemical knowledge. School schedules, syllabi, and procedures of control limited the freedom of the teacher.

## A Captive Public

Textbooks, as objects tailored for specific targets, are defined by their publics. Their contents, format and size, typographic and iconographic features, and even the author's and editor's experiences were meant to supply the demands of a targeted reader according to his or her previous knowledge, age, and cognitive capacities, as well as to the didactic requirements, methods, and practices of an institution. In stark contrast with the standard references to heterogeneous publics found in most eighteenth-century chemistry texts, the didactic books published during the first decades of the nineteenth century contained more and more precise indications about the reader they targeted. The covers and forewords were the usual places to define the ideal public that editors and authors had in mind when writing, manufacturing, and selling a textbook. They offer clues about the horizon of expectations of the intended readers, their background, institutional context, and professional perspectives.

Chemistry textbooks in early-nineteenth-century France were meant for two types of students: medical and pharmaceutical students, and the pupils of the secondary education schools. The latter group disappeared under the Restoration when chemistry was banished from secondary teaching; hence there was a sudden drop in textbook production (followed by a second boom of production in the middle of the century, when chemistry was reintroduced at the secondary level). In fact, the two groups overlapped, since most of the pupils who attended the chemistry courses in the lycées were future students of medicine. After the reorganization of medical and pharmaceutical studies in 1803, chemistry became a propaedeutic subject for both curricula. An examination in chemistry was one of the six compulsory exams that every candidate for the doctoral degree in medicine had to pass. The French Revolution opened two possible ways to become "maître en pharmacie": either the traditional eight years of practical training at the apothecary office, or three years of practice followed by three years of study in one of the pharmacy schools in which chemistry played a major role in the curriculum.[17]

In addition to the students of the medical faculties or the pharmacy schools, students who attended the innumerable private preparatory courses were a favorite target for textbooks. These courses, already established in the eighteenth century, proliferated during the first half of the nineteenth century, especially when the grade of baccalauréat ès sciences became compulsory to enter the Faculty of Medicine. In Paris, the Faculty of Medicine hosted and supported these private courses. The director, J. Tyrat, delivered his course in daily lectures of four hours over a period of two months. Students who attended the private lessons could use the physics cabinet and the chemistry laboratory available on the site. The contents of the courses were strictly adjusted to the exam. Teachers were supposed to define their questions in the same terms as "MM. les examinateurs" did, and to encourage their pupils to answer them in the manner required for the exam.[18]

Thus future medicine and pharmacy students became the main target of the chemistry textbooks published during the first decades of the nineteenth century. During the second third of the century, however, the readership diversified. As chemistry was reintegrated in secondary education as a compulsory subject, there was a dramatic increase in textbook production. From 1830 on, dozens and dozens of texts for the students of the collèges royaux (as secondary institutions formerly named écoles centrales and then lycées were rechristened under the Restoration) were published, and in the late 1840s the first best-sellers were reprinted. Simultaneously a large variety of "manuels" and "aide-mémoire" specifically intended for the preparation of the baccalauréat flooded the market.[19] In 1833, Guizot's law, reorganizing primary education, created an intermediate cycle intended to cover the gap between primary and secondary studies.

Among other subjects, the official syllabus included "notions of physical and natural sciences applicable to daily uses." Consequently, chemistry became part of the teaching in the écoles normales primaires, where future teachers were trained.

In addition to educational policies and official syllabi, local technological institutes interested in the applications of chemistry to arts and industry also contributed to enlarge the public audience for chemistry. Starting in the 1830s, they organized evening lectures and Sunday courses for training specialized workers.[20] Simultaneously, private initiatives led to the creation of teaching centers such as the Conservatoire national des arts et métiers and the Ecole centrale des arts et manufactures of Paris. These specific audiences of workers and manufacturers prompted very successful textbooks such as the *Leçons de chimie élémentaire faites le dimanche, à l'Ecole municipale de Rouen* by Jean Pierre Louis Girardin, reissued eight times between 1835 and 1889, or the eight-volume *Traité de chimie, appliquée aux arts* by Jean Baptiste Dumas, based on the courses delivered initially at the Athénée of Paris and later at the Ecole centrale des arts et manufactures.[21] Chemistry textbooks for technological education formed a heterogeneous group that reflected the diversity of publics interested in the applications of chemistry. For instance, Anselme Payen (1795–1871), a chemistry professor at the Conservatoire national des arts et métiers, published different versions of his course on industrial chemistry in order to reach a variety of audiences. After the first version intended for the gens du monde, he targeted the students of the écoles préparatoires aux professions industrielles and those of the écoles d'art et manufactures, ending up with a volume for "manufacturers and farmers."[22] In contrast to Payen's pragmatic vision of the capacities and interests of his potential readers, Dumas adopted a more idealistic view in the preface to his treatise. He assumed that it would be accessible to a new generation of readers trained in the secondary collèges royaux, and able to apply the theoretical knowledge of chemistry to arts and industry.[23]

As a result of this strategy of diversification, the audience of chemistry textbooks was enlarged. However, as already mentioned, the most striking changes were less quantitative than qualitative. First, the public was captive and subject to teaching and reading practices determined by the spaces and the didactic methods of the time. Second, it was segmented into various layers according to the levels and types of teaching. For each category of book, authors and editors forged a demand, a curiosity, or an urgent and imperious necessity for filling lacunae. Every textbook appeared as supplying a market demand. In 1828, A. Manavit, a professor at the Collège royal in Toulouse, who published one of the first manuals for this kind of institution, presented his text as "an intermediate genre between big treatises and abrégés whose value hardly exceeds that of a table of contents." His textbook was specifically conceived to "help the pupils attending the elementary course of physics and chemistry which the author of this précis is in charge of at the Collège royal of Toulouse, in order to smooth a number of difficulties in their baccalauréat examination, and to allow them to fruitfully attend further courses at the universities."[24] Audiences, we claimed at the beginning of this section, defined textbooks; but it is also true that textbooks helped create and stabilize a diversity of audiences.

### Heroic Authors vs. Obscure Writers of Best-Sellers

Who wrote chemistry textbooks, and why? It is easy to answer the question for the late eighteenth century and the early nineteenth century. The biographies of Lavoisier, Chaptal, and Fourcroy are well known, and their motivations to write textbooks are easy to understand in view of the scientific changes prompted by the chemical revolution and the institutional changes caused by the French Revolution. It is somewhat more difficult to understand the motivations of Louis Jacques Thenard, Mateu Orfila, or Jean Baptiste Dumas, whose books were reissued periodically and used as references for the teaching of chemistry during the first half of the nineteenth century. The task, however, becomes considerably more difficult when we turn to authors such as Mathurin Jacques Brisson, Pierre Jacotot, Jean Baptiste Jumelin, Roch Théogène Guérin, Edmond Jean Joseph Langlebert, Jean Louis Lassaigne, Henry Debray, Eugène Desmarest, Pierre Paul Deherain, and Nicolas Deguin. Some of them authored successful textbooks used by several generations of students. Their names were certainly familiar to nineteenth-century French people, even if they have not been passed down to posterity. They skipped out of history despite the various traces of their publications in French libraries. Biographical sources to study this community of obscure chemistry writers are available, but their activity has been disregarded by historians of science.

The image of textbooks as mere instruments of transmission of knowledge reduces the role of their authors to an exercise in transcribing, summarizing, or adapting knowledge that was already organized. As we will argue below, writing a textbook was much more than passively transmitting pre-established knowledge. Instead, textbook authors acted as mediators among the actors that converged in the production of textbooks. As authors they engaged in the creation of didactic tools adapted to certain categories of readers and to the material conditions and requirements of teaching institutions. Not only did they have to be in agreement with the official norms and regulations that ruled the public educational system; since books are material objects, they also had to adjust to printing technologies and to publishers' editorial policies. From this angle, the profiles of those obscure people who devoted part of their time to textbook writing looks more interesting. Were they motivated by scientific views or by career, financial, political, or ideological interests? What were their backgrounds, their institutional positions, and their professional activities in research and teaching?

Prosopographic study suggests three periods in the evolution of the biographical profile of textbooks writers. In the first period (1789–1808), the authors aimed their books at the new teaching institutions created during the Revolution. The writers in this first generation were the most heterogeneous. Their contrasting backgrounds, professional activities, and involvements in academic research reflects the diversity of situations in which chemistry was cultivated during the second half of the eighteenth century. However, they shared a common feature: almost all of them had been recruited as teachers for the new educational institutions of the République. They had been trained during the Ancien Régime. Their backgrounds ranged from theology in the case of Pierre Jacotot (1755–1821), to military training in the case of Pierre Auguste Adet (1763–1834), to medicine or pharmacy in the cases of Edmé Jean Baptiste Bouillon-Lagrange (1764–1844) and Jean Baptiste Jumelin (1745–1807). Their textbooks were published late in their careers and lives. Some of them, such as Mathurin Jacques Brisson (1723–1806) and Jacotot, had many years of teaching experience in the Ancien Régime institutions. Others, such as Adet and Bouillon-Lagrange, were recruited because of their political engagement or their research achievements.

During the first two decades of the nineteenth century a radical change took place in the biographical profile of chemistry textbook authors. Authors were mainly young medical doctors (and sometimes pharmacists) who published in an earlier stage of their professional careers. Most of them taught students of the private preparatory courses for the entrance exam in medical faculties or medical students preparing chemistry exams at the Faculty of Medicine, or candidates for the degree of maître en pharmacie. For more than twenty years these students learned the foundations and the applications of chemistry through the texts written by authors such as Mateu Orfila, Jean Louis Lassaigne, and Julia de Fontenelle, members of the Société de Chimie Médicale, editors or active collaborators of the *Journal de Chimie Médicale*. They published their works in the publishing house of Nicolas Crochard, associated with the Paris Medical Faculty. In contrast with the heterogeneous group of the previous generation, authors and publishers of this period belonged to the medical milieu and shared common backgrounds, professional interests, and institutional contexts.

During the third period, the 1830s and the 1840s, the influence of the medical milieu was diluted with the revival of chemistry in French secondary education. A community of authors emerged with a new standard biographical profile. They had been educated in the same institutions in which they became teachers. Most often they published their textbooks early in their careers. Textbooks were neither mature works nor the culmination of scientific or teaching activities.[25] The uniform biographical profile of authors of secondary-school textbooks contrasts with the variety of positions that we find among authors of textbooks for primary and technical education. They ranged from secondary teachers to engineers formed at the Ecole polytechnique who wrote small elementary and practical manuals at the end of their careers, including doctors such as Adolphe Dupasquier (1793–1848) and industrialists without university degrees such as Anselme Payen (1795–1871).

### Publishers: Between the Government and the Market

In nineteenth-century France, education gradually became a state monopoly. The various governments attempted to control the contents of science teaching through various systems.[26] During the Revolution and the Empire, textbooks were the main instrument of control.[27] A committee of experts appointed by the government selected the "official manuals," preferably a single one that would guarantee uniform contents in all public and private institutions. The attempts to impose a unique text were especially strict and persistent for secondary education, which was a major concern for all governments. A commission formed by Laplace, Monge, and Lacroix was in charge of designing the syllabus for secondary school and choosing the official textbooks in 1802. Finding no adequate book on the market, they decided to commission a chemistry textbook from the "citizen Adet."[28]

As several attempts to impose an "official manual" failed, a new system was adopted consisting of the publication of a list of authorized textbooks that teachers must use. This system began in 1809 when secondary education was integrated into the Université impériale, prescribing that "all kinds of lectures will be shaped after the classic or elementary printed books, according to the statement attached to the present regulations." Moreover, the "grand-maître" had the right to commission new books when the texts available were judged inadequate. Teachers were free to select the book of their choice from the list, and their decision had to be publicized at the beginning of the course. The list of recommended chemistry texts included Lavoisier's *Elements of Chemistry*, Fourcroy's *Chemical Philosophy*, and Berthollet's *Statique chimique*, and extracts of them could be read in the class by the teacher.[29] Similar lists were published in the following years, and some of them were extremely restrictive. In 1814 the only recommended text was "la chimie de Thenard," which was assumed to be the only one able to relate the most advanced state of chemistry.[30] The reform of 1821 was a turning point in the control of secondary education.[31] Textbooks had to adapt their contents to the official syllabus, but teachers retained the freedom to chose their textbooks. It became the publishers' responsibility to ensure that the textbooks' contents were properly adapted to the mandated syllabus. With this reform, publishers acted as a controller or censor in the name of the government.

As publishers played the crucial role in the process of textbook production, the number of textbooks published increased. More generally, chemistry textbooks followed the general trends of the book market. First, as textbooks in general became one of the main pillars of the book market during the second third of the nineteenth century, they were clearly recognized as an independent editorial genre. Second, our study shows a remarkable process of specialization.[32] In the catalogues of printers and booksellers of the early decades of the nineteenth century, chemistry textbooks were close neighbors of big treatises, specialized monographs, dictionaries, and scientific journals. For instance, Berard and Klosterman, booksellers of the Ecole polytechnique, published the five editions of the *Manuel de chimie* by Bouillon-Lagrange, an important textbook initially written for the students of the Ecole polytechnique and then slightly modified to be used at the Ecole de pharmacie. In addition to a long list of books intended for engineering schools, Berard and Klosterman's catalogue also included prestigious scientific journals such as the *Annales de chimie* and the *Bulletin scientifique* of the Société Philomatique, along with a translation of Thomson's *System of Chemistry*.[33] Nicholas Crochard, another publisher who specialized in medicine and who printed the *Annales de chimie* in 1814, had in his catalogue the two most important chemistry textbooks—Thenard's *Traité de chimie élémentaire* and Orfila's *Eléments de chimie appliquée à la médecine et aux arts*—as well as several manuals intended to aid students in preparing for the Baccalauréat.

Specialization occurred in the 1830s when publishing houses such as Victor Masson gave priority to textbooks in their catalogues. Between 1840 and 1845, Fortin-Masson et Cie published numerous chemistry textbooks such as the *Leçons de statique chimique des êtres organisés* by Dumas and Boussingault, Fresenius's and Gerhardt's treatises of chemical analysis, and several translations of Liebig's textbooks. When Victor Masson took over, textbooks for secondary education prevailed so much that Masson reissued most of the textbooks in the middle of the century, such as Regnault's *Cours élémentaire de chimie* and the various versions of Fremy's and Pelouze's chemistry courses. Louis Hachette, founded in 1826, published the first chemistry textbook for secondary education in 1828 and quickly gathered most of the textbooks for primary and secondary education. Thus, in a few decades the production of textbooks became the monopoly of a few publishing companies. At the same time there was a centralization of most publications in the capital, which prompted the decline of publishing houses in the provinces.

Publishing contracts are privileged sources with which to learn about the negotiation between authors and publishers and better understand the role played by the former.

Unfortunately these scarce and dispersed sources are difficult to find.[34] From a contract signed in 1825 by Béchet jeune and Labé, publishers and booksellers of the Paris Medical Faculty, with Jean Baptiste Dumas for the publication of his future treatise of chemistry, we may infer that publishers introduced in the contract several clauses concerning the structure of the text, its distribution in different volumes, formats and sizes, and so on. In this case, the text had to be "divided into two distinct treatises, one of general chemistry, the other of chemistry purely applied to arts and agriculture, each one comprising four 700 to 800 page volumes in 8 [octavo]; in case this division occurred, each edition of both treatises would be issued in 4,000 copies and [the author] would be paid 7,500 francs for each volume."[35]

This was not a negligible income for the young Dumas, who in a letter to Liebig angrily complained about his "difficult existence" in Paris with an annual salary of just 2,000 francs from his post as "répétiteur at the Ecole Polytechnique."[36] Nor was the economic profit of textbook writing irrelevant for the young Orfila. At a time when his economic resources were limited to his private lectures, he received 5,000 francs from Nicolas Crochard for the first edition of the *Toxicologie générale* in 1813.[37] Similar profits were also relevant for many other authors, such as Nicolas Deguin, who was a secondary-school teacher (with an annual salary of 2,000 francs) and later a professor at a provincial Faculty of Science (between 4,000 and 5,000 francs) during the 1840s. By the middle of the nineteenth century, more than 30,000 copies of Deguin's book on physics had been sold. Deguin also published several editions of his textbook on elemental chemistry.[38] Even for the Parisian mandarins, whose practice of "cumul" gave them larger incomes, the significant money that could be made by writing textbooks was probably an important inducement.[39]

Publishers not only encouraged authors to write textbooks by offering a substantial economic income; they also shaped the material characteristics of chemistry textbooks. In the hands of a few Parisian publishing companies, endowed with power by the official system of control, the material aspects of textbooks quickly differentiated them from other kinds of publications. Publishers highlighted their favorite authors and designers while making all efforts to reduce the costs. As a result of this commercial strategy, small formats with a limited number of pages became the standard during the first half of the nineteenth century, with a specific size for each teaching level. Secondary education textbooks turned into single volumes in octavo format with no more than 500 pages. Only textbooks for technical and superior education continued to be published in two or more volumes. The smaller formats in duodecimo and sextodecimo format, with no more than 200 pages, were reserved for the small hands of primary-school children.

By the mid 1830s, new printing techniques adopted by the big publishing houses changed the material presentation of textbooks and transformed them into a clearly distinguishable type of publication. Such formal changes also had important consequences for textbooks' content. Typographic innovations allowed a hierarchy of information printed on a page, with small characters for technical descriptions or experimental procedures alternating with bigger and bold ones for titles and main ideas. Another result of new engraving techniques was that pictures formerly gathered at the end of the volumes were integrated within the body of the text, and their quality improved. With more detailed though cheaper engravings, textbooks incorporated an increasing amount of visual material, including more realistic representations of instruments, chemical reactions, physical phenomena, industrial processes, and natural landscapes. New images changed not only the way in which experiments, instruments, and phenomena were represented but also the way they were described and explained. In the beginning of the nineteenth century, textbooks had routinely featured long and detailed descriptions of experiments intended to "mettre les élèves dans le cas de les répéter toutes," as Thenard put it.[40] Over time these gave way to short texts giving a few technical details of a chemical reaction expressed analytically by an equation, and including pictures of experimental devices in realistic engravings.[41]

## Textbook Creativity

The common view of textbooks as uncreative and passive vehicles of knowledge conveys not only a particular image of the divorce between research and teaching activities in science but also a static picture of scientific literature. Many eighteenth-century chemists included original research in their textbooks and claimed that they had to perform additional experiments in order to write them.[42] In the nineteenth century, textbooks rarely reported original or new substantial experiments. Yet even in this later period, writing a textbook implied making decisions about structure and contents. This was rarely a personal decision of writers. Rather it was the result of negotiations among a group of actors, including the targeted audiences, publishers, and printers. To discuss the issue of creativity in textbooks, we therefore must abandon the clichés about scientific discoveries as "eureka moments"; it will do us little good to fall back on heroic narratives such as those constructed by nineteenth-century chemist-historians and their followers. By "creativity" we do not even necessarily mean discoveries or scientific innovations. Rather we have in mind any original interpretation of scientific phenomena expressed through writing and teaching, whether or not this original view was accepted and became part of "normal science." We explore the features of textbook creativity through a closer examination of two important issues debated in nineteenth-century chemistry: chemical classifications and atomic theory.

### Looking for an Order of Things

Concerning classifications, historians of chemistry implicitly assume that modern classifications emerged through two founding events: Lavoisier's *Elements of Chemistry* and Mendeleev's periodic system.[43] Both episodes have a common feature: they are among the rare cases in traditional historical narratives in which a creative moment is clearly associated with a pedagogical practice, that is, the act of writing a textbook. Yet *all* textbook authors—not only Lavoisier and Mendeleev—had to deal with classifications of chemical substances in order to organize the chapters of their books. It is therefore legitimate to discuss the creative power of textbooks over a broader sample of textbooks than the two illustrious cases. We shall focus on French textbooks published between these two alleged "founding events."

Lavoisier's *Traité* was not the first textbook to include a plea for the "simple-to-complex" order and the "known-to-unknown" didactic structure. A number of eighteenth-century textbook writers, such as Antoine Baumé, claimed that they moved "from the simple to the compound and from the compound to the more compound."[44] This order, whether called analytic or synthetic, had prevailed in the exposition of chemistry for a few decades before Lavoisier resumed it along with the redefinition of the notion of element as a simple substance. Moreover, Lavoisier's assumption of an identity between the two principles—from simple to complex and from known to unknown—is also found in mid-eighteenth-century textbooks.[45] In following an old textbook tradition, Lavoisier nevertheless increased the importance of chemical analysis in nomenclature and classification. Substances formerly classified as compound, such as metals, turned out to be elementary substances (and vice versa), while an increasing bulk of data concerning elementary analysis became available at the end of the eighteenth century.

In 1800, Antoine de Fourcroy claimed that chemistry would soon be emancipated from natural history thanks to a chemical classification based on the nature and proportion of the constituent principles.[46] According to Fourcroy, the autonomy of chemistry involved the passage from external, visual, physical features to internal, compositional organizing principles, or, as Gaston Bachelard later commented, the replacement of "immediate physical analogies" by "rectified chemical ones."[47] In fact, the change was gradual as the analytical order did not subvert more traditional arrangements. Moreover, the simple-to-complex compositional principle did not solve all the problems related to chemical classifications. For instance, three authors who claimed to follow this principle (Chaptal, Lavoisier, and Fourcroy) wrote very different textbooks indeed during the late eighteenth century.[48] Several choices were available to writers. Many chemical substances were "equally" elementary (or compound). The classification of chemical elements, whose number dramatically increased during the first half of the nineteenth century, required additional criteria. And so did the sequence of the much more numerous chemical compounds, which largely depended on the arrangement of elements.

More criteria meant more problems of congruency. Jacques Thenard—the most influential chemistry textbook author during the first third of the nineteenth century—explicitly employed three organizing criteria: from simple to complex, from known to unknown, and chemical analogies.[49] Although generally accepted by chemists as almost identical, the simple-to-complex order and the known-to-unknown principle raised dilemmas. Admittedly, simple substances should be taught first, although they were largely unknown to beginners. Water, air, and other compound bodies had to be introduced much later.[50]

Compatibility between chemical properties and chemical composition raised more vexing problems. Early-nineteenth-century authors implicitly assumed the existence of a straightforward relationship between elemental composition and chemical properties. In this case, groups based on chemical composition would be congruent with the groups of chemical analogues. But unforeseen divergences occasionally emerged. For instance, Thenard adopted the affinity for oxygen as the leading organizing principle. He applied it not only to simple bodies—non-metals and metals—but to plant and animal principles as well. Relying on Lavoisier's theory of acidity, he regarded acid substances as binary compounds of oxygen and another non-metallic element. Alkaline substances were considered to be binary metallic oxides. Since Thenard's treatise was officially a model for all French textbooks during the first third of the nineteenth century, his classifications were largely adopted, and his classification of metals into six groups survived until the end of the nineteenth century.[51]

In the early nineteenth century, most of the known substances fit Thenard's general schema. However, the discovery of new elements and the "exaggeration of the role of oxygen"—as Ferdinand Hoefer described Thenard's classifications—raised more and more inconsistencies during the 1820s and the 1830s.[52] Ammonia, hydracids, new metals, alkaloids and other newly discovered compounds hardly fit in the scheme.[53] In the sixth and last edition of his textbook, published in 1835–36, Thenard attempted to introduce major changes in his classification in order to face these problems. The institutional context encouraged such changes since a new generation of textbooks was under way for the reintroduction of chemistry in secondary schools.

The younger generation of teacher-authors returned to an older debate between natural scientists: artificial versus natural classifications of chemical elements. In artificial classifications, instantiated by Thenard's choice of the affinities for oxygen or Berzelius's choice of electrical charge, elements were arranged according to one single property, one single character; in the latter, exemplified by Louis-André Marie Ampère's essay published as early as 1816, all of the elements' properties and characteristics had to be taken into account.[54] Ampère criticized the overestimation of oxygen in chemical classifications and suggested a natural classification based on "all the characters of the bodies," so that groups of substances with "the most numerous and essential analogies" might be created.[55] Ampère thought that the "natural order" could be unveiled by the "natural method." Ampère offered a circular arrangement of chemical elements in which properties changed gradually from one group to another, and he emphasized the similarities between bodies placed at the edge of the chain.[56] Ampère distinguished three major families ("gazolytes," "leucolytes," and "chroïcolytes"), which were further subdivided into 15 genres.[57] Ampère's essay raised no enthusiasm among the French chemical community in 1816. It was "rediscovered" in the 1830s and the 1840s when the new generation of textbook writers became more and more concerned about classifications. The debate reached a climax in 1845 when Ferdinand Hoefer (1811–1878), a chemist and historian of chemistry, published a small book about this issue, in which no fewer than eight different chemical classifications were presented.[58]

Although Ampère's essay became the standard reference for the advocates of natural classifications, it was never a model. A first important attempt to introduce natural classifications in secondary-school chemistry textbooks was made by Gay-Lussac's répétiteur at the Ecole polytechnique, César Mathurin Despretz (1791–1863).[59] Despretz was also a teacher of physics in the prestigious Collège Henri IV. In 1828–29 he authored a *Traité de chimie* in which he claimed to adopt a "completely new" order in chemistry while following "what had been done in botanic and natural history." Instead of "arranging the bodies by their affinity for oxygen, hydrogen or other bodies," he made families by "placing together those that had most resemblances."[60] Yet Despretz disagreed with Ampère about the choice of classificatory criteria and chose "more chemical criteria, so to speak."[61] In fact, Despretz's arrangement of metals was more similar to Thenard's artificial classification than to Ampère's natural one.[62] Relying on his eight years of teaching experience, Despretz bragged about the didactic advantages of his natural classification, which conveyed "general ideas" and "relationships that were not shown by artificial classifications."[63]

Despretz was no exception in the 1830s and the 1840s. Most textbook authors did not copy or passively adopt a classification. Rather they discussed the issue, compared various systems, and tried to create their own natural classification. As a result, a variety of systems were found in French chemistry textbooks during the 1830s and the 1840s, coinciding with the diversification of audiences and the emergence of new authors and publishers. Textbook authors were free to decide on the arrangement and contents of their textbooks, since the regularly published official syllabus only defined guidelines for lectures. The 1828 syllabus for the collèges royaux, partly written by Thenard, did not give any indication about the order of exposition.[64] The 1837 syllabus for the baccalauréat-ès-sciences physiques included an arrangement of non-metals along Thenard's artificial classification but no recommendation for the classification of metals.[65] The syllabus for the baccalauréat-ès-lettres (1840) and the classe de philosophie (1843) was even more elusive.[66] Only with the 1852 deep reform of secondary schools did the official syllabus prescribe a definite classification.[67]

The 1852 reform was mainly influenced by Jean Baptiste Dumas, who had developed a concern for classification. Dumas first attempted a natural classification in his 1826 work on chemical atomism. By using "atomic volume," heat capacity, and "modes of combination," Dumas aimed to create a natural classification in which substances whose "molecules had similar properties" were grouped together. This kind of classification, he claimed, would make the study of chemistry easier and, at the same time, would lead to "the discovery of new compounds" by displaying "fair analogies."[68] In 1828, when he wrote the first volume of this textbook, Dumas again praised natural classifications. He arranged non-metals according to the number of atoms of these elements that combined with hydrogen.[69] In subsequent years, other chemist-authors followed Dumas's attempts to create natural classifications by using chemical atomism. Alexandre Baudrimont (1806–1880), a disciple of Ampère and a staunch advocate of atomism, ventured a natural classification in his *Traité de chimie*.[70] Baudrimont's natural classification was founded on his concept of "isodynamic bodies" ("corps isodynamiques"), defined as substances which might replace each other in chemical compounds.[71] The result was a complex system in which an element could be included in two or more of fourteen series.[72] The future of chemistry, however, seemed to lie in natural classifications based on atomic properties. Even the leader of artificial classifications, Jacques Thenard, wrote in his *Philosophie chimique* (1836) that "if ever chemistry possesses a natural classification, it will be grounded on isomorphism."[73] Thenard praised Dumas's classification of non-metals according to "nature, proportions and mode of condensation of their combinations with hydrogen" as "generic characters," and he mentioned other characters, such as electrical or heat conductivity and atomic weight, which were also employed as "specific properties."[74] Taking into account "isomorphism" and "chemical properties of bodies," Hoefer proposed a classification with eleven natural families of elements. In 1841, Hoefer claimed that the time for artificial classification was over and that even the old dichotomy between metals and non-metals should be rejected because it was not "rigorously scientific."[75] He even pretended that Thenard's artificial classification was "generally rejected."[76]

Despite Hoefer's optimistic remarks, natural classifications did not win the battle. While Dumas's natural classification of non-metallic elements was adopted by a majority of textbook writers, Thenard's artificial classification of metals, amended by Henri Victor Regnault (1810–1878) in 1836, was still largely used.[77] Dumas himself, who offered some hints about a natural classification of metals in the second volume of his textbook,[78] finally retreated and adopted an artificial classification slightly different from Thenard's in the third volume dealing with metals.[79] According to Favrot, a private chemistry teacher and a "préparateur" at the School of Mines, Thenard's artificial classification of metals should be employed in textbooks because it was based on "easy-to-learn characters." Ampère's classification, on the contrary, did not have any "salient character" that could be employed to distinguish one group from another.[80]

The same concern was shared by the author of one of the last French attempts at natural classification. Adolphe Dupasquier (1793–1848), a teacher of chemistry at the Ecole de la Martinière in Lyons who published the first volume of a *Traité élémentaire de chimie industrielle* in 1844, regarded natural classification as the most logical one ("la plus logique").[81] Yet he also emphasized its difficulties. On the one hand, several substances were still not known well enough to establish their chemical analogies. On the other hand, Dupasquier thought that it was "an almost insensible passage from the properties of one body to another." All bodies were related by "a general link" ("une liaison générale"), which made a perfect grouping impossible; none could satisfy all points of view.[82] All attempts at natural classification having failed, Dupasquier claimed that time had come for a compromise combining natural and artificial classifications.[83]

A hybrid natural-artificial classification was institutionalized by the 1852 official syllabus decreed by Minister Fortoul after the bifurcation of the two sections "sciences et lettres" in secondary education. The new syllabus, written by Dumas and other scientists, recommended "the classification of non-metallic bodies in four natural families" while metals were arranged according to Thenard's artificial classification.[84] Although this hybrid system was presented as an imperfect and temporary solution, it prevailed until the end of the nineteenth century, as the most successful textbooks published in the 1850s went through more than twenty editions.[85] Thus the impetus that had fostered the search for chemical classifications in French textbooks for two decades was stunted in the middle of the nineteenth century. This stagnation resulted mainly from the imposition of the official syllabus, which froze a temporary system and drastically reduced teachers' and authors' margin of freedom. Yet this does not mean that the creativity of textbooks had been exhausted. Rather, authors displaced their interest to other issues, over which they exercised their power of decision.

### Atoms and Molecules in Textbooks

Did the contemporary controversy that divided the chemical community between atomists and equivalentists in the middle of the nineteenth century offer a more favorable space for individual choices? In view of the usual accounts of this famous controversy by historians of nineteenth-century chemistry, the situation would seem hopeless. It is usually assumed that the word "atom" was banished from chemistry in the wake of Dumas's disappointment with Avogadro's hypothesis. This standard view is based on Dumas's dramatic lecture delivered at the Collège de France in 1836 in which he said "If I had my way, I should erase the word 'atom' from science, in the firm belief that it goes beyond the realm of experiment; and never in chemistry must we go beyond the realm of experiment."[86] Indeed Dumas was influential enough: French chemists rejected the system of atomic weights and adopted equivalent weights, which were supposedly theory-free since they did not imply any commitment to a speculative theory of indivisible elementary particles.

The shift from atoms to equivalents is usually described as a retreat and attributed to two major national features. First, the overarching influence of Auguste Comte's positivism in France is said to have encouraged a prejudice against all hypotheses and theories about the structure of matter while encouraging a narrow Baconianism. Second, the centralized and mandarin organization of science in France, which allowed chemists like Dumas to accumulate various teaching positions in Paris while gaining political power, is said to have encouraged authoritarianism and consequently conservatism. The disastrous effects of the conjunction of this philosophical influence and institutional context is illustrated by Marcellin Berthelot in the second half of the nineteenth century. A staunch opponent of atomism, he became powerful enough to control the French educational system and to ban atomic notation and structural formulas from French teaching until his death in 1906.[87] In this scenario teachers and textbook authors had no choice but to adopt the equivalentist language imposed both by scientific authority and political power. They are portrayed as compliant servants or devoted followers of their mandarin academic chiefs, following their changing views on atomic theory that were supposedly imposed by syllabus control. Whereas the 1828 official syllabus, establishing the contents of scientific courses at the collèges royaux, included for the first time an explicit reference to atomic theory, the next official syllabus for the baccaluréat-ès-sciences, published in 1837—just one year after Dumas's famous lectures at the Collège de France—included a chapter on "proportional numbers" but none on atomic theory.[88]

This interpretation can certainly be supported by a quick glimpse at the forewords of many textbooks published in the middle of the nineteenth century. There one can find many anti-metaphysical claims: all speculative hypotheses should be avoided, only firmly established statements should be exposed to young pupils, and so on. In the name of a didactic imperative, many textbook writers favored a fetishist cult of matters-of-fact. Pelouze and Frémy, for instance, explained their choice in the following terms: "It is mainly a teaching book that we intended to write. It is thus conceivable that we had to admit only what can be called the most positive and verified part of science, namely the facts, and to reject what is only conjecture."[89] However, such statements—spreading the mythical image of experimental knowledge uncontaminated by hypotheses—were mainly rhetorical stances. A closer examination of the contents of a large number of textbooks over several decades invites revision of the standard account of nineteenth-century chemical atomism.

To begin with, French textbooks quickly adopted the atomic theory introduced by Dalton in the early nineteenth century and made increasing use of atomic language for describing chemical reactions. In fact chemical reactions had been described in molecular terms long before Dalton's atomic theory. The chapters on affinity forces, written in the eighteenth century, employed the words "integrant molecules" and "constituent molecules." Thus, unsurprisingly, Daltonian "physical" atoms were not regarded as a novelty by French chemistry textbook authors, who superimposed Daltonian atoms upon the traditional corpuscularism inherited from eighteenth-century chemistry in the first chapters of their books. The problem was not the physical existence of atoms, which was taken for granted, but rather how to know the number of atoms that combined in a chemical compound. Thenard expressed it as follows in 1827: "It seems undoubtedly demonstrated that all combinations proceed atom by atom and that they ordinarily take place between some of them. But how many? This is what we cannot say."[90]

The second question—how many atoms were in a compound—was closely related to the determination of atomic weights and its consequences in chemical formulas.[91] It forms what Alan Rocke calls "chemical atomism." Orfila and Thenard, for example, regarded atomic weights and the laws of chemical proportions as a useful tool for chemical analysis. Consequently they introduced these topics in their books as early as the 1810s. At the same time, as they recognized the various problems related to the determination of atomic weights, they gradually introduced the new methods derived from Dulong and Petit's law, Mitscherlich's isomorphism, and Avogadro-Ampère's hypothesis. Later in the 1820s, Berzelius's formulas reinvigorated atomism in French chemistry textbooks. Berzelian formulas were employed in order to convey chemical binary composition, to explain chemical reactions, and to classify chemical substances.[92] As a result, atoms gained space not only in the first introductory chapters but also in the chapters describing chemical properties of bodies, that is, in the most important part of nineteenth-century chemistry textbooks. Moreover, for the first time, a sharp distinction between equivalent and atomic weights was introduced in textbooks. While these words were employed with almost the same meaning as in the first editions, now separate tables of equivalents and atomic weights were included in Orfila's and Thenard's textbooks. Thus atomism was well established in textbooks by the mid 1830s, so that it deserved a full chapter in many textbooks.

Second, it is true that in the late 1830s and the early 1840s a significant number of French chemists eliminated the word "atom" from their textbooks. But did this mean the banishment of "atomic theory"? A glance at textbooks published during that period offers a more complex panorama. The first chapters still conveyed a picture of chemical reactions in terms of "molecules" (or "atoms") under the influence of chemical affinity forces. Afterwards, they included several chapters on "proportional numbers," "equivalent weights," or, in some cases, even under the title of "atomic theory." There were, however, substantial changes, which might be illustrated by paying attention to several editions of Mateu Orfila's textbook, whose seventh edition appeared in 1843. In that edition, Orfila erased the chapter on atomic theory, which had been introduced in the mid 1830s. He dramatically reduced the number of pages of his chapter on laws of chemical composition and on equivalents. Entire paragraphs were removed, such as those dealing with Gay-Lussac's law and its application to chemical calculations,[93] or with the advantages of equivalent weights in chemistry.[94] The introduction was reduced from 35 pages to six.[95] Orfila retained only one table with equivalent weights, without even mentioning that they differed from atomic weights. Finally, he included a brief description of Berzelian formulas and made a more extensive use of them than in previous editions. However, Orfila pointed out that these symbols should not be confused with "what is called atomic theory," "an hypothesis" with which he claimed he would never deal.[96] In the descriptive chapters, the paragraphs on chemical composition were rewritten in order to use exclusively chemical equivalent weights and formulas.

These changes did not reveal any positivistic or anti-metaphysical bias, since Orfila and a number of other textbook writers still employed a corpuscularian approach in their textbooks. Chemical reactions were still described in terms of particles (whether called atoms or molecules) endowed with affinity forces. For instance, a very popular book published by A. Bouchardat in 1842, which went through three editions, assumed that bodies were made up of small parts that "could be named molecules, atoms or particles." Bouchardat distinguished two kinds of molecules that he called "atomes intégrants" and "atomes constituants."[97] Similar terms were used by Grosourdy (1838–39), Guérin (1840), and Favrot (1841), whose books were intended for medical and pharmaceutical students. However, Jean Lassaigne, a chemistry lecturer at the Alfort Veterinary School, continued using "molécules intégrantes" and "molécules constituantes" while he largely employed atomic theory in his textbook during the 1840s. In technology-oriented lectures at the Conservatoire des arts et métiers, Anselme Payen distinguished between "molécules intégrantes ou atomes physiques" and "molécules constituantes ou atomes chimiques" during the early 1830s.[98] In contrast, Jean Jacques Colin, chemistry lecturer at the military school of Saint-Cyr, preferred the word "molecule" and developed a theory of his own. In 1841, when he was writing up the third edition of his textbook on chemistry, Colin published a booklet titled *Elementary considerations about chemical proportions, equivalents and atoms, being an introduction to the study of chemistry*.[99] It was an attempt to write an introduction to chemistry similar to Liebig's *Introduction à l'étude de la chimie*, but more adapted to "French ideas." The main result of this adaptative effort was the replacement of the word "atom" by the word "molecule" in order to avoid self-contradictory expressions such as "half-atom" or "an atom and a half." Colin strongly claimed that "chemical phenomena take place between molecules," and he employed the term "molecular weight" instead of "atomic weight." While acknowledging the differences between equivalent and atomic weights, Colin argued that there was no real problem because it was easy to convert the former values into the latter by simple arithmetic operations. In any case, he employed atomistic formulas in several chapters of his 1845 textbook.[100] In short, textbook writers followed the general trend of skepticism about calculations of the number of atoms in molecules, but they did not advocate agnosticism about the corpuscular structure of matter.

Rather than a "retreat" from atoms to equivalents, these examples show a wide spectrum of uses of the notions of "atoms" and "molecules." Rather than choose between two alternative matter theories, textbook writers appropriated some elements of atomic theory in their own rather creative ways. As a result, a broad range of situations could be found during the late 1830s and the early 1840s. Sometimes "atom" and "molecule" were treated as synonyms or quasi-synonyms, whereas Orfila preferred "atoms" in the early decades and "molecules" in the 1840s. It would be difficult to account for such changes by invoking any positivistic or anti-metaphysical tendency, since corpuscularian images still underlay the description of chemical reactions. It would be even more difficult to invoke an authoritarian order coming from top officials when there was such a striking lack of discipline in the use of the basic concepts.

Third, after the first international conference of chemists (held in Karlsruhe in 1860) prompted a general adoption of atomic weights and structural formulas in Europe, France (where Berthelot reigned supreme in the Academy of Sciences) was the last fortress of equivalentism. Yet there were committed atomists gathered around Adolphe Wurtz, an Alsatian chemist and professor at the Faculty of Medicine, who established a kind of research school on the model of Liebig's research school in Giessen. While they had to publish their research papers in the official language of equivalents, a number of French atomists took the opportunity to write textbooks to spread their atomistic convictions. Wurtz himself did not conceal his proselytic intentions when he published the second edition of his *Leçons élémentaires de chimie moderne* in 1871.[101] Simultaneously his disciple Alfred Naquet published a textbook titled *Principes de chimie fondée sur les théories modernes*. According to its publisher, the entire press run of Naquet's textbook sold out in a year and a half, and the book was translated into English, German, and Russian.[102] In the 1870s, Edouard Grimaux published a booklet dedicated entirely to the atomic theory, in addition to delivering lectures on the same topic at the Ecole polytechnique.[103] Paul Schützenberger, another student of Wurtz who became a professor at the Collège de France, opened his monumental *Traité de chimie générale* with words of revenge against Dumas's famous banishment of atoms. Thus, despite their usually uncontroversial character, textbooks did not remain outside the theoretical controversy about chemical atomism. Rather, they provided spaces of freedom for committed atomists who could not use the language of their choice in their research publications.

## Conclusions

The emergence of textbooks as a distinctive genre has been analyzed in this essay as a negotiated process involving four main actors: the captive public of students, the authors, the publishers, and the French government. All of them cooperated to stabilize the concept of textbooks as material and commercial objects transmitting scientific knowledge for didactic purposes. Given their respective natures and functions, each mediated between the others. The public, which was the raison d'être of the whole enterprise, was only a virtual actor, since the ideal public defined by authors and publishers differed from the students who actually made use of textbooks. The French government defined the rules of the game, at least for primary- and secondary-school textbooks. The publishers defended the commercial interests of their private companies while negotiating with "public instruction." The authors were the only individual actors. In the beginning of the nineteenth century they derived their authority from their own scientific achievements, guaranteeing the validity of their claims by themselves. By contrast, in the middle of the nineteenth century the reliability of their writings derived from a chemical community represented by a few illustrious names and titles of publications. Their authority mainly relied on their teaching experience in a specific institution for a specific audience. Remarkably, the chemical community was never directly involved in the enterprise, since authors did not necessarily represent it. Yet it was always present behind the scenes, providing validation for scientific statements and exercising political power through some of its leaders.

The examples analyzed above show to what extent writing a textbook remained a creative activity. First, all didactic exposition of a science requires selecting and defining items and then organizing them into a coherent discourse. Second, even when the official syllabus prescribed the contents, it only provided the table of contents; the authors had to supply their own chemical narrative. The fact that we found a diversity of expositions of chemistry in the same period with very different views even of the basic concepts of the discipline means that there was no orthodoxy. Even in the French institutional framework that favored dogmas, chemical textbooks never became mere catechisms. Although textbooks were presented by their authors and publishers as spaces devoid of personal opinions and public controversies, they retained a degree of originality since they had to make decisions about ongoing debates. In fact, from a more pragmatic perspective, originality was no exception. Rather, it was the general rule—especially because textbooks had to claim originality either by addressing new audiences or by adopting novel didactic approaches in order to succeed in the market.

Textbooks no longer can be viewed as dogmatic and conservative vehicles of normal science. On the one hand, they are windows revealing new actors in the history of science: audiences, publishers, printers, and the silent crowd of unknown authors participated in the construction of science. On the other hand, this window sheds new light on the relation between teaching and research practices: textbooks played an important role in discipline-building and in creating theories. The price to be paid for opening these new avenues for the history of science is to study textbooks as objects of research for their own sake. The window glass is not transparent. Textbooks are intrinsically complex because of their multiple identities. As material objects, they are one of the various "paper tools" employed in many scientific practices. As commercial items, they are subjected to the constraints of the market. As educational tools, they are shaped and reshaped by local or national traditions and their ethnic, religious, and political roots. Textbooks can also be considered instruments of professionalization and mediators between scientific communities and the society at large. For all these reasons, textbooks offer historians a rich resource for studying the production of new scientific knowledge and new generations of scientific practitioners.

## Notes

1. John H. Brooke, "Introduction: The study of chemical textbooks," in *Communicating Chemistry: Textbooks and their Audiences, 1789–1939*, ed. Anders Lundgren and Bernadette Bensaude-Vincent (Science History Publications, 2000), 1–18, on 1.

2. See Bernadette Bensaude-Vincent, "Mendeleev's periodic system of chemical elements," *British Journal for the History of Science* 19 (1986): 3–17; Nathan M. Brooks, "Dimitrii L. Mendeleev's *Principles of Chemistry* and the periodic law of the elements," in Lundgren and Bensaude-Vincent, *Communicating Chemistry*, 295–311; Michael D. Gordin, The Ordered Society and Its Enemies: D.I. Mendeleev and the Russian Empire, 1861–1905 (PhD dissertation, Harvard University, 2001), chapter 2.

3. Owen Hannaway, *The Chemists and the Word: The Didactic Origins of Chemistry* (Johns Hopkins University Press, 1975).

4. These and other questions were discussed in a workshop on "Chemical textbooks and their audiences" as part of a larger European program on the "Evolution of chemistry (1789–1939)." The workshop, which was held in Uppsala in 1996 with contributions from all over Europe, revealed a wide variety of textbook traditions with distinct national characteristics. The proceedings were published in Lundgren and Bensaude-Vincent, *Communicating Chemistry*.

5. Auguste Comte, *Cours de philosophie positive* (Hermann, 1975 [1830]), 2nd Leçon, volume 1, 50–51.

6. Gaston Bachelard, *La formation de l'esprit scientifique* (Vrin, 1972 [1938]), 24–28.

7. Thomas Kuhn, *The Structure of Scientific Revolutions* (University of Chicago Press, 1962); idem, *The Essential Tension* (University of Chicago Press, 1977).

8. Kathryn Olesko, *Physics as a Calling: Discipline and Practice in the Königsberg Seminar for Physics* (Cornell University Press, 1991), 15–16.

9. On the emergence of popular science as an editorial genre in various national contexts, see Bernadette Bensaude-Vincent and Anne Rasmussen, *La science populaire dans la presse et l'edition: XIXe et XXe siècles* (CNRS, 1997). On encyclopedias, see Richard Yeo, *Encyclopaedic Visions: Scientific Dictionaries and Enlightenment Culture* (Cambridge University Press, 2001). Another interesting study on an early-modern genre of scientific literature is provided by William Eamon, *Science and the Secrets of Nature: Books of Secrets in Medieval and Early Modern Culture* (Princeton University Press, 1994).

10. Thomas H. Broman, "J. C. Reil and the 'journalization' of physiology," in *The Literary Structure of Scientific Argument*, ed. Peter Dear (University of Pennsylvania Press, 1991), 13–72.

11. See James A. Secord, *Victorian Sensation: The Extraordinary Publication, Reception, and Secret Authorship of* Vestiges of Natural History of Creation (University of Chicago Press, 2000), esp. chapter 2.

12. Alain Choppin, *Les Manuels Scolaires: Histoire et Actualité* (Hachette, 1992); idem, "L'Histoire des manuels scolaires: Une approche globale," *Histoire de l'éducation* 9 (1980): 1–25.

13. For a review of recent literature on chemistry in French education, see the bibliography quoted in Bernadette Bensaude-Vincent, Antonio García-Belmar, and José R. Bertomeu-Sánchez, *La Naissance d'une science des manuels (1789–1852)* (Editions des Archives Contemporaines, 2003).

14. For more information about chemistry in medical and pharmaceutical curricula, see José R. Bertomeu-Sánchez and Antonio García-Belmar, "Mateu Orfila's *Eléments de chimie médicale* and the debate about chemistry applied to medicine during the early XIXth century in France," *Ambix* 47 (2000): 1–28.

15. Janis Langins, "The decline of chemistry at the *École polytechnique* (1794–1805)," *Ambix* 28 (1981): 1–19; idem, *La République avait besoin de savants: Les débuts de l'École polytechnique* (Belin, 1987); Bruno Belhoste, Amy Dahan Dalmedico, and Antoine Picon, *La Formation polytechnicienne, 1794–1994* (Dunod, 1994); Bruno Belhoste, *La Formation d'une technocratie: L'Ecole polytechnique et ses elèves de la Révolution au Second Empire* (Belin, 2003).

16. Claudette Balpe, "L'enseignement des sciences physiques: Naissance d'un corps professoral (fin XVIIIe-fin XIXe siècle)," *Histoire de l'éducation* 73 (1997): 49–85.

17. See Bertomeu-Sánchez and García-Belmar, "Mateu Orfila's *Eléments*."

18. J. Tyrat, *Nouveau manuel complet et méthodique des aspirants au Baccalauréat ès Sciences* (J. Delalaine, 1837).

19. During the 1830s and the 1840s, the term "manuel" was associated with texts that reduced the content of a chemistry course to a series of basic notions summarized in brief paragraphs so that students would learn them by heart and repeat them in the exams. These and other distortions introduced by the Baccalauréat in teaching methods were criticized by chemistry teachers and authors who considered the manuals to be the epitome of vicious educative methods.

20. Thérèse Charmasson, Anne-Marie Lelorrain, and Yannick Ripa, *L'Enseignement technique de la Révolution à nos jours* (Economica–Institut National de Recherche Pédagogique, 1987).

21. J. P. L. Girardin, *Leçon de chimie élémentaire, faites le dimanche à l'Ecole municipale de Rouen* (Rouvier, 1836–37); J. B. Dumas, *Traité de chimie appliquée aux arts* (Béchet jeune, 1828–1846).

22. A. Payen, *Cours de chimie élémentaire et industrielle destiné aux gens du monde* (Thomine, 1832).

23. Dumas, *Traité de chimie appliquée aux arts*, volume 1, viii.

24. F. M. J. Malaguti, *Leçons élémentaires de chimie* (Dezobry et E. Magdeleine, 1852), volume 1, préface.

25. Balpe, "L'enseignement des sciences physiques."

26. Choppin, *Les Manuels scolaires;* idem, "Le cadre législatif et réglamentaire des manuels scolaires, I: De la Révolution à 1939," *Histoire de l'éducation* 29 (1986): 21–58; Alain Choppin and Martine Clinkspoor, *Les Manuels scolaires en France,* volume 4, *Textes officiels* (Institut National de Recherche Pédagogique, 1993).

27. The use of an "official textbook" as an instrument for educational, political, ideological, or religious control has been a common practice used in many different contexts and periods. See Karl Hall's essay in this volume, in which he refers to textbooks in the post-revolutionary Soviet Union.

28. See *Travail de la Commission chargée pour les mathématiques de désigner les livres classiques à l'usage de tous les lycées,* as quoted in Bruno Belhoste, *Les Sciences dans l'enseignement secondaire en France,* volume 1, *1789–1914* (Institut National de Recherche Pédagogique, 1995), 78–80.

29. Ibid., 83–86.

30. *Règlement des études des lycées et collèges,* September 28, 1814, in Belhoste, *Les Sciences dans l'enseignement secondaire en France,* volume 1, 88–90.

31. Bruno Belhoste, "Les caractères généraux de l'enseignement secondaire scientifique de la fin de l'Ancien Régime à la première guerre mondiale," *Histoire de l'éducation* 41 (1989): 3–45, on 34–35; idem, *Les Sciences dans l'enseignement secondaire en France,* volume 1, 103.

32. Valérie Tesniere, "L'édition universitaire," in *Histoire de l'édition française,* ed. Henri Jean Martin and Roger Chartier (Promodis, 1983–1986), volume 3, 221–222; F. Barbier, "Une production multipliée," in ibid; Y. Mollier, "Histoire de la lecture, histoire de l'édition," in *Histoire de la lecture: Un bilan des recherches,* ed. Roger Chartier (Institut Mémoires de l'Edition Contemporaine, 1995).

33. Thomas Thomson, *Système de chimie* (Veuve Bernard, 1809).

34. See Secord, *Victorian Sensation,* chapter 4, for an outstanding study of the role of publishers in marketing a popular scientific text in nineteenth-century Britain.

35. Contract signed by Thenard, Dumas, and Bechet in Paris, April 30, 1849 (folder "Thenard," Archives de l'Académie des Sciences, Paris).

36. Dumas to Liebig, November or December 1831, as quoted in Alan Rocke, *Nationalizing Science: Adolphe Wurtz and the Battle for French Chemistry* (MIT Press, 2001), 56–57. Dumas complained about his "place de 2000 francs pour tout potage."

37. M. G. Chapel d'Espinasoux, "La Jeunesse d'Orfila: Fragment d'une autobiographie inédite publié par . . . ," *Revue hebdomadaire* 22 (1914): 615–634, and 23 (1914): 86–113; quotation on 97.

38. Archives Nationales de France (Paris), F17/20540.

39. Gay-Lussac's total salary was 45,400 francs at the beginning of the 1830s. See Maurice Crosland, *Gay-Lussac, 1778–1850* (Belin, 1991), 318–319. On "cumul" see also Rocke, *Nationalizing Science,* passim.

40. L. J. Thenard, *Traité de chimie élémentaire, théorique et pratique* (Crochard, 1813–1816), 4 volumes, in volume 1, ii–iii.

41. For an excellent discussion of how engraving techniques shaped a scientific discipline and teaching, see Klaus Hentschel, *Mapping the Spectrum: Techniques of Visual Representation in Research and Teaching* (Oxford University Press, 2002).

42. Andrew Warwick's study of mid-nineteenth-century mathematical education in Cambridge, England offers an additional example of how the boundaries between "original scientific research" and education are extremely diffuse, in constant transformation; the flow of knowledge between research and teaching is far from unidirectional. According to Warwick's study, original research results in mathematical physics were often announced as assigned problems on the Cambridge Mathematics Tripos exam. See Andrew Warwick, *Masters of Theory: Cambridge and the Rise of Mathematical Physics* (University of Chicago Press, 2003).

43. For a broader discussion of this topic and additional bibliography see José-Rámon Bertomeu-Sánchez, Antonio García-Belmar, and Bernadette Bensaude-Vincent, "Looking for an order of things: Textbooks and classifications in nineteenth century France," *Ambix* 49 (2002): 227–251.

44. A. Baume, *Chymie éxpérimentale et raisonnée* (P. F. Didot le jeune, 1773), volume 1, xii–xiv; quotation on xiv.

45. Piere-Joseph Macquer, *Elémens de chymie théorique* (J.T. Hérissant, 1753), xvi–xvii.

46. Antoine-François Fourcroy, *Système des connaisances chimiques* (Baudouin, 1800), volume 1, xxxiii–xxxv.

47. Bachelard, *La Formation de l'esprit scientifique*, 29–39.

48. Bernadette Bensaude-Vincent, "A view of the chemical revolution through contemporary textbooks: Lavoisier, Fourcroy, and Chaptal," *British Journal for the History of Science* 23 (1990): 435–460.

49. Thenard, *Traité de chimie élémentaire*, volume 1, i–ii.

50. See, for instance, Berzelius's criticisms against the order from simple to complex in J. J. Berzelius, *Traité de chimie* (Firmin-Didot, 1845–1850), volume 1, 2. See also Jean Louis Lassaigne, *Abrégé élémentaire de chimie* (Bechet jeune, 1829), volume 1, 69.

51. See, for instance, M. J. B. Orfila, *Elémens de chimie médicale* (Crochard, 1817); J. S. E. Julia de Fontenelle, *Manuel de chimie médicale* (Béchet, 1824); Lassaigne, *Abrégé élémentaire de chimie*. Thenard's metal classification was employed even in the early twentieth century in Troost's popular chemistry textbook: cf. Louis Troost, *Traité élémentaire de chimie*, eighth edition (G. Masson, 1885), 397–399; ibid., fifteenth edition (G. Masson, 1910), 382–384. For a description of Thenard's classification, see J. R. Bertomeu-Sánchez and A. García-Belmar, "Mateu Orfila y las clasificaciones químicas," *Cronos* 2 (2003): 3–35; Bertomeu-Sánchez, García-Belmar, and Bensaude-Vincent, "Looking for an order of things."

52. Ferdinand Hoefer, *Nomenclature et classifications chimique* (J. B. Baillière, 1845), 55.

53. For a discussion of these issues, see José-Ramón Bertomeu-Sánchez and Antonio García-Belmar, "Mateu Orfila y las clasificaciones químicas," *Cronos* 2 (1999): 3–35.

54. A. M. Ampère, "Essai d'une classification naturelle pour les corps simples," *Annales de Chimie* 1 (1816): 295–308; 373–394;1–32; 105–125; quotation on 296. On Ampère's earlier drafts of his classification, see C. L. Dowland-Pillinguer, "A Chemist Full of Bold and Ingenious Ideas": The Chemical Philosophy of A.M. Ampère (1775–1836) (PhD dissertation, Cambridge University, 1988). For a presentation of the dilemma between artificial and natural classification see Bertomeu-Sánchez et al., "Looking for an order of things."

55. Ampère, "Essai d'une classification naturelle," 297.

56. Ibid.: "deux corps placés aux deux extrémités de la chaîne formée par tous les autres se rapprochent et s'unissent mutuellement par des caractères communs."

57. Ibid., 120–125. For Jussieu's ideas on continuity and classification in natural history, see Peter F. Stevens, *The Development of Biological Systematics: Antoine-Laurent de Jussieu, Nature, and the Natural System* (Columbia University Press, 1994). For a recent review of the topic, see *Spaces of Classification*, ed. Ursula Klein (Max-Planck-Institut for Wissenschaftsgeschichte preprint, 2003).

58. Hoefer, *Nomenclature et classifications chimique.*

59. According to his biographer, Gay-Lussac became a "powerful patron of Despretz during all the phases of his academic life": Crosland, *Gay-Lussac, 1778–1850*, 344–345. Cf. *Discours de M. Becquerel prononcé aux funérailles de M. Despretz . . . Le mardi 17 Mars 1863* (Firmin Didot, 1863).

60. C. M.Despretz, *Elémens de chimie théorique et pratique* (Méquignon-Marvis, 1829–30), volume 1, 1.

61. Ibid., volume 1, iv.

62. The "chromoïdes" group was closely similar to Thenard's fourth group. Cf. Thenard, *Traité de chimie élémentaire*, fifth edition (Crochard, 1827), volume 1, 320–322. The only exception was arsenic, which Thenard placed in the non-metallic group. Moreover, except for zirconium, the "alluminoïdes" group was identical to Thenard's first group.

63. Ibid.

64. "Programme du cours des sciences physiques pour les élèves des deux années de philosophie, 1 avril 1828," in Belhoste, *Les Sciences dans l'enseignement secondaire en France*, volume 1, 115–121.

65. "Programme bac-ès-sciences mathématiques et physiques, 3 février 1837," in Belhoste, *Les Sciences dans l'enseignement secondaire en France*, volume 1, 139–147. The program for chemistry is in ibid., 142–144.

66. "14 Juillet 1840, Questions de mathématiques et de physique au baccalauréat ès lettres," in Belhoste, *Les Sciences dans l'enseignement secondaire en France*, volume 1, 152–161; "Questions de physique et de chimie," in ibid., 156–159. See in particular question 40: "Exposer les propriétés qui distinguent l'hydrogène, le carbone, le phosphore, le soufre, le chlore, l'iode. 24 février 1843.

Programmes de physique et de chimie en classe de philosophie," in ibid., 194–198 (chemistry on 197–198).

67. "30 août 1852, Nouveaux programmes de l'enseignement scientique," in ibid., 273–301, on 276–277 and 294–296.

68. J. B. Dumas, "Mémoire sur quelques points de la théorie atomique," *Annales de Chimie et de Physique* 33 (1826): 337–391, on 340. In the 1820s, Dumas published his studies on the "atomic volume," which he calculated as the ratio between atomic weight and density. See A. Le Royer and J. B. Dumas, "Essai sur le volume de l'atome des corps," *Journal de Physique de Chimie et de Histoire Naturelle* 92 (1824): 402–411. See also T. M. Cole, "Early atomic specultations of Marc Antonie Gaudin: Avogradro's hypothesis and the periodic system," *Isis* 66 (1975): 335–336. On Dumas's other works on chemical classification see John Van Spronsen, *The Periodic System of the Chemical Elements: A History of the First Hundred Years* (Elsevier, 1969), 74–75 and 85–87.

69. Dumas, *Traité de chimie appliquée aux arts*, volume 1, lxxivv. The classification was: 1st genre: Hydrogène; 2nd genre: F, Cl, Br, I; 3rd genre: Se, S, Appendice: O; 4th genre: P, As, Appendice: N; 5th genre: B, Si, Appendice.

70. Alexandre Edouard Baudrimont, *Introduction à l'étude de la chimie par la théorie atomique* (Colas, 1833); idem, *Traité de chimie générale et expériméntale* (J.-B. Ballière, 1844–1846).

71. Baudrimont, *Introduction*, 53–54.

72. Baudrimont, *Traité de chimie générale et expériméntale,* volume 1, 297. A similar approach had been discussed by Baudrimont in his *Traité élémentaire d'histoire naturelle* (H. Cousin, 1839), 47.

73. Thenard, *Traité de chimie élémentaire,* sixth edition (Crochard, 1834–1836), volume 5, 443.

74. Ibid., volume 5, 517–518.

75. Ferdinand Hoefer, *Eléments de chimie minérale* (Dezobry et E. Magdeleine, 1841), 45. He compared this chemical dichotomy with the old botanic groups: arbres vs. herbes.

76. Ibid., 5. Thenard's classification "quoique ingénieuse, est incomplète; et commence à être généralement abandonnée."

77. V. Regnault, "Recherches relatives à l'action de la vapeur d'eau à une haute température sur les métaux et sur les sulfures métalliques: Essai d'une nouvelle classification des métaux d'après leur degré d'oxidabilité," *Annales de Chimie* 62 (1836): 337–388.

78. Dumas, *Traité de chimie appliquée aux arts,* volume 2 (1830), 39–42. Dumas mentioned a "first family" with "two sections": 1: potassium, sodium, and lithium; and 2: barium, strontium, and calcium.

79. Ibid., volume 2, 67–68. Dumas discussed his classification in a chapter on "the action of oxygen on metals," and he even mentioned that his six sections were "borrowed from M. Thenard."

80. C. Favrot, *Traité élémentaire de physique, chimie, toxicologie et pharmacie* (Béchet jeune et Labé, 1841), 361.

81. Alphonse Dupasquier, *Traité élémentaire de chimie industrielle* (C.Savy jeune, 1844). Dupasquier died in 1848, and the second volume was not published. See Amédée Bonet, *Eloge d'Alphonse Dupasquie* (Imprimerie de Léon Boitel, 1849), 29–30.

82. Dupasquier, *Traité élémentaire de chimie industrielle*, volume 1, 60.

83. Ibid., volume 1, 65–66.

84. "30 août 1852, Nouveaux programmes de l'enseignement scientique," in Belhoste, *Les Sciences dans l'enseignement secondaire en France*, volume 1, 273–301, on 276–277 and 294–296.

85. L. Nekoval-Chikhaoui, *Diffusion de la classification périodique de Mendeleiev en France entre 1869 et 1934* (Thèse Université de Paris IX, 1994), 85. The author claimed that many textbook writers affirmed the adoption of "la classification de Dumas": 1: F, Cl, Br, I; 2: O, S, Se, Te; 3: N, P, As; and 4: C, Si, B.

86. Jean Baptiste Dumas, *Leçons sur la philosophie chimique* (Cultures et civilisations reprint, 1972 [1837]), 246; quotation translated by Robert Fox, *The Caloric Theory of Gases from Lavoisier to Regnault* (Clarendon, 1971), 291.

87. For reviews of the recent literature, see Rocke, *Nationalizing Science*; José-Ramón Bertomeu-Sánchez and Antonio García-Belmar, "Atoms in French chemistry textbooks during the first half of the nineteenth-century," *Nuncius* 19 (2004), no. 1: 77–119.

88. "Programme du cours des sciences physiques pour les élèves des deux années de philosophie, 1 avril 1828," in Belhoste, *Les Sciences dans l'enseignement secondaire en France*, volume 1, 115–121; "Programme des baccalauréats ès sciences mathématiques et ès sciences physiques, 3 février 1837," in ibid., 139–147: "lois suivant lesquelles les corps se combinent; nombres proportionnels."

89. J. Pelouze and E. Frémy, *Cours de chimie générale* (V. Masson, 1848), volume 1, 1.

90. Thenard, *Traité de chimie élémentaire*, fifth edition (1827), volume 1, 173–174.

91. For instance, following a simplified description of the case of water, 1 part of hydrogen reacts with 8 parts of oxygen to create 9 parts of water. The formula HO implies that oxygen has an atomic weight of 8 (taking hydrogen as having unit atomic weight), whereas assuming the formula $H_2O$ implies 16 as the atomic weight for oxygen.

92. On the multiple meanings of Berzelian formulas as paper tools, see Ursula Klein, *Experiments, Models, Paper Tools: Cultures of Organic Chemistry in the Nineteenth Century* (Stanford University Press, 2003).

93. Cf. M. Orfila, *Elémens de chimie, appliquée à la mèdicine et aux arts*, sixth edition (Crochard, 1835–36), volume 1, 27–29; idem, *Elémens de chimie*, seventh edition (Fortin, Masson, et Cie, 1843), volume 1, 11–12.

94. Cf. M. Orfila, *Elémens de chimie*, sixth edition (1835–36), volume 1, 36–40; idem, *Elémens de chimie*, seventh edition (1843), volume 1, 14–16.

95. Cf. M. Orfila, *Elémens de chimie,* sixth edition (1835–36), volume 1, 23–58; idem, *Elémens de chimie,* seventh edition (1843), volume 1, 11–17.

96. M. Orfila, *Elémens de chimie,* seventh edition (1843), volume 1, 17.

97. A. Bouchardat, *Cours des sciences physiques* (G.Baillière, 1842), 2. The second edition appeared in 1845 and the third in 1848.

98. A. Payen, *Cours de chimie élémentaire et industrielle* (Thomine, 1832), volume 1, 50.

99. Jean-Jacques Colin, *Considérations élémentaires sur les proportions chimiques* (Chez Gauthier Laguionie, 1841).

100. Jean-Jacques Colin, *Cours de chimie,* third edition (Gaultier-Languionie, 1841), 24–25; ibid., fourth edition (1845), 23–24.

101. Adolphe Wurtz, *Leçons élémentaires de chimie moderne* (Masson, 1867–68), advertisement; ibid., second edition (1871).

102. Alfred Naquet, *Principes de chimie fondés sur les théories modernes* (F. Savy, 1865).

103. Edouard Grimaux, *Chimie inorganique élémentaire* (Victor Masson, 1874); idem, *Introduction à l'étude de la chimie, théories et notations chimiques* (Dunod, 1883).

# 8 "Think Less about Foundations": A Short Course on Landau and Lifshitz's *Course of Theoretical Physics*

**Karl Hall**

Writing a textbook is not a simple matter.
—Joseph Stalin, 1950

Several weeks after the theoretical physicist Lev Landau nearly lost his life in an automobile accident outside Moscow in January 1962, an article titled "A fundamental work" appeared in the pages of the Soviet Union's national newspaper, *Izvestiia*.[1] The author, V. L. Ginzburg, was one of the many distraught colleagues to stand vigil at Landau's bedside during his arduous rehabilitation.[2] It was the annual cycle of nominations for the highest Soviet civilian honor, the Lenin Prize, that provided the formal occasion for Ginzburg's piece, which made no mention of Landau's plight. Rather, in accordance with prescribed ritual, Ginzburg was contributing a public testimonial in favor of the two nominees: Landau and his closest collaborator, Evgenii Lifshitz.

The nomination was not for their scientific research, however. The object of Ginzburg's praise was a set of textbooks known as the *Course of Theoretical Physics*, which he claimed had codified the modern style of theoretical physics like no other work. "Just like painting, fiction, or poetry," wrote Ginzburg, "the science and scientific literature of every era are distinguished not only by their content, but also possess their own characteristic form, their style." The depth and originality of Landau and Lifshitz's *Course*, he emphasized, lay in its achievement of a contemporary synthesis without straining toward unification (*unifikatsiia*). Scientific peers and government committee were in accord with Ginzburg: later that year, the two men won the Lenin Prize, the first time it was awarded for physics pedagogy. A year that had begun tragically was then capped with the award of the Nobel Prize to Landau for his work on the theory of condensed matter.

The pedagogical laurels are all the more striking in view of the inauspicious beginnings of the *Course*. The first volume of the series to appear in print, *Statistical Physics*, came out in both Russian and English in 1938, just before Landau lost a year in the custody of the secret police. The next three volumes appeared in wartime, with no prospects for translation. By the time of Stalin's death, in 1953, two-thirds of the *Course* had already been composed in isolation from international colleagues, during the peak years of Soviet autarky. There was little reason to expect that it would become perhaps the century's best seller in the field. And yet, if one takes the totals for all editions—complete translations in all major European languages and Japanese, individual volumes in ten other languages, and half a dozen new editions—it is safe to say that a million volumes of the *Course* have been sold in all corners of the scientific world.[3] (One anecdotal indicator of the esteem the *Course* continues to enjoy is that the reserve shelves once set aside for the Marxist-Leninist canon at the Academy of Sciences Library in St. Petersburg now display the volumes of Landau and Lifshitz.[4]) This is surely an astounding record for such a specialized set of textbooks.[5]

The *Course of Theoretical Physics* has become a quintessentially international text, but that does not mean that it is generic. Many top theorists today see themselves as having attained a certain maturity of outlook once they are prepared to entertain strong opinions about how any one of the volumes of Landau and Lifshitz ought to be rewritten for a contemporary audience. Yet the *Course* is now such a massive "fact on the ground" that any efforts to historicize it must run counter to the sheer inertia of its presentness. What I want to study is the historical processes by which this set of teaching texts could itself become canonical, and, in turn, what it did and did not teach young theorists to take for granted in the canon of physics. Moreover, I will look at how the *Course* violated certain generic conventions of its own day in subtle ways, and in doing so showed students how learned skills foster productive ideas, how norms spark novelty. To do this, we must inquire after the peculiarly Soviet origins of this quintessentially international textbook during its lengthy process of composition and dissemination, and pay close attention to the subtler factors that enabled it to reach its intended global audience.

If I seem to exaggerate the social role of this rather dry set of texts, it is precisely because other collateral institutions that might have underwritten the disciplinary status of theoretical physics were largely absent when Landau began his career in the 1920s. The small cohort of Soviet theorists did not work in institutes modeled on those of the Central European universities, where this kind of "pure" science had enjoyed its greatest support. To the contrary, the need for specialized theorists was repeatedly questioned by both insiders and outsiders to the physics profession, for whom the taint of theoretical abstraction ill served the applied agendas of Soviet physical research. Indeed, before he moved to Kharkov in 1932, Landau and his friend George Gamow failed rather conspicuously in a bid to create an institute of theoretical physics in

Leningrad.[6] When Landau first started administering the Theoretical Minimum (a series of nine exams that he devised covering mathematical tools along with all the major domains of theoretical physics) and planning the *Course*, advanced degrees (abolished since 1918) were not yet a formal part of the professionalization process.[7] He began training theorists when he was still employed by the People's Commissariat of Heavy Industry, and early in their careers some of his closest associates had to make do teaching physics to students in places like the Tanning Industry Institute. Not one to mince words, Landau once bluntly informed a senior Soviet experimentalist that "physicists cannot double as engineers."[8] The *Course* would eventually help dignify a distinct social identity for the theorist that explicitly blocked off the kinds of functionalist expectations that had dominated the early Soviet period.

The authors achieved this by creating a "new tradition" (as Stalin frequently urged the new Soviet service elite to do) while relying discreetly on carefully selected calculating techniques and principles, for the most part from a disciplinary past in which Russian physicists had played only a minor role.[9] I will argue that the authors employed interpretive conventions in their texts that unintentionally mirrored in intriguing ways some of the conventions frequently encountered in socialist realist literature. The devices they used to disrupt generic expectations within subfields (e.g., mechanics, electrodynamics, statistical physics) served at the same time to establish the authenticity of their own authorial voice in contradistinction to contemporary textual practices. It is in this way that certain historically contingent pedagogical techniques have come to be identified with a resolutely international style. My contention is that the *Course* was never "always already" international in any transcendent sense—despite the manifest intentions of its authors—but rather was made so by historically identifiable means. In that sense, we are all Soviets now.

The authors of the *Course of Theoretical Physics* crafted a series that would be universal without being fundamental, yet at the same time it was to be briskly normative rather than exhaustive in its methods. As Landau once counseled an aspiring theorist, "one has to think less about foundations. The chief thing you need to master is the work technique, and understanding of subtleties itself will come later."[10] Though it was scarcely alone, the *Course* has perhaps done more than any other textbook in the second half of the twentieth century to destabilize conventional notions about how the physics canon is made. For Landau and Lifshitz, canonical textbook depictions of the past discovery of first principles as triumphs of resolution en route to the present were of little use in helping students develop a proper sense of contemporary problem-solving priorities. Such depictions misleadingly valorized exceptional individuals and "foundational" investigations, while distracting talented but mortal students from mastering a broader set of practices that could still take them to the first ranks if they were but sufficiently disciplined in their training.

## An End to Discussions

Every Soviet text could be called upon to perform tutorial functions within the ongoing Bolshevik efforts to remake political and cultural life. Books indeed did extensive metaphorical service as weapons for dispersing the enemies of the Bolshevik Enlightenment project. The "bourgeois" professor, religion, drunkenness, ignorance, and illiteracy were all easy targets.[11] To Bolshevik partisans, the sciences offered inexhaustible sources of ammunition for these battles, and a richly variegated popular literature on science sprang up in the 1920s and the 1930s.[12] Broadly speaking, cultural experimentation under the new regime thrived on dissolving barriers between political, economic, social, scientific, and cultural realms, in search of a novel "socialist culture," especially when combined with Stalin's violent drive for party unity in the 1930s. This explosion in cultural production (including science popularization), however, inevitably led to functional differentiation rather than consolidation, meaning that at the broadest level, the initial situation around 1930 was more one of fractured competencies than monolithic repression.

Once party leaders were forced to realize that a few "red" experts and a lot of technology were insufficient to realize their millenarian projects, however, they increased pressure on "specialists" of all stripes to cultivate the next generation of science and engineering cadres. This meant more training, and thus more textbooks. But what kind of textbooks? Even orthodox venues like the journal *The Book and the Proletarian Revolution* did not seek to dissolve all boundaries, arguing that "there is no need for the textbook to be transformed into a multi-purpose workbook on physics, mechanics, and social studies."[13] While still subject to party scrutiny, the core function of the science text was not subject to all of the same interpretive strictures that Soviet literary texts were, since literature was regarded as a superstructural epiphenomenon whose social function had to be consciously imposed, while the essential functionality of the science text merely had to be channeled to the proper ends.

Scientists thus occupied a defensive but defensible position in this contentious curricular process, since both proscription and prescription were much less easily achieved by party and government authorities in scientific culture than in literary culture, and few scientists were eager to remove the remaining professional barriers to their enactment. Landau the textbook author was thus acutely aware of the strengths and vulnerabilities of his position at a time when the Communist Party had begun to consume itself in its struggle for revolutionary unity. Yet despite the different "rules" for science texts and literary works, Landau's texts do bear some relation to the newly dominant set of interpretive strategies known as socialist realism, which requires a few words of explanation before we turn to Landau's own corpus.

During the 1930s Stalin's allies and subordinates increasingly strove to generate seamless accounts of the Soviet Union's purported social and technological progress toward socialism realized. Countless memoirs and critical monographs have documented the impact of these campaigns on literature and the arts. Spurred on by Stalin's reported remark that writers should become "engineers of human souls" (not Soviet Mary Shelleys, but the literary equivalent of hydroelectric dam builders), delegates to the First Congress of Soviet Writers in 1934 were enjoined to pay constant heed to socialist realist themes.[14] One of the so-called fellow travelers of early Soviet literature, Isaac Babel, gave a notorious speech at the Congress condemning worn and outdated conventions as the enemy of correct literary style, effectively rejecting the elaborate textures of his own prose. Reminding his audience of the virtues of concision, Babel urged them to "look how Stalin forges his speech, how terse his words are, parsimonious, of such full musculature, how they esteem the reader."[15] In this ritual of self-abasement, Babel plainly illustrated socialist realism's call for uniformity of diction, parsimony of expression, and strong didactic purpose in all forms of literature.[16]

Partly a backlash against the profusion of radical literary styles in the 1920s, socialist realism has often been dismissed in retrospect as cramped neoclassicism wholly attuned to ideological imperatives.[17] It was conservative in that it selectively embraced as its models the "best" (realist) traditions in world literature (e.g., Émile Zola) along with the Russian classics, and it was neoclassical in that it seemed to rely on a strict hierarchy of conventions that echoed the early novel form in emphasizing restraint, stasis, clarity, and rigidity. The "boy meets tractor" plots, the district party secretaries offering the hero moral instruction, and so forth all seem to testify to an exceedingly stale genre in which aesthetics is "polluted" by ideology. But it is too easy to simply dismiss this seemingly jumbled style as inauthentic. I would follow the literary scholar Greg Carleton in insisting that the interpretive conventions of socialist realism cannot be understood adequately with reference to standard notions of genre.[18] We all bring a strong set of generic expectations to our reading of fiction, poetry, art, history, and even scientific texts. The kind of "truth-telling" we ask of fiction, for example, usually bears little relation to the "truth-telling" we expect of history; how we frame our analyses of a given text depends crucially on what genre we assume it belongs to. It is the willful confusion of these genres that is most troubling about socialist realism, whose histories so often veered into the fabulous, and whose fictions so often stood in as authoritative portrayals of what ought to have been (and what was immanent). These practices are readily evident in the 1938 publication of the *History of the All-Union Communist Party (Bolsheviks): Short Course*, easily the most important textbook of its day.[19]

In his reading of the notorious odes by Soviet writers to the building of the White Sea canal, Carleton rightly argues that the topoi (i.e., recurrent motifs, rhetorical commonplaces) of socialist realism functioned as transgeneric rhetorical models that provided the primary focus for modes of representation and interpretation. Within the socialist realist text, topoi had "precedence" over extra-textual reality; one did not have to inquire after their ontological status to know that this or that device (a foundry floor; a stern-eyed chekist; a prisoner in a corrective labor camp; ritual encounters between father and son figures invoking the Great Family of Soviet experience) performed a necessary discursive role.[20] Thus a character's evolving psychological complexity—so crucial to standard Western notions of narrative development—was here rendered irrelevant, since the characters achieved meaningful status by performing certain readily identifiable behaviors that marked their places within the (often complex) semantic structure of the text as a whole. Their job was to instantiate a grander process of "overcoming" (Nature, The Past, The West) that could just as well have referents in non-fiction as fiction: the same function was being performed in either case. In the extreme example analyzed by Carleton, a seeming jumble of lyric, biography, fictional dramatization, photograph, and caricature could be reconciled precisely because no one genre was the sole transmitter of the "reality" of the canal. In socialist realism the reader was always supposed to engage at the level of topoi *before* making any determinations of genre.

If the reification of topoi was ultimately more important to the "authenticity" of a text than its adherence to any single genre, "then a fictional or hypothetical character and a real person can stand side by side and reinforce each other without palpable tension."[21] It was not that genre boundaries were erased in practice, but that our conventional epistemological grounds for relying upon them as interpretive cues had little bearing on the socialist realist text's own cohesiveness. A disconcerting consequence of reliance upon topoi at the expense of generic unity is the absence of conventional cause-and-effect sequencing: frame trumps narrative, and worrisomely upends our ability to track the standard markers of historical texts. In the *Short Course* these textual strategies were used (paraphrasing Carleton) to cover, obviate, blur, and otherwise make irrelevant the criteria for discerning actual from putative or desired reality. My claim is that similar literary technologies were also adopted by Landau and Lifshitz for the benefit of a select readership perfectly distinct from the "mass" reader of conventional socialist realist texts. It is not my intention to conflate ontological categories in comparing these disparate texts. I instead will demonstrate that certain (mathematical) semantic precepts were employed by the authors of the *Course of Theoretical Physics* to destabilize operative contemporary textbook genres in order to link the series together more effectively.

Landau missed (and would have dismissed) the initial fanfare over the *Short Course*: he was sitting in jail at the time.[22] In the meantime, the head of the Soviet government, Viacheslav Molotov, signaled to all Soviet scholars that they would do well to emulate the authors of this textbook. His lengthy hectoring of educational officials at a Kremlin gathering in 1938 contained a prominent call for more and better university-level textbooks.[23] Though this posturing was largely setting the stage for the *Short Course* and would have its strongest impact on the social sciences, natural scientists could not miss its significance.[24] Newspaper editorials faulted the Academy of Sciences for not participating in textbook composition, and textbook publication was declared "a matter of state importance."[25] Expanding on Molotov's themes, the newly appointed top official for higher education—originally trained as a chemist—carefully stressed the need for more textbooks in all the science disciplines.[26]

The message was not lost on Soviet physicists.[27] The Academy of Sciences physics journal, though practiced at brief editorial paeans to the politics of the day, took the much more unusual measure of reprinting the relevant speeches of Stalin and Molotov in their entirety.[28] In due course an obscure scholar complained about the lack of a synoptic theoretical physics survey in Russian, notwithstanding the more specialized efforts of Jakov Frenkel, Igor Tamm, and Vladimir Fock, top theorists who were prey to philosophical idealism in the eyes of this critic.[29] In physics as in other disciplines, who would set down in print the "new traditions" had truly become a pressing concern for scientists. The stage was set.

## For a Truly Scientific Soviet Book

By the time Landau and Lifshitz set to work writing the *Course*, in the 1930s, history had become a liability for many Soviet physics textbooks in print since 1917.[30] The first Soviet theorists had only disdain for tsarist-era holdovers, many of whom were burdened by what they regarded as an indiscriminately expansive nineteenth-century conception of the physics discipline in which a wide variety of machines, applications, and bulk phenomena that were increasingly becoming the province of engineers were still appealed to as the proper domain of physics, yet without benefit of any common explanatory framework.[31] It was a series of developments in the 1930s that gradually led physicists to attach greater importance to the writing of textbooks. The first was the retreat from radical curricular reforms such as the "brigade" method of instruction implemented in 1931.[32] A poorly conceived attempt to collectivize student training and concentrate instruction in a laboratory setting, the brigade methods constituted an end run around the standard teaching mechanism of lecture and textbook in combination. Quickly abandoned in spirit and formally discarded several years later, brigade training served to emphasize that neither a prescribed classroom dynamic nor an admirable emphasis on practical skills was adequate to inculcate students with professional expertise. The return to more conventional methods then redoubled concerns about the suitability of available textbooks.

The appearance of texts like A. K. Timiriazev's *Introduction to Theoretical Physics* in 1933 was also a grave concern to theorists.[33] This was a work hostile to relativity theory and much of quantum theory, but its author, a professor at Moscow University and the most prominent party member among physicists at the time, advertised it as the only textbook of modern physics consistent with dialectical materialism.[34] Except for the philosophically tendentious introduction and conclusion, it was in fact an entirely conventional and technically respectable text that happened to be more than a generation out of date by the time of its publication. Although "Soviet" for superficial reasons, its contents fell squarely within an older textbook genre, relying in particular on the techniques of classical kinetic theory. An impatient Vladimir Fock, the best-known Soviet contributor to quantum theory at the time, harshly castigated the author for both its scientific inadequacies and its misleading polemics, and called for a "truly scientific Soviet book" in response, exhorting theorists to take up cudgels and pens for the good of the profession.[35]

Whether or not Soviet science publishers found Fock's criticisms discomfiting, they had not demonstrated any disposition to reject manuscripts during the quota fever of the first Five-Year Plan.[36] For much of the 1930s chaos reigned among the publishers of science texts, whose operations were constantly subject to reorganization, breakup, and consolidation with every shift in priority of the planning system.[37] Poets and writers may have faced relentless censorship, historians may have revised their manuscripts with committees looking over their shoulders, but for years to come it was more often the case that the hasty output of the beleaguered Soviet presses would far outstrip the ideological ambitions of state and party officials when it came to science textbooks. The editorial publishing council of the Academy of Sciences, for example, generated a lengthy list of policy resolutions in 1935 whose bureaucratic conventionality was marred only by a small item buried in subcategory (b) of resolution 4: "establish the most thorough control over the scientific, ideological and literary composition (*oformlenie*) of submitted manuscripts."[38] More than a decade later, there was still little evidence that the corresponding resources or personnel had been cultivated to attain these nominal goals. Editors with advanced science degrees and considerable experience were paid 2½ times less than junior staffers at the Foreign Literature Publishing house, leaving science publishers continually shorthanded.[39] The more specialized the content of the text, the less likely that it would be subject to ideologically motivated prior censorship.[40] The real danger lay in the common exercise of publicly condemning "errors" after publication, and physicists could not help but ponder strategies to forestall potential critics.

These adverse circumstances helped sharpen the collective awareness of the physicists regarding the larger ramifications of the textbook shortage, which occasioned many a discussion in the Academy of Sciences.[41] Invoking a familiar phrase from recent government campaigns for reform at the elementary and secondary levels,[42] Landau's former teacher, Frenkel, suggested a competition to create standardized textbooks (*stabil'nye uchebniki*) at all undergraduate levels, perhaps obtaining subvention funds for that purpose from the state's Higher School Committee (KVSh).[43] As was so often the case, however, actions lagged behind words, and no concrete institutional initiatives were pursued.[44] It was more often the case that enterprising individual authors took the initiative in composing "truly scientific Soviet" textbooks, not least because publication royalties provided one of the few ways for theorists to supplement their modest incomes at a time when many experimentalists engaged in consulting work for state industry.[45]

### As Is Natural in the Atmosphere of Leningrad

Frenkel himself had long been taking the initiative in writing the most innovative physics textbooks of the early Soviet era.[46] Frenkel's first students in Petrograd in the early 1920s entered the classroom expecting displays of amber, fur, and sparks, but instead encountered lectures on electrodynamics with no analogue anywhere in the literature.[47] Once he resolved to try his hand at a textbook based on his lectures, Frenkel was undeterred by the initial reluctance of Soviet publishers to support him.[48] He took the simple expedient of turning to Springer-Verlag in Berlin to publish the first volume of his *Lehrbuch der Elektrodynamik* in 1926.[49] His ambitions were not lost on his first European readers. They were immediately struck by the text's unusual structure, which they were quick to attribute to its revolutionary setting; such breaks with the standard genre were "natural in the atmosphere of Leningrad," explained one reviewer of Frenkel's textbook in *Nature*.[50]

What made Frenkel's *Lehrbuch* so unusual was that it offered an unapologetically logico-deductive rather than historical-inductive approach to its subject.[51] In retrospect, one could summon equally weighty methodological arguments for either approach and write off the choice between the two as "merely" stylistic, but that would be missing the point. It was the historical-inductive framework that defined the genre of the European physics textbook at this historical moment.[52] Among other things, Frenkel upset generic expectations by taking electric dipoles rather than single charged particles as the basis for his presentation. He had also argued in the professional literature that a structureless elementary electron removed one aspect of the classical self-energy problem, and incorporated this in his textbook. Couched in terms of the locality assumption, Frenkel's point electron became one of the central concepts of quantum field theory.[53] To young students not yet versed in these issues, such assumptions would come naturally in the course of studying his *Lehrbuch*, and Frenkel consistently pursued this field-theoretic approach throughout his career.[54]

The composition of the Landau and Lifshitz *Course* would in turn subtly reflect both the positive and negative lessons learned from these and other peers working to transform Soviet physics instruction. In postwar editions Frenkel wavered on the pedagogical utility of adopting the principle of relativity as a starting point for microscopic electrodynamics. Landau and Lifshitz adopted this tactic for their textbook on the *Classical Theory of Fields* (completed at the end of 1939), and it remains in place today.[55] Not until well into the postwar era did other textbooks venture this approach.[56] While not exactly axiomatic in flavor, *Field Theory* was nonetheless very nearly self-contained, since the reader was only assumed to have mastered certain basic methods from analytical mechanics. Foremost among these was the principle of least action, which Landau and Lifshitz used in concert with the principle of relativity to present Maxwell's equations as the proper basis for developing electrodynamics in tensor form. This did lead to certain incongruities, since relativistic particle motion had to be a direct consequence of these beginning postulates, but only the Lorentz interval qualified as an appropriate invariant to satisfy the principle of least action. If the reader were to inquire after the physical basis for the requirement of Lorentz invariance, however, their only recourse would be to return to the opening paragraphs of the textbook: "Experiment shows that . . . all the laws of nature are identical in all inertial reference systems."[57] (Experiment also provided a value for the velocity of light, *c*.) Even less satisfying was the ad hoc introduction of the appropriate potential term in the Lagrangian in order to "derive" Maxwell's equations in their proper form.[58]

Then as now, textbooks claiming Maxwell's equations as their starting point were uncommon. In large part this is because the logico-deductive mode of organization has proven less heuristically accessible than historico-inductive approaches built up from simpler phenomenological laws like those of Faraday and Coulomb. These standard approaches made plausibility arguments for Maxwell's theory before formally establishing that all the phenomenological laws could indeed be derived from Maxwell.[59] Landau and Lifshitz's deliberate adoption of this unusual organizational style reinforced their readers' sense that they were encountering a distinctive and authoritative mode of discourse.[60]

There was methodology to this madness. Basic concepts of electrodynamics like the motion of a charged particle in a Coulomb field were used in problem examples as an excuse to introduce some of the basic notions of particle scattering. This was but one of many instances in which the parts of the *Course of Theoretical Physics* were explicitly coordinated with the larger goal of training the vanguard of theory. "Although the scope of the book is kept rigidly within the bounds of classical physics," commented Léon Rosenfeld on the English edition of *Field Theory*, "most of these problems have been chosen with a view to future use in quantum theory."[61] Later, in non-relativistic *Quantum Mechanics*, scattering problems were consolidated in a single chapter in preparation for subsequent exposure to quantum field theory. Similarly, an important consequence of incorporating general relativity in a text on field theory was that it enabled Landau and Lifshitz to edge the physics community away from an essentially geometrical conception of gravity to a more dynamical and unabashedly field-theoretic approach, as David Kaiser has demonstrated for the successive postwar editions of *Classical Theory of Fields*.[62] The kind of methodological unity Landau aimed for in his own research could thus be instilled in a new generation of theoretical physicists.

Before considering further volumes, I should stress the care with which Landau negotiated the publication of the *Course* as a whole. Frenkel would be harshly attacked by Soviet "patriots" for publishing several of his textbooks abroad before issuing Russian editions.[63] Except for *Statistical Physics*, whose Russian edition was delayed by Landau's jailing, every volume of the *Course* appeared first in Russian, though it was always written with an international audience in mind. While still in Kharkov, Landau had used his foreign colleagues as intermediaries to begin putting out feelers abroad about simultaneous publication of the future *Course of Theoretical Physics*.[64] In England he found an audience eager "to give Russia its place in our academic studies, and Russian science and learning its proper international status."[65] Oxford's R. H. Fowler quickly lent his approval to "what might prove to be a first class standard work on theoretical physics."[66] Some initial confusion about the nature of the project was rather revealing, with Fowler counseling against publishing Landau's "works to date in five volumes," and suggesting instead "a book in English which contained perhaps the more important parts of his Russian works, so long as the book so composed formed something of the monograph type."[67] Landau was after something "of a much more elementary character" than an encyclopedic monograph of his own work, his friend Rudolf Peierls assured the editors: "It is to be a textbook for students."[68] This series would ensure, however, that all of Landau's major contributions to physics were memorialized in suitable pedagogic form.

**Topos and Disciplinary Topography**

The first volume in sequence of the *Course of Theoretical Physics* was *Mechanics*, the traditional subject at the core of any physics curriculum. Landau assigned much of the assembly of a full mechanics text to Lev Piatigorskii, the lone party member of his theory group in Kharkov. Their collaborative text was completed the month of Landau's arrest in 1938, an event that delayed its appearance until after his rehabilitation.[69]

The eight-paragraph introduction to *Mechanics* is the closest Landau ever came to reflecting on the tasks of theoretical physics in methodological terms, and it is not hard to see why he opted to remove these remarks from subsequent editions.[70] The positivist thrust of the decidedly off-the-cuff remarks is clear, but the portions that could be construed as empiricist are partially counterbalanced by reference to "logically closed theories" that "never lose their value." But of course the real point of the introduction was not philosophical (and entwined with *Erkenntnistheorie*); it was pedagogical (and tied to the sociology of Soviet physics at the time). The authors told students from the outset to be conscious of disciplinary boundaries, and rather than focus on the status of foundational theories as the source of disciplinary identity, they concentrated instead on approximation methods as potential sinks for the loss of same.

In *Mechanics* one finds many of the signs of a quasi-logico-deductive approach to the teaching of physics that were explicitly acknowledged only in later editions. "As with the other branches of theoretical physics," Landau and Lifshitz would subsequently admit, "our exposition makes no use of the historical approach."[71] Galilean and Newtonian mechanics were of little concern in their own right, since the principle of least action and conservation laws formed the starting point for discussing all equations of motion. Landau and Piatigorskii stressed their own approach as a move away from formal computation. Calculating mere numbers was a matter for mathematical physics: "Once the equations taking into account only the essential factors are obtained, the task of theoretical physics, strictly speaking, basically comes to an end." This was not an attempt at an ideal Cartesian mechanics, but a choice of conventions and an epistemological attitude, for Landau consistently maintained that classical mechanics could be treated as a logically closed system only insofar as theorists knew just what quantities they were approximating. It would be a waste of time and energy to calculate too precisely.[72]

Textbooks like those of Landau and Lifshitz, as well as the influential *Classical Mechanics* of the American physicist Herbert Goldstein somewhat later, represented in part an attempt to break with the traditions of the Cambridge Mathematical Tripos as handed down in classic texts like E. T. Whittaker's *Analytical Dynamics*.[73] Whittaker's mastery of the subject matter was such that he could scarcely bring himself to exclude any available theorem or technique, occasionally obscuring physical problems in his attention to mathematical completeness.[74] Where Whittaker chose to derive Hamilton's principle from Lagrange's equations (the historical order), Landau thought it more important to show that Lagrange's equations follow from Hamilton's principle, since this opens up the royal road to field theory, the vital working tool of the modern theorist.[75] More than mere principle was at stake here, for in this manner Landau also made his bid to reclaim mechanics for physics faculties (as opposed to mechanico-mathematical faculties) in the Soviet university.[76]

At the same time, the approach taken in *Mechanics* scarcely resembled the axiomatics celebrated by Landau's old-regime colleagues in the 1920s.[77] Landau did not even bother to distinguish Hamilton's (integral) principle from the slightly less general principle of least action, since he was probably more concerned with establishing the appeal of the latter as a useful form for making the transition to optics and to quantum mechanics.[78] Landau's *Mechanics* began with the principle of least action, asserting that "this principle expresses the law of motion of any mechanical system." There was no attempt to mention nonholonomic or dissipative systems as appropriate to the domain of mechanics, but seldom subject to the Hamiltonian formulation.[79] The authors also took their time introducing the concept of force, and once they did so, they did not start with forces depending on velocity. As Rosenfeld remarked upon reading further in the *Course of Theoretical Physics*, Landau had been fiddling with this 'violon d'Ingres' since his days in Copenhagen, in the "belief that one could encompass the whole of physics in a monumental 'principle of least action.'"[80] Even Goldstein did not lend such primacy to the principle of least action, while most contemporary mechanics textbooks followed more conventional historical presentations.[81]

Fock registered his displeasure with what he regarded as the "slipshod" notions of mathematical rigor employed in *Mechanics*, fearing they would corrupt the physical arguments.[82] On the other hand, Frenkel was one of several colleagues over the years to protest the "formal nature" of Landau's methods.[83] For his part, Landau gave every indication that he felt he had steered the most effective pedagogical course. What he and Lifshitz did with the principle of least action is a crucial example of their textual strategy for the *Course* as a whole. They sought to shift the student reader's interpretive process to a topical plane that would precede any generic assumptions about how a classical mechanics text or an electrodynamics text or a quantum mechanics text taken alone should "naturally" frame its respective problem domain. In this fashion problems with no obvious empirical relation to each other could be more readily rendered plausible objects of study by invoking the topos of the least action principle. The unwieldy empirical narrative that could link the two phenomena is absent, but the association by means of the topos nonetheless makes the technique's application "authentic" much more quickly in the eyes of the student. Breaking generic assumptions and binding via topoi: Landau and Lifshitz showed that socialist realist didactic strategies were neither limited to literature and the arts, nor were they always used merely to propagate "orthodox motifs."

Making reference to topoi served to defeat the specialist's presumption that each phenomenal domain has its own best methods (some more empirical than others), which would otherwise make it that much more difficult to develop points of linkage between any two domains, because their respective techniques might not be immediately reconcilable within a strict hierarchy of commonly accepted procedural principles. The mature theorist might look at this and insist that the unity of the phenomena is logically entailed in the original proof of the principle of least action, so long as you have established that the given system is a conservative one. That may be so, but as a practical matter this had little to do with how students learned to apply the principle to concrete problems, and Landau's aim was to make the technique transferable in a more efficient fashion within the overall context of the *Course*. For Landau, the whole point of broad theoretical training was to make it possible to employ "remote analogies" in the solution of the specialized problem at hand.[84]

In the 1930s, theoretical fluid dynamics was probably the most developed subject that had the least overlap with the agendas of modern theoretical physics, thanks to its heavy reliance on nonlinear differential equations and empirical methods. Just as they had with *Mechanics*, Landau and Lifshitz used the volume *Fluid Mechanics* to redraw or blur boundaries.[85] Landau and Lifshitz made the first attempt in the literature to reassert the methodological unification physics could offer to suitably chosen aspects of fluid mechanics—with the added domestic bonus of attributing important developments to Russians (N. E. Zhukovskii and A. N. Kolmogorov) who did not figure in canonical Western textbooks.[86] It was well understood by Landau's Soviet contemporaries that he was engaged in selectively reclaiming conceptual and institutional territory for the discipline of theoretical physics, and could not be seriously faulted if that overriding imperative occasionally led to overly compressed explication of major concepts, abuse of phrases like "it is apparent," or unevenness of presentation.[87] There was no pretense that the fundamental equations of hydrodynamics could be subsumed within atomic theory, only the promise that consistent invocation of established principles could prevent one man's phenomenology from becoming another man's poison.

## The Order Parameter (and the Culture of Language)

Landau is best known for his contributions to the theory of condensed matter, so it should come as no surprise that the most influential volume in the *Course* would become *Statistical Physics* (now in two volumes). It was the first in the series to be completed, and was submitted to the State Publishing House for Technical-Theoretical Literature immediately upon Landau's arrival at Peter Kapitza's Institute of Physical Problems in February 1937.[88] David Shoenberg, a friend and onetime research student of Kapitza's at the Mond Laboratory in Cambridge, had already undertaken to translate the manuscript into English. As proofs from the Clarendon Press at Oxford began arriving over the months that followed, he had occasion to draw Landau into extended debate over the structure of the text. "There were many points in *Statistical Physics* which were stated as self-evident but which I could not properly understand," Shoenberg later recalled, and his demands for clarifications from Landau helped him "improve the presentation in the translated version."[89]

These improvements in the first of the volumes of *Course of Theoretical Physics* to be published also marked the first steps in the assimilation of the *Course* by a wider audience. Subtle differences between the original text and the translation readily indicate that this process immediately began effacing the historical specificity of *Statistical Physics*, which was composed in a distinctive voice well suited to its setting. Not unlike many a party boss in other arenas of Soviet culture, the authors breezily assumed an air of authority in diverse matters that seldom made any distinction between personal tastes and judgments and the reigning consensus among physicists. Neutrality, temperamental leanings toward methodological agnosticism, the patient wait-and-see attitude of the optimistic empiricist were not stylistic options in the discourse of these theorists.

Landau and Lifshitz opted for a unified presentation of thermodynamics and classical statistical mechanics using the work of Josiah Willard Gibbs as their starting point. In the briefest of prefaces, the English reader was immediately alerted that "no attempt has been made at mathematical rigour in the treatment, since this is anyhow illusory in theoretical physics, but we have instead tried to make clear the fundamental physical assumptions on which the results are based." Without further preliminaries, the authors then launched into explanations of probability, statistical distributions, phase space, and entropy in the opening pages of the text. The Russian preface evidenced a sharper tone, however. It took the opportunity to dismiss at the outset those physicists who regarded statistical physics as one of the less mathematically rigorous domains of physics. Landau and Lifshitz saw no point in imposing unwarranted rigor upon statistical physics, which had been rendered a "logically connected and harmonious system" by the work of Gibbs. They faulted the authors of standard texts on the topic for introducing his methods only as the culmination of the admittedly invaluable efforts of Clausius, Maxwell, and Boltzmann, rather than as the fundamental point of departure pedagogically.[90] For the two Soviet theorists, this internal harmony meant that any additional considerations could be jettisoned.

There is an additional subtext to the first edition that many non-Soviet readers would have missed. Landau and Lifshitz informed their readers that the theory of liquids would not be presented because it depended on molecular interactions that could only be defined on a case-by-case base, without offering qualitative conclusions applicable to all liquids. Immobile tools were useless to the disciplinary topography of the *Course*. They also served notice that in their treatment of the relation between solids and liquids, they would unfortunately be faced with strange assertions from certain quarters about there being no fundamental differences between liquids and crystals. For Landau, symmetry considerations alone dictated a sharp phase transition, and he brooked no opposition in the matter.[91] Landau and Lifshitz realized that in essence every science textbook faced vetting for its attention to the "culture of language" (*iazykovaia kul'tura*).[92] Evidence of convoluted "chancery language" in university textbooks was grounds for caustic criticism, and physics was no exception.[93] When Landau's students in the 1930s perceived him to be speaking another 'language' than they were used to from the university, they were also testifying to his ongoing efforts to create a uniquely authoritative discourse for theoretical physics.[94]

Aside from a brief reference to the size of the universe, the only experimental numerical value mentioned in *Statistical Physics* is Boltzmann's constant, as an amazed reviewer noted, impressed if not entirely approving that Landau and Lifshitz "have kept within their chosen territory with an inflexibility that is unusual."[95] Indeed, much of the latter half of the book stuck to the thermodynamic approach, when many of the topics (chemical potentials, phase equilibria, surface phenomena) begged for statistical treatment. Statistical considerations were not inserted until the second edition (1958), though phase transitions were crucially excepted from these modifications. Even when the emphasis was on thermodynamics, the text's presentation remained unrepentantly deductive in form. An early reader judged it "well worthwhile, being sound and presenting the matter in a systematic way from a somewhat unusual point of view."[96] A truly distinctive program of pedagogy had been launched.

The contrast with two equally prominent contemporary texts could not have been greater.[97] R. H. Fowler's *Statistical Mechanics* was notable for its encyclopedic coverage and idiosyncratic choice of method. While acknowledging that Gibbs's method had "the advantage in logical precision," Fowler explicitly avoided any talk of "deeper theory." At the other extreme, Richard Tolman's *Principles of Statistical Mechanics* gave careful attention to a host of rigorous proofs passed over in silence by the Soviet theorists. This same laconic style of presentation would characterize the second, postwar edition, when Landau and Lifshitz finally conceded the utility of incorporating quantum statistics in a unified treatment.[98] The effect achieved was still something other than axiomatic, but rather made the impression that "no holds are barred," as George Uhlenbeck put it.[99] Landau and Lifshitz happily invoked relativity theory, second quantization, group theory, and, in general, whatever theoretical tool would serve their physical ends in the context of statistical physics.

By far the most important of these tools was the order parameter, a calculating device invented by Landau that effectively blurred the boundary between microscopic and macroscopic while opening up the possibility of unified descriptions of highly disparate phenomena. Though more restricted in scope than the principle of least action, this qualified as another central topos marking the cohesion of the *Course*, one that became crucial to the research agendas of numerous young Soviet theorists. Unlike Fowler, however, Landau and Lifshitz maintained that all of these eclectic procedures were, indeed, "deeper theory" in the best sense. The result did have its appeal. "The mathematical and physical deductions," wrote one reviewer, "are kept free from too high a degree of exactness, a fact which makes the book very pleasant to read."[100] To his mind Landau and Lifshitz had performed a signal service in offering a modernized treatment of the notoriously opaque writings of Gibbs.[101]

### Quantum Mechanics without Foundations

Quantum mechanics was in most respects the most modern subject matter included in the *Course of Theoretical Physics*, so renouncing a historico-inductive presentation in their textbook was rather less likely to distinguish Landau and Lifshitz's *Quantum Mechanics* from its competitors. In pragmatic terms, even among those physicists on the Continent who wholeheartedly endorsed probabilistic physics and regarded measurement interaction problems as central to quantum theory, Niels Bohr's epistemological concerns were more often ignored in favor of problem techniques aimed as much at the experimentalist as the theorist, as witness Arnold Sommerfeld's influential wave-mechanics supplement to *Atombau und Spektrallinien*.[102] Landau and Lifshitz thus had many more counterparts in Europe and America who were equally eager to launch their textbooks from first principles, or at the least to claim an explicit operational foundation that identified their work as distinctly modern.[103] The Soviet authors nonetheless had some unusual reasons for minimizing the historical trappings of the theory, while at the same time paying careful attention to parallels with classical physics encountered elsewhere in the *Course*.[104]

Though in preparation before the war, Landau and Lifshitz's *Quantum Mechanics* did not appear until 1948. It dispensed with even cursory nods toward Bohr's complementarity, and shared with Frenkel an unwillingness to attribute great significance to issues of observer interference when discussing the uncertainty principle. The introduction to the Russian edition (conveniently excised from the 1958 English edition) makes clear their position on quantum mechanics as a tool of instruction rather than a site for mathematical and foundational disputes:

> We cannot but remark that in many quantum mechanics courses the exposition has been substantially complicated in comparison with the original works. While such complication is usually argued for in terms of generality and rigor, upon careful examination one can easily observe that both are in actuality often illusory to such a degree that a significant fraction of "rigorous" theorems are mistaken. Insofar as this kind of complication of exposition seems to us completely unjustified, we have striven, on the contrary, for all possible simplicity and much of the time we have returned to the original works.[105]

Returning to the original works did not mean adopting a historico-inductive organization, however. Save for a cursory nod to measurement problems in the first chapter, instruments and questions of state preparation played no part in the text. Canonical experiments like those of Davisson and Germer or Stern and Gerlach got no mention at all. At the same time, Landau and Lifshitz largely dispensed with mathematical preliminaries, introducing operators for physical quantities from the outset. Consistent with their desire to train theorists, they also dwelt at length on group theory, keeping the discussion as physical as possible by presenting the topic in crystallographic terms.

Pick up a copy of the first English edition of *Quantum Mechanics* (1958) and you will find that it has many of the minimal historical appurtenances one would expect for an intermediate-level student text. In many instances these modest historical cues were added by the translators, and are not necessarily found in the original Russian edition. Landau and Lifshitz consciously avoided philosophical and political "danger zones" associated with quantum theory. Their Russian introduction began with the uncertainty principle, but with Soviet memories of Nazi aggression still fresh in the late 1940s, the authors found it advisable to broach the subject without mentioning Germany's most famous physicist by name or actually labeling Heisenberg's famous principle as such. (The term "uncertainty relations" is used freely later in the text, e.g., §14.) Neither did

Bohr's name grace the abbreviated discussion of the measurement problem.[106] According to the current editor of the *Course*, more detailed comments were withheld until the next edition in the knowledge that they could have drawn philosophical criticism.[107]

What is perhaps most striking about both *Quantum Mechanics* in particular and the *Course of Theoretical Physics* in general is the ongoing balance between completeness and pragmatism, between universal scope and local utility. This was all in keeping with the "strongly pragmatic character" of Landau's Theoretical Minimum.[108] To achieve that delicate balance, Landau and Lifshitz engaged in a great deal of idealization, making no pretense that nascent experimentalists would receive any guidance about how to proceed to the next stage of specialization in their laboratory endeavors. This quasi-closed system deliberately avoided paradox (logical or empirical), notwithstanding the unintuitive nature of quantum mechanics. "You don't get any dead-end problems there," claimed one of Landau's disciples of the *Course*, and with good reason.[109] A loosely defined but vigorously enforced internal consistency was paramount in the world of the Theoretical Minimum, an attitude that did not go unnoticed by outsiders. One senior Soviet textbook writer put it this way: "There is a widespread view among theorists that when presenting quantum mechanics it is generally advisable to renounce the historical succession and construct the presentation purely logically, in more or less dogmatic fashion."[110] For this Landau and Lifshitz were unapologetic. Their methods bespoke a strong authorial drive to call up a new social order among physicists, one in which theorists could speak a common language amid the fragmentation of experimental practice. This was no retreat into "high theory," nor was it (to sustain the literary parallel using a contemporary term of opprobrium) "avant-gardism." Landau and his disciples maintained healthy working dialogues with selected experimentalists, and carefully prioritized their choice of exemplary problems accordingly. But their revision of the teaching canon shared a certain socialist realist abhorrence of "mere" naturalism/empiricism as a species of provincialism that required continual correction. The energetic, unceasing revision of the *Course* itself, a process tightly controlled by its authorial collective, might even be said to be exemplary in a positive sense of the collective editorial practices publicly celebrated (and frequently abused behind the scenes) in socialist realist literature.[111]

## The Theoretical Physics Approach to Life

When first published, immediately after World War II, *Sovetskaia kniga* (*The Soviet Book*) quickly became the most prestigious general review of books in the Soviet Union, and it was the venue in which Landau published the only book review of his career. In a few terse paragraphs, he confessed to being a firm believer in "systematization" and praised textbooks written "very clearly and without any extraneous complications, which, unfortunately, one often encounters in books on issues in theoretical physics."[112] In *Sovetskaia kniga*, however, such "extraneous complications" meant that reviews of books on physics topics always came after the feature review in the sequence of topical reviews that followed. Thus it was in early 1950 that Academician Landau and Professor Lifshitz found the second edition of *Field Theory* criticized by associates of Fock in a review immediately adjacent to a fawning feature review of a book on science in the Stalin era, the official tribute by the Academy of Sciences to the Soviet leader on his seventieth birthday.[113]

This accidental juxtaposition of stalinist hyperbole and *Field Theory* (general theory of relativity included at no extra charge, in just 350 pages) reflects the larger tensions in Landau's Soviet world. Once a victim of repression, he had only scorn thereafter for the "plumbers, carpenters, and joiners" in Stalin's Kremlin.[114] But the forced political cynicism of his mature years should not obscure the youthful motives that licensed such an ambitious pedagogical enterprise. He detested the class-based access to scientific careers that he saw in Great Britain, mocked the proponents of Aryan physics in Germany, and dismissed the imprecations of Bolshevizing philosopher-courtiers, but he did not regard any of these as threats to the integrity of science as an international enterprise. There was not a moment before the age of 30 when he doubted that ideology was already on his side. In those days it seemed clear to him that the Soviet Union stood poised to become the most effective patron of modern science on the world scene.[115] If the international audience for the *Course* was struck by its unusual authorial voice, this was not because Landau had set out to articulate a uniquely Soviet scientific style (as against "Western" science). That would have been a false distinction for him. His consistent ambition was to demonstrate that Soviet physics could show itself the most modern among the many international competitors for the title.

Landau soon came to the conclusion that theorists as a group needed to be more pragmatic about articulating autonomous research agendas during the wait for the final theory combining quantum mechanics, special relativity, and gravitation, since experimentalists were not behaving as if they detected the impending end to the narrative of modern physics. The *Course* helped achieve that aim. The first generation of Soviet theorists showed frequent impatience with the ambiguity of history, where "history" for theorists was first and foremost the periodization of the discipline tacitly imposed by experimental traditions and explicitly enforced by Marxist philosophers. In their textbooks, Landau and Lifshitz rejected received notions of the conceptual evolution of the discipline in favor of innovative pedagogical structures. For these men textbooks could serve in part to establish new theoretical traditions that were independent of a Marxian version of physics as a catalogue of instrumental techniques and empirical feats largely determined by historical circumstance and heroic individuals in service to class interests. Such mundane empiricist profanations would have no part in the proper elevation of the culture of theory.

It would be a mistake to think that the Soviet system offered only obstacles to the realization of this massive project. The award of the Lenin Prize in 1962 was in keeping with the many material incentives for producing scientific work whose international glory redounded, however indirectly, to the Soviet Union. There were social incentives as well. The writing of textbooks was strongly linked to the formation of distinctive Soviet scientific research schools, and the inimitably cosmopolitan *Course* nicely served that parochial purpose at the same time. As an example of attitudes prevalent in the late Stalin era, consider this statement by a prominent university educator: "A textbook cannot be constructed via compilation of someone else's thoughts, accomplishments and scientific results, it has to be the sum total of great scientific and methodological work carried out by one or another scientific school."[116] Textbooks, in other words, did not simply codify the mopping-up process of normal science in Kuhnian fashion—they were markers of its innovative potential, and distinctive discourse reinforced that presumption.[117] Moreover, in a period when the use of conventional historical narrative risked endorsing "bourgeois" or "reactionary" members of the physics canon, the "school" as a coherent social unit became by far the most accepted domestic means to legitimate the standing of Soviet theorists in comparisons with the likes of J. J. Thomson, Arnold Sommerfeld, or Enrico Fermi. The *Course* sealed an already strong case for Landau's inclusion in this elite group.[118]

Certainly his students regarded the *Course* as the most important indicator of the school in sociological terms.[119] These same students, by and large, were also sensitive to the ways in which the widespread adoption of the textbooks had the potential to dilute the identity of the Landau school.[120] As successive volumes of the *Course* slowly codified his disciplinary standards, however, Landau himself showed no desire to distinguish the explicit from the tacit and administered the Theoretical Minimum exams using problems straight from the textbooks, year after year. It did not disturb him that prospective examinees would consult recent examinees about his limited repertoire of exam questions. His closest colleagues objected, but Landau always insisted that if students had read all the problems, worked through the solutions that followed, and understood them, then that was enough for him. The students had only paper and pencil, and the knowledge that Landau would peer over their shoulders every twenty minutes or so to ascertain *how* they were solving the problems.[121] The *Course* would constitute the school through identification of practitioners who had mastered an elite set of skills rather than a sacred body of knowledge.

Looking back on the lengthy enterprise, one of Landau's first students unsurprisingly thought the *Course* warranted praise as a revolution in theoretical physics pedagogy.[122] But of course the pedagogical relevance of this elegant toolkit further depends on what students *did* with their acquired skills, and how they subsequently went about solving actual research problems.[123] Whether or not one calls it revolutionary, the *Course* did indeed help structure the ways in which Landau's students framed original problems on their own. Though it was not common practice to cite textbooks in peer-reviewed articles, students of Landau who had passed the Theoretical Minimum often referred to the available volumes from the *Course* in their early publications.[124]

This practice was more than a gesture of deference, and rarely derivative (e.g., "following equation 15.3 of volume 5. . ."). One of the genre-destabilizing aspects of the *Course* was its inclusion of problem solutions in the text, solutions notorious to students everywhere for their intentionally quirky compactness. Mastery of these set problem-solving techniques carried over directly to the choice of independent problems for research.[125] Take the case of R. G. Arkhipov, who is of interest precisely because he followed a career path fairly distant from Landau's ambit after passing the Theoretical Minimum. In one of his first papers, Arkhipov cited both *Statistical Physics* and *Fluid Mechanics*, and in the latter case he explicitly referred the reader to problem 2 of §60 and problem 3 of §61, each suitably reformulated (not simply reproduced) for his present purposes.[126] Such examples could be multiplied. Landau's last official graduate student cut his teeth professionally by identifying mistakes in these exemplary problems and refining the problems for later editions, earning the right to co-author the volume on relativistic quantum mechanics that Landau had long postponed.[127] After the death of Lifshitz in 1985, he eventually inherited the responsibilities of general editor of the *Course*.

To complete the Theoretical Minimum demonstrated more than proof of technical competence, however. Successful students were also initiated into the teacher's peculiar moral economy. The applicant would leave his notebooks and volumes from the *Course* in the entryway of Landau's apartment, and proceed upstairs to a small room, alone with paper and pencil. Passing one of the later exams in the sequence would often spur Landau to confide in the initiate, broaching broader subjects that could not always be reconciled with conventional Soviet political discourse and social mores in the 1950s—subjects that would once merely have been "mainstream radical" in Landau's university days. In particular, Landau liked to impress upon the young theorist how he ought to comport himself, and even when he should marry (at age 30).[128] In essence the *Course* served as a professional distillation of a larger system of group values that Landau liked to call the "theoretical physics approach to life."[129]

Landau's texts project an obsession with imposing a synoptic vision of order upon the external world. I do not mean simply the ontological striving for explanatory unity that perhaps motivates every physicist, but a rather more skeptical attitude in which theoretical closure is an epistemological construct purposely imposed on representations of natural phenomena whose fundamental intractability Landau readily acknowledged in his own daily research. In keeping with the *Course*'s cohesion at the level of topos rather than genre, one might say that Landau taught a picaresque physics, with brilliant leaps from one domain to the next in episodic fashion, but without sustained development of a single 'plot' running from the subatomic to the macroscopic realm.[130] This was not a pedagogical program meant to prepare the way for unification à la Einstein or Dirac, for the first generation of theorists trained in the wake of the revolution in quantum mechanics did not share their elders' optimism. Thus, the *Course* represented a signal accomplishment to Landau's contemporaries. In 1959 George Uhlenbeck commented approvingly that the *Course* of Landau and Lifshitz "is in the great tradition, and is the only one attempted by physicists of the present generation."[131] The fact that this remains true more than forty years later has done nothing to diminish the physics community's esteem for the ambitious scope of the project.

Clearing away the empirical debris of history in order to get at the most effective methods for practicing theory in the here-and-now did still have its critics. Not least among them was that good Marxist and erstwhile Bohr collaborator Léon Rosenfeld, who suggested that a more effective pedagogical dialectic would involve an inductive approach on the first pass through the physics, followed by the deductive path of Landau and Lifshitz for analysis of "more delicate problems" on the return journey:

> For this return journey the advanced student could find no better guide than our authors. But for his first introduction to the true spirit of theoretical physics he will still have to rely on the old masters, Abraham and Sommerfeld, who (albeit unwittingly) display that fine sense for dialectics which is so strangely lacking even in the best productions of our Russian friends.[132]

In a way, Rosenfeld was missing the point, since one of the advantages of adopting the deductive mode for Landau and Lifshitz was the ease with which they could ignore "the old masters" of the profession. Landau, as a young man, is said to have proclaimed: "You can't find out anything new from thick books. Thick books are the graveyard where the ideas of the past are interred."[133] It is evident from the *Course of Theoretical Physics* that Landau maintained a highly instrumental conception of texts throughout his career. More than a generation after its conception, the *Course* would swell into several thousand pages of creative output that powerfully testified to Landau's conscious efforts to build a novel and sustainable culture for Soviet theoretical physics, an endeavor that strikes us as anything but nihilistic. The project was originally launched by that impulse to clear the ground in professional terms, however, and it was sustained by the same need to avoid the ambiguities and impositions of history, "the most serious science" to so many of his Soviet contemporaries.[134]

## Notes

1. V. Ginzburg, "Fundamental'nyi trud," *Izvestiia*, January 28, 1962, p. 4. Imagine Steven Weinberg getting similar coverage in the *New York Times*, not for *Dreams of a Final Theory* or *The First Three Minutes*, but for his three-volume *Quantum Theory of Fields*. That improbability suggests just how strong the public linkage between authorial persona, pedagogical production, and scientific achievement (*dostizhenie*) was in the Soviet Union.

2. The extensive efforts to rehabilitate Landau are recounted in A. Dorozynski, *The Man They Wouldn't Let Die* (Macmillan, 1965).

3. M. I. Kaganov, "Entsiklopediia teoreticheskoi fiziki," *Uspekhi fizicheskikh nauk* 145 (1985): 349, reprinted in *Vospominaniia o L. D. Landau*, ed. I. M. Khalatnikov (Nauka, 1988), 316–322.

4. My thanks to Michael Gordin for drawing this to my attention.

5. This is all the more remarkable insofar as it is not an introductory series. Though I have no data for Pergamon, this figure seems a safe extrapolation from the print runs of the Russian editions (e.g., the second edition of *Quantum Mechanics* was printed in 70,000 copies, and the third edition of *Statistical Physics* in 57,000 copies; both books are now in their fifth editions). Published reminiscences by Landau's colleagues indicate that new editions exhausted their print runs in short order.

6. See G. E. Gorelik and G. A. Savina, "G. A. Gamov . . . zamestitel' direktora FIANa," *Priroda* no. 8 (1993): 82–90; K. Hall, Purely Practical Revolutionaries: A History of Stalinist Theoretical Physics, 1928–1941 (PhD dissertation, Harvard University, 1999), chapter 2.

7. Even after higher degrees were reinstated in 1934, the Theoretical Minimum remained as a prestigious informal standard, free from meddling by the state's Committee on Higher Education. Apparently the reputation of the Theoretical Minimum was such that one man who had no intention of becoming a theoretical physicist passed the exams as a matter of intellectual prestige before turning to a career in the life sciences.

8. Arkhiv Rossiiskoi Akademii Nauk (hereafter ARAN), St. Petersburg branch, f. 2, op. 1 (1936), d. 8, l. 116. This is an unpublished remark from the familiar March 1936 session of the Academy of Sciences.

9. I. V. Stalin, "Rech' na prieme rabotnikov vysshei shkoly" (May 17, 1938), *Pravda*, May 19, 1938, reprinted in *Sochineniia*, 1 [14], ed. Robert H. McNeal (Hoover Institution Press, 1967), 275–278.

10. From a January 1958 letter quoted in Maiia Bessarab, *Landau: Stranitsy zhizhni*, second edition (Moskovskii rabochii, 1978), 116.

11. See, for example, a cartoon in which the State Publishing House, in the form of a cannon, fires books at these same targets. "Kniga—oruzhie proletariata," *Pravda*, May 21, 1929.

12. James T. Andrews, *Science for the masses: The Bolshevik state, public science, and the popular imagination in Soviet Russia, 1917–1934* (Texas A&M University Press, 2003).

13. N. Suvorov, "K voprosu ob uchebnikakh po fizike," *Kniga i proletarskaia revoliutsiia* no. 3 (1933): 76–78, on 76.

14. Among others, A. A. Zhdanov attributed the remark to Stalin in his speech at the Congress, "Sovetskaia literatura—samaia ideinaia, samaia peredovaia literatura v mire," *Izvestiia*, August 20, 1934. On earlier informal uses of the phrase by Stalin, as well as the broader context for the metaphor, see David Joravsky, *Russian Psychology: A Critical History* (Blackwell, 1989), 329.

15. I. Babel, "Poshlost'—vot vrag," *Izvestiia*, August 26, 1934. Babel was subsequently repressed.

16. For example, the art historian Mathew Cullerne Bown describes an editorial from *Artists Brigade* in 1932 that invoked the use of "vivid, concrete, figurative language" in which the "empirically observed fact" would be developed into a "socially significant generalization." Matthew Cullerne Bown, *Socialist Realist Painting* (Yale University Press, 1998), 125.

17. The extensive scholarly literature on socialist realism does not bear recitation here. For a classic study of its uses in stalinist culture-building, see Katerina Clark, *The Soviet Novel: History as Ritual* (University of Chicago Press, 1981). A more recent collection of astute reinterpretations may be found in Thomas Lahusen and Evgeny Dobrenko, eds., *Socialist Realism Without Shores* (Duke University Press, 1997).

18. Greg Carleton, "Genre in Socialist Realism," *Slavic Review* 53 (1994): 992–1009.

19. For an early English edition of this ubiquitous book, see *History of the Communist Party of the Soviet Union (Bolsheviks): Short Course* (International Publishers, 1939). On the composition and use of the *Short Course*, see especially David Brandenberger, *National Bolshevism: Stalinist mass culture and the formation of modern Russian national identity, 1931–1956* (Harvard University Press, 2002).

20. Carleton ("Genre in Socialist Realism") makes explicit the connection here with the uses of topoi in medieval literature, e.g., lions and tigers as obligatory components of the proper "landscape scene."

21. Carleton, "Genre in Socialist Realism," 1003.

22. G. E. Gorelik, "'Moia antisovetskaia deiatel'nost' . . .' Odin god iz zhizni L. D. Landau," *Priroda* no. 11 (1991): 93–104; "The Top-Secret Life of Lev Landau," *Scientific American* (1997), August: 72–77.

23. His summons to active composition was seconded by one of the few student speakers. V. Molotov, "O vysshei shkole," *Izvestiia*, May 20, 1938; E. M. Sergeev, *Moskovskii universitet—vzgliad skvoz' gody* (Izd. Moskovskogo universiteta, 1992), 53–54.

24. Lest there be any doubt, three party-minded scientists (including the young mathematics phenom S. L. Sobolev) quickly patched together a quotation-laced editorial echo of the official sentiments, giving due priority to Molotov's call for textbooks. See S. Sobolev, Kh. Koshtoiants, and B. Kuznetsov, "Za peredovuiu nauku stalinskoi epokhi!" *Pravda*, May 22, 1938.

25. See the editorials "Nauka i praktika," *Izvestiia*, May 11, 1938; "Vypusk uchebnikov—delo gosudarstvennoi vazhnosti," *Izvestiia*, August 27, 1938.

26. S. Kaftanov, "Vysshaia shkola i podgotovka sovetskikh spetsialistov," *Bol'shevik* no. 16 (1938), August 15: 14, 19–20.

27. Nor was the message lost on geographers and mathematicians in the Academy, who hastened to make their own contributions. See A. A Grigor'ev, "O stabil'nykh uchebnikakh po geografii dlia srednei shkoly," *Vestnik Akademii nauk* no. 7–8 (1938): 61; A. A. Lusternik, "Rabota sotrudnikov Matematicheskogo instituta Akademii Nauk SSSR nad uchebnikami," *VAN* no. 7–8 (1938): 67.

28. See *Izvestiia Akademii nauk* no. 3 (1938), 259 ff. Stalin's was the speech cited in note 9.

29. N. M. Dubitskii, "Za Sovetskii uchebnik po teoreticheskoi fizike," *Sovetskaia nauka* no. 5 (1939): 131–149. Note the implicit acknowledgment that Timiriazev's textbook was inadequate to the task. (See note 33.)

30. Reasons for the eventual demise of prominent pre-revolutionary textbooks like those of Eikhenval'd, Gol'dgammer, and Mikhelson are detailed in chapter 13 of Hall, Purely Practical Revolutionaries.

31. The primary offender was O. D. Khvol'son, *Kurs fiziki*, 5 volumes (Gosizdat, 1923). A last 'supplementary volume' devoted to physics since 1914 appeared in 1926, with no pretense of integrating its subject matter with the previous volumes; cf. Khvol'son, *Kurs fiziki. Dopolnitel'nyi tom. Fizika 1914–1926 gg.* (Gosizdat, 1926). From its beginnings in 1897, Khvol'son's textbook went through a total of seven editions, the last appearing in 1933.

32. Narkompros RSFSR, *Nepreryvnaia proizvodstvennaia praktika v industrial'no-tekhnicheskikh VUZakh: Metodicheskie materialy* (Gosizdat, 1930); A. Iashin, "Laboratornyi metod," *Krasnoe studenchestvo* no. 23 (1931): 12; Brigadier Vaits, "Laboratorno-brigadnyi metod v Leningrade," *Krasnoe studenchestvo* no. 26–27 (1931): 18–21; Peter Konecny, *Builders and deserters: Students, state, and community in Leningrad, 1917–1941* (McGill University Press, 1999), 152–154.

33. A. K. Timiriazev, *Vvedenie v teoreticheskuiu fiziku* (GTTI, 1933).

34. For evidence of Timiriazev's hostility to the early efforts of Tamm and Frenkel (not to mention Dirac), see Timiriazev, "Volna idealizma v sovremennoi fizike na Zapade i u nas," *Pod znamenem marksizma* no. 5 (1933): 94–123.

35. V. A. Fock, "Za podlinno nauchnuiu sovetskuiu knigu," *Sotsialisticheskaia revoliutsiia i nauka* no. 3 (1934): 132–136. Igor Tamm later echoed Fock's concerns, appalled that Timiriazev's textbook served as a vehicle for propagating N. P. Kasterin's mathematically untenable representation of electrodynamics; cf. I. E. Tamm, "O rabotakh N. P. Kasterina po elektrodinamike i smezhnym voprosam," *Izvestiia Akademii nauk* no. 3 (1937): 437.

36. From 1928 to 1931 alone, the total publication volume of titles in the exact sciences nearly tripled, while the volume for technology and agriculture increased twice again as much. A. I. Nazarov, *Kniga v sovetskom obshchestve* (Nauka, 1964), 143. The first Five-Year Plan had originally called for a 257% increase in total output for the exact sciences. Cf. S. T. Sbitnikov and B. D. Udintsev, "Piatiletnii perspektivnyi plan pechati SSSR," *Na knizhnom fronte* no. 39–41 (1929): 26.

37. In the most prominent instance, ONTI was first abolished in 1934 and replaced by a Unified Scientific-Technical Publishing House, which then became the State Unified Scientific-Technical Publishing House in 1938, but was promptly split up late in 1939 in favor of a reformed ONTI, lasting this time until 1963. See Nazarov, *Kniga v sovetskom obshchestve*, 142.

38. "Zasedanie Redaktsionno-izdatel'skogo soveta AN SSSR," March 27, 1935, ARAN, f. 454, op. 2, d. 5, l. 11. Though S. I. Vavilov was a member of the committee, he was not present at this meeting.

39. "Ob izdanii fiziko-matematicheskoi literatury," an appeal from publishing official L. Grachev to Party Central Committee secretary A. A. Zhdanov, undated (received August 11, 1947), Gosudarstvennyi Arkhiv Rossiiskoi Federatsii, f. 4851, op. 1, d. 332, ll. 131–136.

40. In terms of institutions and personnel, the formal bureaucratic obstacles on the route to publication were considerably greater in the 1970s and the early 1980s than in the 1930s and the 1940s, even if the political atmosphere was much less hostile and arbitrary.

41. See, e.g., "Protokoly zasedaniia prezidiuma i plenuma fizicheskoi gruppy," May 23, 1936, ARAN, f. 437, op. 1, d. 30, l. 9, which includes an active discussion of the need for physics texts.

42. On earlier measures to institute standardized textbooks, see the Central Committee resolution "Ob uchebnikakh dlia nachal'noi i srednei shkoly" and the accompanying editorial "Dat' shkole khoroshii, stabil'nyi uchebnik!" *Pravda*, February 13, 1933; the joint Sovnarkom and Central Committee resolution, "Ob izdanii i prodazhe uchebnikov dlia nachal'noi, nepolnoi srednei i srednei shkoly," *Pravda*, August 8, 1935. An early article which explicitly condemns the lack of an introductory Soviet physics text at the college level is A. Pinkevich, "Zametki o vysshei shkole," *Izvestiia*, July 21, 1934.

43. "Materialy sessii Fizicheskoi gruppy Akademii Nauk SSSR po voprosam prepodavaniia fiziki vo VTUZakh," *Izvestiia Akademii Nauk* no. 1 (1937): 29.

44. It did not help matters that the KVSh leadership was arrested and charged with "wrecking" not long thereafter. The continued absence of an acceptable introductory physics textbook remained a source for editorial complaint—e.g., "Za khoroshii sovetskii uchebnik," *Tekhnicheskaia kniga* no. 6 (1940): 3–6.

45. On royalties as a motive for textbook writing, see Frenkel to A. Landé, January 28, 1929, folder 20, Nachlass Alfred Landé, Staatsbibliothek Preussischer Kulturbesitz zu Berlin. Writing physics textbooks was also the principle source of income for Andrei Sakharov's father; cf. *Memoirs*, tr. R. Lourie (Knopf, 1990), 13–14.

46. I. E. Tamm's *Fundamentals of the Theory of Electricity* should also be mentioned, since it was the first Soviet text on electrodynamics for a Russian audience, and helped set the mark for modern

pedagogical standards incorporating the full apparatus of Maxwell's theory: Tamm, *Osnovy teorii elektrichestva*, volume1 (Gosizdat, 1929). For details on Tamm's text, see chapter 13 of Hall, Purely Practical Revolutionaries.

47. A. I. Lur'e, "Vospominaniia," in *Ia. I. Frenkel': Vospominaniia, pis'ma, dokumenty*, second edition, V. Ia. Frenkel', comp. (Nauka, 1986), 57.

48. V. F. Frenkel', *Iakov Il'ich Frenkel'* (Nauka, 1966), 182–183, 146–147. Gosizdat initially turned down Frenkel's text, citing limited readership and the current paper shortage.

49. J. Frenkel, *Lehrbuch der Elektrodynamik*, I: *Allgemeine Mechanik der Elektrizität* (Springer-Verlag, 1926).

50. E. E. F. d'Albe, review of the first volume of *Lehrbuch der Elektrodynamik*, *Nature* 119 (1927), June 11: 851. Dispensing with historical developments also caught the attention of Pascual Jordan, who cited Frenkel's focus on the logical structure of the theory as one reason to study the new text despite the dominance of Föppl's and Abraham's superb texts in the German literature; see his review in *Die Naturwissenschaften* 15 (1927): 292–293. Compare also H. Backhaus's review in *Zeits. techn. Phys.* 10 (1929): 108.

51. "*Von der historischen Entwicklung der Elektrizitätslehre ist hier ganz abgesehen*," wrote Frenkel in the forward (italics in original); see also V. Frenkel', *Ia. I. Frenkel'*, 183–185.

52. For electrodynamics, Frenkel's inductively oriented Russian predecessors were: I. I. Borgman, *Osnovaniia ucheniia ob elektricheskikh i magnitnykh iavleniiakh*, third edition (St. Petersburg, 1914); A. A. Eikhenval'd, *Teoreticheskaia fizika*, part I: *Elektrichestvo* (Gosizdat, 1926); cf. A. Eichenwald, *Vorlesungen über Elektrizität* (Springer, 1928). In his study of French interwar physics, Dominique Pestre has documented a similar tendency toward historical, inductive, and exhaustive presentations of material in the leading textbooks of the late Third Republic. Dominique Pestre, *Physique et physiciens en France 1918–1940* (Éditions des Archives Contemporaines, 1984), chapter 2. Further individual English and German examples of the same tendency are cited below.

53. Cf. S. S. Schweber, *QED and the Men Who Made It* (Princeton University Press, 1994), 87.

54. Frenkel, "Zamechaniia k kvantovo-polevoi teorii materii," *Uspekhi fizicheskikh nauk* 42 (1950): 69–75; "Korpuskuliarnyi aspekt materii," *Uspekhi fizicheskikh nauk* 44 (1951): 110–116.

55. L. Landau and E. Lifshitz, *Teoriia polia* (GTTL, 1941). A much-revised second edition, issued in 1948, formed the basis for the first English edition, *Classical Theory of Fields*, tr. Morton Hamermesh (Addison-Wesley, 1951). Much of the original manuscript had been assembled when both men were still in Kharkov, and Lifshitz apparently continued work on it while Landau was in prison from April 1938 to April 1939. Lifshitz to Fock, May 12, 1939, ARAN StP, f. 1034, op. 1, d. 516, l. 12.

56. An early American example of this approach is Robert B. Leighton, *Principles of Modern Physics* (McGraw-Hill, 1959), which is more modest in its pedagogical aims than *Classical Theory of Fields*.

57. Landau and Lifshitz, *Teoriia polia*, 9.

58. These points are raised in L. Rosenfeld's review of *Classical Theory of Fields, Proc. Phys. Soc. London* 65A (1952): 567–568.

59. In his course of lectures in Munich, Arnold Sommerfeld offered the most famous exception to this rule, though these lectures did not appear in the form of a textbook until after World War II. An erstwhile student of Sommerfeld's, Julius Stratton at MIT, likewise invoked Maxwell from the beginning of his course, but did not take the additional unusual step with Sommerfeld of incorporating the special theory of relativity in his treatment, much less use it as the starting point after the fashion of the Soviet theorists. Arnold Sommerfeld, *Elektrodynamik* (Geest and Portig, 1949); J. A. Stratton, *Electromagnetic Theory* (McGraw-Hill, 1941). Though Stratton introduces the Lorentz transformation early in his text, he does not pretend to a thorough treatment of the special theory of relativity.

60. Another unusual organizational feature of *Field Theory* was the inclusion of geometrical optics and diffraction, which was justified by describing the former as the short-wavelength approximation of local plane waves, and the latter as an effect of deviations from that approximation.

61. Rosenfeld, review of *Classical Theory of Fields, Proc. Phys. Soc. London* 65A (1952): 567.

62. See David Kaiser, Making Theory: Producing Physics and Physicists in Postwar America (PhD dissertation, Harvard University, 2000), chapter 11. Kaiser's essential arguments, minus the exposition of Landau's role, may be found in his article "A ψ is just a ψ?: Pedagogy, practice, and the reconstitution of general relativity, 1942–1975," *Stud. Hist. Phil. Mod. Phys.* 29B (1998): 321–338.

63. M. A. Kuz'min, "'Osobye' pozitsii rabolepstvuiushchikh professorov," *Vestnik vysshei shkoly* no. 2 (1948): 4–5. Kuz'min worked at the Leningrad Polytechnic Institute, where Frenkel also taught; Frenkel's home institution was the Physico-Technical Institute just across the street.

64. J. G. Crowther to K. Sisam, October 6, 1935, LB 7866, OUP Archives. Enclosed with Crowther's letter was his translation of a letter from Laszlo Tisza, who wrote from Kharkov on Landau's behalf and indicated that Rudolf Peierls had already conveyed OUP's tentative interest in the series.

65. Sisam to Soviet Commercial Attaché in London, April 20, 1944, LB 9158, OUP Archives.

66. Fowler to Sisam, October 10, 1935, LB 7866, OUP Archives.

67. Fowler to Sisam, October 9, 1935, LB 7866, OUP Archives. By the time Fowler wrote again to Sisam the next day, he had been set straight by Peierls.

68. Crowther to Sisam, May 13, 1936, LB 7866, OUP Archives. Crowther was reporting a conversation with Peierls.

69. L. Landau and L. Piatigorskii, *Mekhanika* (GTTL, 1940).

70. Landau's subsequent falling out with Piatigorskii may also have influenced the decision. The reasons are discussed in my dissertation.

71. See the introduction to the English edition in L. Landau and E. Lifshitz, *Mechanics*, tr. J. B. Sykes and J. S. Bell (Pergamon, 1960), vii.

72. Landau and Piatigorskii, *Mekhanika*, 9, 10.

73. Herbert Goldstein, *Classical Mechanics* (Addison-Wesley, 1950); E. T. Whittaker, *A Treatise on the Analytical Dynamics of Particles and Rigid Bodies*, fourth edition (Cambridge University Press, 1937). Whittaker's text became available in a Russian edition in 1937.

74. In Whittaker's work more than 200 pages of text separate the presentation of Lagrange's equations of motion from Hamilton's canonical formulation, while the Poisson brackets are introduced as but one method among many to express contact transformations, without any mention of their utility in identifying constants of motion for physical systems.

75. Landau's growing concern with high-energy physics probably motivated him to expand sections on particle scattering into a separate chapter in later editions. Compare chapter 3 of *Mekhanika* (1940) and chapter 4 of *Mechanics* (1960).

76. A. K. Kikoin, "Kak ia prepodaval v Khar'kovskom universitete," *Vospominaniia o L.D. Landau*, 161.

77. See the paean to axiomatics in V. K. Frederiks and A. A. Fridman, *Osnovy teorii otnositel'nosti* (Academia, 1924), as reprinted in *Einshteinovskii sbornik 1984–1985* (Nauka, 1988), 105. Evidence of Soviet philosophers' hostility to axiomatic tendencies may be found in the review of Frederiks and Friedmann's book by I. Orlov, *Pod znamenem marksizma* no. 7 (1925): 232–234. By way of contrast to Landau's approach to mechanics, cf. G. Hamel, "Die Axiome der Mechanik," in volume 5 of *Handbuch der Physik*, ed. R. Grammel (Springer, 1927), 1–42.

78. With Hamilton's principle the total time is held constant, while the principle of least action takes the total energy as a constant. Strictly speaking, any dynamical system possessing an integral of the energy can be reduced to a lower-order system, and the two principles are then identical for such reduced systems; cf. Whittaker, *Analytical Dynamics*, 246–248.

79. Much later in the text, Landau briefly treats the dissipative function (§34, in the chapter on small oscillations) and rigid bodies in contact (§49). In the second edition the chapter on small oscillations is condensed even further to make way for the newly separate preceding chapter on particle collisions, and the dissipative function is no longer treated.

80. L. Rosenfeld, review of *Classical Theory of Fields*, *Proc. Phys. Soc. London* 65A (1952): 567.

81. Clemens Schaefer, *Mechanik*, 3d ed., volume 1 of *Einführung in die theoretische Physik* (Walter de Gruyter, 1929); William MacMillan, *Statics and the Dynamics of a Particle*, volume 1 of *Theoretical Mechanics* (McGraw-Hill, 1927); Arnold Sommerfeld, *Mechanik*, fourth edition (W. Klemm, 1948); J. C. Slater and N. H. Frank, *Mechanics* (McGraw-Hill, 1947). In the interest of streamlined pedagogy, Sommerfeld and Slater and Frank quickly introduce Lagrange's equations in their texts. By contrast, Schaefer requires more than 200 pages to reach Lagrange, while the painfully thorough MacMillan labors for nearly 350 pages before doing so.

82. V. Fock, review of Landau and Piatigorskii's *Mekhanika*, *Uspekhi fizicheskikh nauk* 28 (1946): 377–383.

83. Frenkel's remark is noted in Vl. P. Vizgin, "Martovskaia sessiia," *Voprosy istorii estestvoznaniia i tekhniki* no. 3 (1991): 55 n. 15.

84. Iu. Kagan, "Davaite voz'mem integral. . ." in *Vospominaniia o L. D. Landau,* 137.

85. L. Landau and E. Lifshitz, *Mekhanika sploshnykh sred* (GTTL, 1944); *Fluid Mechanics* (Addison-Wesley, 1959); *Theory of Elasticity* (Addison-Wesley, 1959).

86. Cf. Ludwig Prandtl, *Essentials of Fluid Dynamics* (Hafner, 1952); Horace Lamb, *Hydrodynamics,* sixth edition (Cambridge University Press, 1932).

87. V. L. Ginzburg, review of *Mekhanika sploshnykh sred, Uspekhi fizicheskikh nauk* 28 (1946): 384–386. A similar appreciation of the disciplinary utility for the theorist of Landau and Lifshitz's *Fluid Dynamics* is expressed in M. J. Lighthill's review in *Proc. Phys. Soc. London* 76 (1960): 586–587.

88. L. Landau and E. Lifshitz, *Statisticheskaia fizika* (GTTL, 1938).

89. D. Shoenberg, "Recollections of Landau," in I. M. Khalatnikov, ed., *Landau: The Physicist and the Man* (Pergamon, 1989), 224–226. For the English translation of *Statisticheskaia fizika,* cf. L. Landau and E. Lifshitz, *Statistical Physics,* tr. D. Shoenberg (Clarendon, 1938).

90. Later influential texts which employed Gibbs as their starting point acknowledged their debt to Landau and Lifshitz, e.g., C. Kittel, *Elementary Statistical Physics* (Wiley, 1958), vi.

91. Landau and Lifshitz, *Statisticheskaia fizika,* 5–6. Soviet readers recognized the target of these remarks: Frenkel, author of a rival theory. For more on this agonistic professional discourse, see chapter 13 of Hall, Purely Practical Revolutionaries.

92. A. Nemilov, "Kakov dolzhen byt' sovetskii estestvenno-nauchnyi uchebnik dlia vysshei shkoly," *Kniga i proletarskaia revoliutsiia* no. 10 (1935): 85.

93. K. I. Bylinskii, "Zametki o iazyke uchebnikov vysshei shkoly," *Vestnik vysshei shkoly* no. 1 (1950): 54–60, esp. p. 57, where the collectively written introductory *Course of Physics* edited by N. D. Papaleksi is chided for bad writing of this sort.

94. Ia. A. Smorodinskii, "Po zakonam pamiati," in *Vospominaniia o L. D. Landau,* 218.

95. The contents of the text were rather more restricted than was implied by the title: cooperative phenomena like plasmas, dipole assemblies, or gravitational systems were not included. The quotation is from R. W. G., review of Landau and Lifshitz's *Statistical Physics, Nature* 142 (1938), October 8: 655–656. The same "rather strange feature" regarding the absence of experiment characterized the much-revised 1958 edition. Cf. G. S. R., review of *Statistical Physics, Proc. Phys. Soc. London* 75 (1960): 327.

96. Fowler to Sisam, April 24, 1937, LB 7866, OUP Archives. Fowler was commenting on a draft translation of the first six chapters.

97. R. H. Fowler, *Statistical Mechanics,* second edition (Cambridge University Press, 1936); Richard C. Tolman, *The Principles of Statistical Mechanics* (Oxford University Pres, 1938).

98. Landau and Lifshitz, *Statistical Physics,* second edition, tr. E. Peierls and R. Peierls (Addison-Wesley, 1958).

99. George E. Uhlenbeck, review of the second edition of *Statistical Physics, American Journal of Physics* 27 (1959), May: 372.

100. W. H., review of Landau and Lifshitz's *Statistical Mechanics, Science Progress* 34 (1939): 160.

101. This did not prevent later Soviet critics from faulting Landau and Lifshitz for slighting the good materialist Boltzmann in their preference for Gibbs. Iakov Terletskii also felt they had been sloppy in their treatment of the Second Law of Thermodynamics, permitting archaic notions of the heat death of the universe to slip in as a result. See Ia. P. Terletskii, "Ob odnoi iz knig akademika L. D. Landau i ego uchenikov," *Voprosy filosofii* no. 5 (1951): 190.

102. A. Sommerfeld, *Atombau und Spektrallinien, Wellenmechanischer Ergänzungsband* (F. Vieweg, 1929), v.

103. Max Jammer, *The Philosophy of Quantum Mechanics* (Wiley, 1974), 59.

104. Early Soviet texts generally stopped short of Born and Jordan's "systematic extension into a logically closed system" of Bohr's concepts; cf. *Elementare Quantenmechanik* (Berlin, 1930), vi. As elsewhere, the wave-mechanics approach proved more popular, especially in the textbooks written by Frenkel. For more detailed comparisons of Western and Soviet editions, see chapter 13 of Hall, Purely Practical Revolutionaries.

105. L. Landau and E. Lifshitz, *Kvantovaia mekhanika*, part I (GTTL, 1948), 9–10. English edition: *Quantum mechanics: Non-relativistic theory* (Addison-Wesley, 1958).

106. §1.1 is titled "Introduction" (Russian edition) rather than "The uncertainty principle" (English edition). Heisenberg and Schrödinger are not named until a footnote on p. 46 (Russian). In contrast to the English edition, Bohr's name is also absent from the first section of the Russian edition. Note also that cross-references to other volumes of the *Course* are sparser in the English edition, since the translations were not completed in the same order as the originals, and the *Course* had still not reached its final form. The Russian edition lacked any index at all (as compared with the skimpy one in the English edition). Curiously, Fritz London is either elided from the Russian edition (§86), or miscredited: Heitler and London is attributed to brother Heinz (§78).

107. Personal communication of L. P. Pitaevskii, August 1, 2002. My own reading of the second edition makes it difficult for me to see what additional controversy could have stemmed from the changes in the second edition, as opposed to what was already potentially controversial in the first edition.

108. A. I. Akhiezer, "Uchitel' i drug," 51–52.

109. As quoted in Anna Livanova, *Landau*, second edition (Znanie, 1983), 91. Livanova does not identify the student.

110. E. V. Shpol'skii, *Atomnaia fizika*, third edition, volume 1 (GTTL, 1950), 9. The first edition of Shpol'skii's textbook appeared the same year as Landau and Lifshitz's *Quantum Mechanics*, and Shpol'skii's implicit criticism is clearly addressed to them, though he readily acknowledged the difference in audience level for their respective texts.

111. Western readers usually regard the taint of the ideological as evidence of the ultimate "provincialism" of socialist realism, but many Soviet authors remained resolutely modernist (and thus well within Western "universalizing" rubrics) in their desire to transcend "local" discourses. To illustrate the loose parallel, consider Maxim Gorky's advice to Fedor Gladkov, author of the continually revised *Cement*, which Gorky thought suffered from stylistic deficiencies including ornamental ingredients, regionalisms, and "vulgarisms": "Your language [of the first—1924—version] will be difficult for the citizen of Pskov, of Viatka, for the inhabitants of the higher and middle Volga, to understand. Here too, like many other contemporary authors, you reduce in an artificial way the scope of influence of your book, of your creation." Quoted in Thomas Lahusen, "Socialist realism in search of its shores: Some historical remarks on the 'historically open aesthetic system of the truthful representation of life'," in Lahusen and Dobrenko, *Socialist Realism Without Shores*, 16.

112. L. Landau, review of S. Z. Belen'kii's *Lavinnye protsessy v kosmicheskikh luchakh*, *Sovetskaia kniga* no. 11 (1948): 24–25.

113. "Kniga o nauke stalinskoi epokhi," *Sovetskaia kniga* no. 2 (1950): 3–18; M. G. Veselov and P. P. Pavinskii, review of Landau and Lifshitz's *Teoriia polia*, second edition, ibid., 19–22. The Academy tribute was *Iosifu Vissarionovichu Stalinu Akademiia Nauk SSSR* (Izd. Akademii nauk, 1949).

114. Landau's 1947 Kremlin characterization is taken from a 1947 secret police report excerpted in Iu. I. Krivonosov, "Landau i Sakharov v 'razrabotkakh' KGB," *Voprosy istorii estestvoznaniia i tekhniki* 3 (1993): 123–132, on 126.

115. L. Landau, "Burzhuaziia i sovremennaia fizika," *Izvestiia*, November 23, 1935.

116. I. I. Artobolevskii, "Uchebnik i nauchnye shkoly," *Vestnik vysshei shkoly* no. 6 (1950): 22–24.

117. M. G. Iaroshevskii, "Logika razvitiia nauki i nauchnaia shkola," in S. R. Mikulinskii, et al., eds., *Shkoly v nauke* (Nauka, 1977), 7–97, on 77.

118. Iu. A. Khramov, *Nauchnye shkoly v fizike* (Kiev: Naukova dumka, 1987), canonizes Joffe, Rozhdestvenskii, Mandelstam, Vavilov, Landau, Tamm, and Kurchatov as creators of schools. Tamm was the only other pure theorist.

119. M. I. Kagan, "Shkola Landau. Chto ia o nei dumaiu . . ." *Priroda* no. 3 (1995): 76–90, and most of the reminiscences in *Vospominaniia o L. D. Landau*.

120. See the discussion in Kathryn M. Olesko, "Tacit knowledge and school formation," in *Research Schools: Historical Reappraisals*, volume 8 of *Osiris*, 2d. ser. (1993): 16–29.

121. Interview with V. L. Pokrovsky, January 19, 2002; A. A. Abrikosov, "O L. D. Landau," in *Vospominaniia o L. D. Landau*, 36; B. L. Ioffe, "Landau's theoretical minimum, Landau's seminar, ITEP in the beginning of the 1950's," http: //arxiv.org/abs/hep-ph/0204295 (also published by World Scientific).

122. Akhiezer, "Uchitel' i drug," in *Vospominaniia o L. D. Landau*, 53.

123. See Martin H. Krieger's *Doing Physics: How physicists take hold of the world* (Indiana University Press, 1992) on the notion of "toolkit."

124. A highly unsystematic survey of the Soviet *Journal of Experimental and Theoretical Physics* (1945–1960) yielded the following papers in short order. The list could no doubt be lengthened considerably, but is limited to early career papers not co-authored with Landau or Lifshitz. Since the relativistic quantum theory volume was not published in Landau's lifetime, his students working solely in particle physics go unrepresented here. Cf. A. A. Abrikosov, "Ob orientatsii elektronnogo momenta galidnogo atoma v molekulakh tipa ClCN i ClCH3," *ZhETF* 19 (1949): 853–854 (*Quantum mechanics*); R. G. Arkhipov and I. M. Khalatnikov, "Rasprostranenie zvuka cherez granitsu mezhdu dvumia sverkhtekuchimi fazami," *ZhETF* 33 (1957): 758–764 (*Mechanics of continuous media*); I. E. Dzyaloshinskii, "On the stability of the phase boundaries between normal and superconducting states," *JETP* 3 (1957): 980–981 (*Statistical physics*); Dzyaloshinskii, "Thermodynamic theory of 'weak' ferromagnetism in antiferromagnetic substances," *JETP* 5 (1957): 1259–1272 (*Statistical physics, Quantum mechanics*); L. P. Gor'kov, "Statsionarnaia konvektsiia v ploskom sloe zhidkosti vblizi kriticheskogo rezhima teploperedachi," *ZhETF* 33 (1957): 402–407 (*Mechanics of continuous media*); L. P. Pitaevskii, "K voprosu ob anomal'nom skin-effekte v infrakrasnoi oblasti," *ZhETF* 34 (1958): 942–946 (*Electrodynamics of continuous media* and *Field theory*).

125. Andrew Warwick shows a similar dynamic at work in the training for the Cambridge Tripos examinations; see his *Masters of Theory: Cambridge and the Rise of Mathematical Physics* (University of Chicago Press, 2003), chapter 1.

126. R. G. Arkhipov, "Neustoichivost' techeniia sverkhtekuchei plenki," *ZhETF* 33 (1957): 116–123. Cf. references 6 and 7 therein.

127. V. B. Berestetskii, E. M. Lifshitz, L. P. Pitaevskii, *Relativistic Quantum Theory* (Pergamon, 1971). Berestetskii had co-authored a 1953 textbook on quantum electrodynamics with Landau student A. I. Akhiezer. It served as the assigned text for the Theoretical Minimum in the interim.

128. Kagan, "Davaite voz'mem integral . . . ," 137.

129. More literally, "theorphysical approach to life." Abrikosov, "O L. D. Landau," 33. Many of the essays in this volume echo this phrase, and invoke Landau's strong "moral" views.

130. My invocation of the picaresque genre follows Sharon Traweek's discussion in *Beamtimes and Lifetimes: The World of High Energy Physicists* (Harvard University Press, 1988), 103–104.

131. George E. Uhlenbeck, review of second edition of *Statistical Physics*, *AJP* 27 (1959), May: 372.

132. L. Rosenfeld, review of *Classical Theory of Fields*, *Proc. Phys. Soc. London* 65A (1952): 568.

133. As recounted by Yuri Rumer, "Stranichki vospominanii o L. D. Landau," in the afterword to Landau and Rumer, *Chto takoe teoriia otnositel'nosti*, third edition (Sovetskaia Rossiia, 1975), 77.

134. ARAN, f. 596, op. 4, d. 4, l. 20. The young experimentalist and party member B. M. Vul so characterized history in response to remarks by Landau's close friend M. P. Bronshtein about the ambiguous role of historical arguments when searching for the ultimate fundamental laws of nature.

# 9 In the "Context of Pedagogy": Teaching Strategy and Theory Change in Quantum Chemistry

**Buhm Soon Park**

The context of pedagogy differs almost as much from the context of justification as it does from the context of discovery.
—Thomas Kuhn, 1977

Theory change in science is quintessentially a social phenomenon. Changes are made when the accumulation of theory choices made by individual scientists reaches a certain point. The question of theory choice is a complex one that has invited a number of historical investigations and philosophical discussions. It is generally accepted that a good theory should be accurate and consistent, have broad scope, be simple, and lead to plentiful new findings. On a par with these "standard," "shared" criteria, Thomas Kuhn placed "individual" criteria—idiosyncratic aspects dependent upon the scientist's personality, training, and intellectual milieu—as factors that influence theory choice. It is now well known that philosophers of science accused him of making theory choice a "matter for mob psychology." Kuhn also evoked "the context of pedagogy," a phrase he coined in responding to the philosophers.[1] Kuhn felt that his critics mistakenly dumped individual criteria into the context of discovery, the area of study that they thought was more amenable to psychological than philosophical analysis, and considered theory choice a problem to be dealt with only in the context of justification. They were misled, Kuhn argued, by science pedagogy or what he called "textbook science," which often obscures the origins of a new theory and distorts the relationship between the theory and its evidentiary support. Should there be anything to be distinguished from the context of justification, Kuhn suggested that it be the context of pedagogy, rather than the context of discovery. Here is an irony: while Kuhn was one of the early historians attentive to the constructive role of teaching and training in theory choice, he was not enthusiastic about studying pedagogy as an historical subject. Rather, he was concerned about the processes of scientific change reconstructed—and possibly distorted—for pedagogical conveniences.[2]

There is no doubt that textbook writers place more emphasis on pedagogical efficacy than historical accuracy in the narration of subject materials. Yet the analysis of pedagogical elements in a textbook may afford a valuable insight into a book's motivation and intention, its social function and intellectual setting, and, furthermore, the dynamics of normal science.[3] In this essay I explore the integral role played by pedagogy in theory change in one of the most notable events in the history of twentieth-century chemistry: the replacement of valence bond theory by molecular orbital theory. I focus in particular on pedagogical strategies developed by British quantum chemists Charles Coulson and Michael Dewar as part of their endeavor to communicate with organic chemists. I show that their teaching strategies were crucial to setting the pace of theory change. At the heart of their pedagogy, I have found, was the translation of the prized tools of valence bond theory—the concept of hybridization and the criterion of maximum overlapping—into the language of molecular orbital theory. This translation was the crucial step for visualizing molecular orbitals in an intuitively correct way to most chemists. I also examine the ways in which Coulson and Dewar dealt with the relationship between the two theories. It is in the context of pedagogy, I argue, that we can see the creative aspect of the transmission of knowledge, the process that may affect the course of theory change in science.

## Characteristics of Theory Change in Quantum Chemistry

The case of theory change in quantum chemistry is indeed complicated. First and foremost, it is debatable whether molecular orbital theory really "replaced" or "overthrew" valence bond theory. A brief sketch of the rivalry between the two theories helps to illustrate this quandary. Both theories were developed in the late 1920s a few years after the advent of quantum mechanics and both underwent further refinements and modifications in subsequent decades. A partisan split soon emerged, with Linus Pauling (California Institute of Technology) becoming the spokesman for valence bond theory, and Robert Mulliken (University of Chicago) and J. E. Lennard-Jones (Cambridge University) leading the molecular orbital camp. In the 1930s, no theory took a preeminent position, and chemists did not fault each other solely based on theoretical commitment. As John Slater (MIT) and J. H. Van Vleck (Harvard) reiterated, the valence bond and molecular orbital theories were two different approximation methods for solving the Schrödinger equation, that is, two different paths toward the real truth.[4]

To judge from the number of research papers and review articles published in the journals of their specialty, the balance among quantum chemists began to tip toward the molecular orbital theory in the early 1950s.[5] This trend continued into the next decade, and as one valence bond theorist conceded in 1968: "What are the possibilities of development and application of valence bond theory in the near future? The valence bond theory has not yet been so extensively applied as the molecular orbital theory. The use of ever bigger computers might remove some of the computational difficulties inherent in valence bond calculations and open up some fields of research that up to now have appeared to be out of practicable range."[6] Indeed, the molecular orbital theory's advantage in computation became more pronounced with the rapid development of computing technology beginning in the early 1950s.[7]

This computational edge, however, was not self-evidently of great interest to non-quantum chemists, whose main concern was not to achieve quantitative accuracy in computations but to interpret molecular properties and chemical reactions qualitatively. If we broaden our attention from those who "developed" quantum-chemical theories to those who "used" them as their research tools, we can see a different aspect of the rivalry between the two theories. The early reception of valence bond theory was quite noticeable in organic and inorganic chemistry, and it was not until the 1950s that the molecular orbital theory began to demonstrate its usefulness in both fields. John Platt, a former colleague of Mulliken at Chicago, made an apt observation in 1966:

> These two theories, the "valence bond theory" or "resonance theory" and the "molecular-orbital theory," are only different ways of applying to molecular structure the ideas and equations of quantum mechanics which were developed for atoms and electrons in the 1920s. But the contention between their adherents has divided the chemical world for a generation. The theories are supposed to be formally identical when all higher-order corrections are included, but in practice they are as different as night and day. . . . The knowledge of the [molecular orbital] theory is spreading, and for fundamental calculations it is now used far more widely than its rival, and simplified versions of it are being taught to college freshmen and even to high school students.[8]

In 1973, David Urch, a chemist at Queen Mary College in London, gave his thought on when the molecular orbital theory came to be a tool for organic and inorganic chemists:

> Little real progress was made in the application of molecular orbital theory until after the end of World War II and then the progress was made in organic rather than inorganic chemistry. Coulson and Dewar demonstrated that molecular orbital theory, even in its most simple form, was widely applicable to organic problems, mostly in aromatic systems, but also heterocyclic and aliphatic systems as well. The success of MO theory in the organic field eventually spurred its reapplication in inorganic chemistry. . . . Finally in the 1960s MO ideas have been applied to main group systems, rationalizing bond lengths and other structural problems in much the same way as was done a decade or more earlier in organic chemistry.[9]

In addition to these personal comments, there is ample evidence for the increasing popularity of molecular orbital theory in the 1950s within the chemical community as a whole. As one organic chemist counted, a "mere handful of papers (approximately 20) in the 1930s was followed by approximately 70 papers in the 1940s, whereas the decade of the fifties just completed has witnessed some 600 papers on this subject."[10] Convenient teaching tools also appeared, such as a "mnemonic device for molecular orbital energies," converting algebraic formulas into simple graphical representations.[11] In 1973, Robert Morrison and Robert Boyd of New York University, authors of a popular textbook in organic chemistry, showed their preference for molecular orbital theory:

> The structure of molecules is best understood through quantum mechanics. Exact quantum mechanical calculations are enormously complicated, and so various methods of approximation have been worked out to simplify the mathematics. The method that is often the most useful for the organic chemist is based on the concept of *molecular orbitals*: orbitals that are centered, not about individual nuclei, but about all the nuclei in the molecule. . . . We cannot learn here how to make quantum mechanical calculations, but we can see what the results of some of these calculations are, and learn a little about how to use them.[12]

Similarly, James Huheey of the University of Maryland proclaimed the supremacy of molecular orbital theory in his 1972 textbook on inorganic chemistry:

> Molecular orbital theory undoubtedly offers us the best current interpretation of the properties of coordination compounds. As we have seen, the valence bond theory in its usual form and the crystal field theory are but simplified versions of the molecular orbital theory in which certain aspects of the latter have been conveniently ignored. Often these omissions are necessary in order that the problem be tractable. It is quite legitimate to use either VB or CFT to predict the qualitative properties of complexes as long as one realizes the inherent limitations. With increasingly accurate wave functions and increased use of high-speed computers to carry out the tedious calculations, it is probable that the molecular orbital theory will assume a preeminent position in the theoretical interpretation of metal complexes.[13]

Thus two different pictures of theory change can be drawn. From the standpoint of quantum chemists, there was no reigning theory to be replaced or overthrown; the valence bond theory had been in contention for about two decades, and then it simply lost favor as the research front moved to problems that dealt with a great amount of computation, problems better handled by the molecular orbital theory. Few quantum chemists, if any, had to undergo a conversion process. Instead, the valence bond theory just failed to attract a new generation of quantum chemists eager to take full advantage of electronic digital computers. For non-quantum chemists, however, a shift had to be made. The valence bond theory was the leading research tool for about three decades, and the molecular orbital theory successfully challenged its rival first in the field of organic chemistry and then in inorganic chemistry. In the classroom, teachers explicitly stated that they would switch gears from the valence bond to the molecular orbital theory, and they adopted or rewrote textbooks introducing the molecular orbital theory favorably. The rivalry between the two theories, once compared to the boat race between Oxford and Cambridge Universities on the Thames,[14] was actually more intense among the general audience of chemistry than the small group of quantum chemists.

The sources of the shift from valence bond theory to molecular orbital theory were numerous and heterogeneous. For instance, there were a variety of contested research areas, ranging from the question of theoretical soundness to the explanation of molecular properties. Theorists were faced with such issues as electron correlation, ionic factors, and relativity corrections as part of their endeavor to improve accuracy in the calculation of bond length and binding energy of a molecule. They also sought to provide a systemic, coherent interpretation of molecular spectra. In contrast, organic chemists were primarily interested in explaining and predicting the stability of chemical compounds, the information needed to propose a reasonable reaction mechanism. The main concern for inorganic chemists was to understand the structure and stability of coordination compounds—molecules usually having transition metals at the core—and to explain their magnetic and spectroscopic properties.

Indeed there was no universal criterion of theory choice that led the way for theory change. Some valued accuracy, others scope and consistency. There was no single crucial experiment, either. Cyclobutadiene, $C_4H_4$, drew considerable attention, since the valence bond and molecular orbital theories offered different predictions about its stability and shape. As Stephen Brush shows, this test case was settled in favor of the molecular orbital theory by 1978.[15] By that time, however, a large number of organic chemists had already embraced this theory. As early as 1962, Andrew D. Liehr, an inorganic chemist at Bell Laboratories, pointed to a growing concern over the excited electronic states of molecules as the main reason for a "rapid fall from favor of the valence bond theory, and a consequent rise in esteem of the molecular orbital and [its variation] crystal field theories."[16] But he did not explain why chemists became interested in the excited states, and it is also unclear whether the growing interest in this research area was a cause or a result of the wider reception of molecular orbital theory.

It thus comes as no surprise that historians have looked beyond the criteria of scientific merits, instead paying particular attention to the role of individual scientists in the processes by which the two theories were received within the chemical community. For the early popularity of valence bond theory, due credit has been given to Pauling, whose shrewd teaching strategy for his chemical audience contrasted sharply with that of Mulliken and the other main architects of molecular orbital theory.[17] Similarly, recent study has recognized Coulson's contribution to the reception of molecular orbital theory. Kostas Gavroglu and Ana Simões see him as the individual who had "as much rhetorical and pedagogical skill as Pauling,"[18] and Brush fittingly describes Coulson's 1952 textbook *Valence* as the book that tried to do for the molecular orbital theory what Pauling's *The Nature of the Chemical Bond* (1939) had done for the valence bond theory.[19] Still unanswered is an important question: how could Coulson convince his chemical audience that molecular orbitals, basically non-localized within a molecule, might be the right tool for describing the chemical bond, which most chemists deemed to be localized between atoms?

### Coulson's Strategy: Visualization of Chemical Bonds with Molecular Orbitals

Charles Alfred Coulson (1910–1974) received his undergraduate and graduate education at Cambridge University, where he became a polymath. Coulson came to Trinity College in 1928 with an Entrance Scholarship in Mathematics. He learned pure and applied mathematics from such distinguished mathematicians as G. H. Hardy and E. Cunningham, and he also took lectures from renowned physicists, including J. J. Thomson, Ernest Rutherford, and Arthur Eddington. He proved his talent in mathematics and physics by winning First Class Honors in the Mathematical Tripos Parts I and II and in the Natural Science Tripos Physics Part II; but his first paper, published before his graduation in 1932, was about meteorology. For his PhD research, he was first attracted to the theoretical physicist Ralph H. Fowler, who had an infectious enthusiasm for quantum mechanics.[20] But he actually worked with John Lennard-Jones, a former student of Fowler's and the first to hold the Plummer Chair of Theoretical Chemistry, which had just been established in 1932. Four years later, Coulson received a doctoral degree for his study on the $H_3^+$ ion and methane,[21] but his intellectual curiosity was not confined to chemistry. After winning a Fellowship at Trinity College, he studied for two years the effects of radiation on bacterial cultures. Coulson continued to move freely across disciplines in his subsequent professional career.

In quantum chemistry, Coulson certainly transcended the rivalry between the valence bond and molecular orbital theories. Although he subscribed to the molecular orbital theory under the influence of Lennard-Jones and endeavored to promote it throughout his life, he also appreciated some merits of the valence bond theory. In his 1937 paper on the hydrogen molecule, for example, Coulson made a critical comparison: the molecular orbital treatment gave a better value for the internuclear distance and the vibration frequency of the molecule, but its energy value was half a volt worse than the valence bond theory's.[22] In other areas, such as the treatment of polar molecules and the consideration of the screening effect of one electron upon another,

Coulson found the molecular orbital theory gave better results. All things considered, he declared that the two theories were "almost equally valid."

Coulson was well aware that, despite the equivalence of the two theories in describing molecular properties, the valence bond theory might be more appealing to chemists. As the name denoted, the valence bond theory addressed questions directly related to valence, one of the fundamental concepts in chemistry: viz., why molecules form at all; why two hydrogen atoms form a stable compound (the hydrogen molecule), whereas two helium atoms do not form a corresponding molecule; why atoms form compounds in definite proportions; and why atoms in a molecule are arranged in a specific way in space. Traditional theories of valence assumed that there were various kinds of chemical bonds connecting atoms, and the valence bond theory kept alive this image of the chemical bond, "localized" between the atoms, in its quantum-mechanical treatment of molecular properties. Pauling, for instance, proposed that the energy of a system should be lowest when two atomic orbitals overlapped as much as possible. Dubbed the criterion of maximum overlapping, this idea allowed a visual understanding of the strength of chemical bonds. Pauling and Slater also independently developed the concept of hybridization with which they could explain why valence is directed in certain directions.[23] For water, which has a bent rather than a linear structure, they found a clue in two unpaired $2p$ orbitals of oxygen, which are shaped like a dumbbell and positioned perpendicular to each other: the combination of these $2p$ orbitals of oxygen and $1s$ orbitals of two hydrogen atoms leads to a bond angle of about 90°. Yet methane needed a different treatment, since the carbon atom in its normal state has only two unpaired $p$ orbitals, not enough to form four bonds with four hydrogen atoms. Slater and Pauling considered an excited state of carbon, where one of the $2s$ electrons jumps to the $2p$ state, thus having four orbitals (one $2s$ and three $2p$) available for the bond formation. But it was still difficult to explain why four bonds have a tetrahedral orientation. Therefore, instead of pure $2s$ and $2p$ orbitals, they employed linear combinations of them ($sp^3$ hybridization) that would give methane the tetrahedral structure. The criterion of maximum overlapping provided a theoretical ground for why this hybrid orbital would give a stronger bond than any other possible hybrids.

In the molecular orbital theory, however, these conceptual tools were not necessary. When the concept of hybridization appeared in 1931, Mulliken described it simply as another way of explaining the deformation of atomic orbitals in a molecule:

The molecules $CH_2$, $CH_3$, and $CH_4$ have respectively the same electron configurations as the oxygen, fluorine and neon atoms. Of course all the orbits [atomic orbitals] are more or less deformed, and the $p$ orbits become differentiated into not more than three sub-forms, as a result of the fact that parts of the positive charge are located in the protons instead of being concentrated in the central nucleus. Or probably, as Pauling and Slater have shown, all the outer orbits are so much modified that we no longer should distinguish 2*s* and 2*p* orbits, but may better think, in $CH_4$, in terms of four new 2-quantum orbit-types, each a sort of hybrid of 2*s* and 2*p*, with 2*p* predominating in the mixture. These new orbit-types are adapted to the probable tetrahedral symmetry of the molecule.[24]

Unlike Slater and Pauling, Mulliken did not characterize the directed valence by the central atom's wave functions projecting out in particular directions. He had quite a different picture of the chemical bond: instead of localized bonds between atoms, he envisioned itinerant electrons wandering from one atom to another in the given molecular structure. These electrons were classified as "bonding," "antibonding," or "nonbonding" according to the way they affected the energy of the system. Because of this departure from the traditional chemical picture of bonds, Mulliken called for the shift from the "ideology of chemistry" to the "ideology of electron configuration" in thinking about valence.[25] It was Van Vleck who first made molecular orbital calculations for the tetrahedral structure of $CH_4$ without using hybridization and the criterion of maximum overlapping.[26] In so doing, he showed that the molecular orbital procedure embodied the essential ideas of directed valence, and that the hybridization of atomic orbitals was not a necessary condition for tetrahedral valence bonds but rather an "artificial assumption."[27]

But Coulson maintained that the molecular orbital theory did not have to abandon a pictorial interpretation of chemical bonds. To prove his point, he first appropriated the conceptual tools developed in the valence bond theory, such as hybridization and maximum overlapping, and gave a mathematical justification for their use in the molecular orbital framework.[28] He then elaborated his view in a 1941 review article titled "Quantum theory of the chemical bond."[29] Coulson explained why he preferred the molecular orbital theory: "We confine ourselves here to . . . the molecular orbital approximation, not because it is the better (neither is satisfactory, and the existence of the two complementary approximations is an indication of our partial failure to solve the problem), but because it is easier to maintain one point of view consistently throughout this account." The consistency he had in mind was the use of "orbitals," which, though mathematically defined as one-electron wave functions, could be used for a rough picture of charge distribution in both atoms and molecules. It is particularly noteworthy that Coulson visualized the process of molecular formation by juxtaposing atomic and molecular orbitals as a sort of before-and-after illustration. He began with the example of $H_2$.[30] Figure 9.1 represents each electron's molecular orbital, $\Psi = \psi_a + \psi_b$, which is symmetrical about the line *AB*. It is also clear from probability functions that the electrons in this bond are closely localized in the region between *A* and *B* in a shape resembling a sausage. Hence Coulson called it a "sausage-type" orbital.[31]

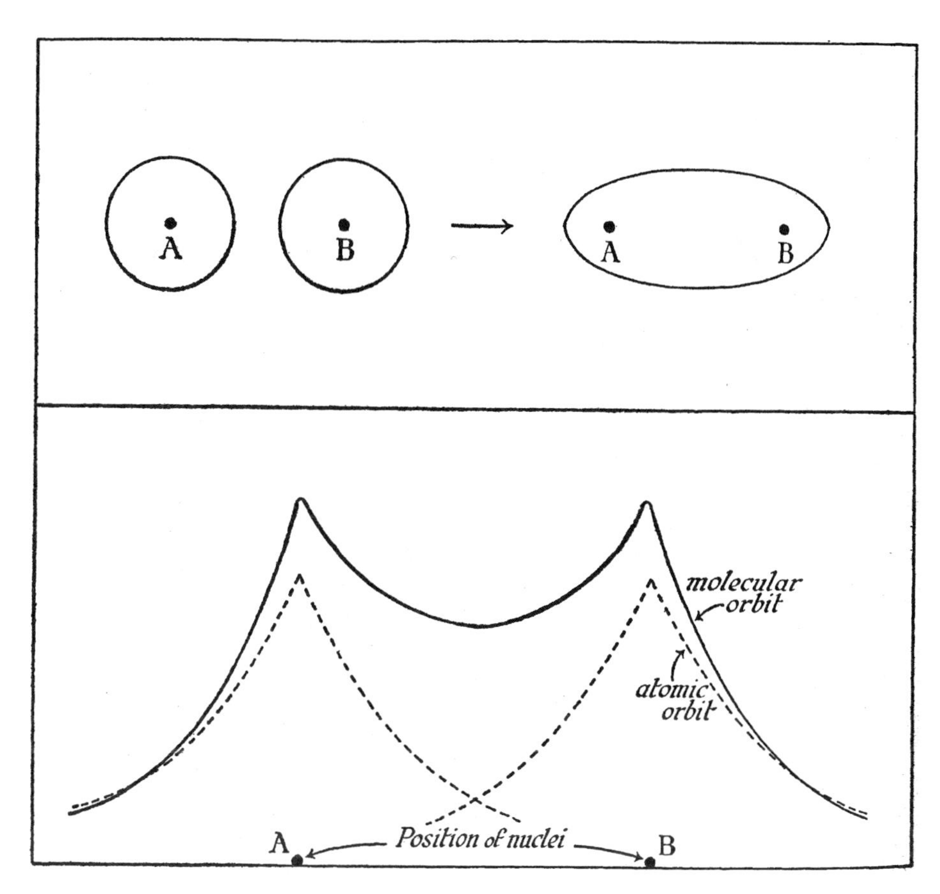


**Figure 9.1**
Coulson's visual explanation of the formation of $H_2$. Its molecular orbital is built from atomic orbitals (above), and the corresponding probability functions are given along the axis of $H_2$ (below). Source: Charles A. Coulson, "Quantum theory of the chemical bond," *Proceedings of the Royal Society of Edinburgh* 61 (1941): 115–139, on 120.

By the same token, Coulson described single bonds in polyatomic molecules as directed, localized, and electron-paired. The water molecule, for instance, has two "sausage" bonds, each of which is composed of oxygen's 2*p* orbital and hydrogen's 1*s* orbital. The angle between the bonds, about 105°, could be accounted for by the dumb-bell-shape 2*p* orbitals of oxygen and the repulsion between the two hydrogen atoms. (See figure 9.2.)

Coulson explained the pyramidal structure of ammonia in a similar way. For the tetrahedral structure of methane, however, he drew on the criterion of maximum

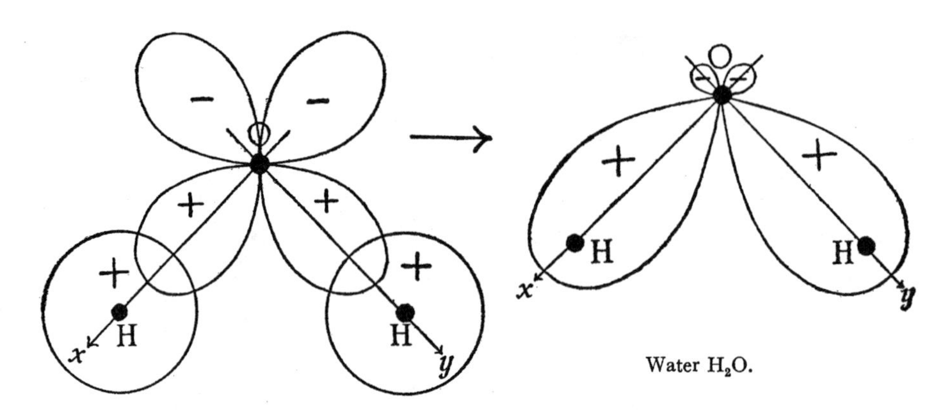


**Figure 9.2**
The formation of the molecular orbital in $H_2O$. Source: Charles A. Coulson, "Quantum theory of the chemical bond," *Proceedings of the Royal Society of Edinburgh* 61 (1941): 115–139, on 122.

overlapping and the hybridization concept. He explained: "If we suppose that the four unpaired electrons in Carbon are one each in the four [hybridized] tetrahedral orbits . . . , it can form four equivalent bonds, and the Criterion of Maximum Overlapping shows that these are directed to the four corners of a regular tetrahedron, making angles of 109°28' with each other. In this way the familiar facts of stereochemistry in organic compounds involving only single bonds receive a natural explanation."[32]

For unsaturated molecules, which have double or triple bonds, Coulson provided an ingenious visual explanation by taking ideas from both valence bond and molecular orbital theories. On the one hand, Pauling and Slater construed the double bond of ethylene ($H_2C = CH_2$) as two equivalent single bonds: carbon atoms are hybridized tetrahedrally and then linked together by the pairing of two sets of hybridized orbitals. This yielded a picture of two tetrahedrons sharing an edge (see figure 9.3); note that this picture was based on the $sp^3$ hybridization, not the $sp^2$ (trigonal) one.[33]

On the other hand, molecular orbital theorists offered two different versions of the double bond. Hückel interpreted it as composed of two different kinds of bonds, σ and π, without considering the ways in which the localized C-H bonds might affect the double bond. Mulliken, by contrast, considered that there was no localized bond and instead drew the picture of all valence electrons of ethylene moving around the whole nuclear framework. He classified types of molecular orbitals with group theory and estimated the relative order of the energies of various molecular orbitals by empirical rules, not by theoretical calculations.[34] Coulson basically followed Hückel's line of thought, but grafted the concept of hybridization onto it. For him, each carbon atom in ethyl-

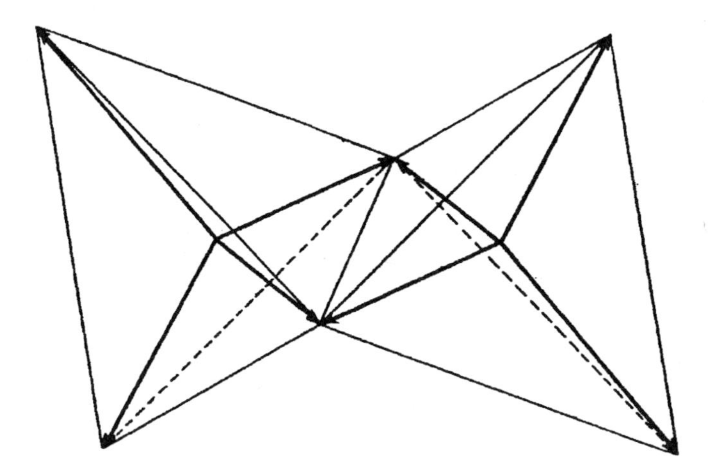

**Figure 9.3**
Pauling's 1931 schematic drawing of ethylene. Note that the double bond of this molecule is described here as composed of two equivalent single bonds, rather than two different kinds of bonds, σ and π. Source: Linus Pauling, "The nature of the chemical bond," *Journal of the American Chemical Society* 53 (1931): 1367–1400, on 1379.

ene is hybridized trigonally ($sp^2$ hybridization), and the hybridized orbitals make "sausage" bonds with orbitals from two hydrogen atoms and the other carbon atom. These are σ bonds in Hückel's terminology. Then dumbbell-shaped $2p_z$ orbitals from carbon atoms are paired together to form another kind of bond. Coulson said: "If we arrange the two $CH_2$ planes so that they are coincident, the two dumb-bells are pointing parallel, and this is the direction in which they overlap most."[35] He called the resulting molecular orbital "double-streamers," which is the "π bond" according to Hückel. Coulson then went on to interpret the triple bond in a similar way using "sausage bonds" and "double-streamers." (See figure 9.4.)

So far, Coulson's visualization of chemical bonds kept the idea of localized bonds intact. The greatest success of his exposition, however, came with his departure from this idea in dealing with the problem of benzene. In fact, the unique structure and reactivity of benzene had long defied the application of the three-way classification of bonds—single, double, and triple bonds—as evidenced by Kekulé's oscillation model in the late nineteenth century. Pauling and his collaborator George Wheland overcame this difficulty by explaining that five canonical structures "resonate" with one another in the valence bond treatment. Yet the concept of resonance, although convenient, caused much confusion and misunderstanding amongst chemists. Hückel also developed a molecular orbital treatment of benzene, but there were very few chemists who

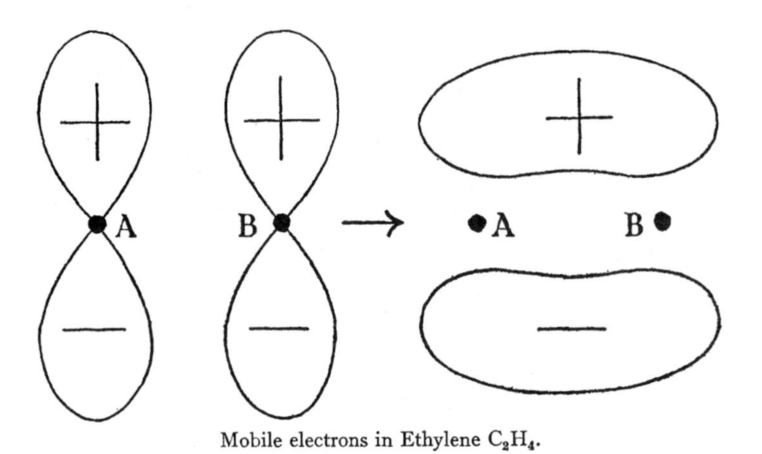


**Figure 9.4**
The formation of "double-streamers" in ethylene. Source: Charles A. Coulson, "Quantum theory of the chemical bond," *Proceedings of the Royal Society of Edinburgh* 61 (1941): 115–139, on 123.

could follow his sophisticated mathematical array. Coulson revived Hückel's method and explained it with a plethora of visual imagery. According to Coulson, the benzene bond is formed in three stages: first, each carbon atom is hybridized into the trigonal state, forming "sausage" bonds with adjacent carbon atoms and hydrogen; second, all the remaining six unpaired $2p_z$ electrons are located in parallel, perpendicular to the plane of the hexagon; and third, when these electrons are allowed to interact, they become free to migrate from one nucleus to another. Hence they are "non-localized," "mobile" electrons.[36] Coulson compared this with the double bond: "The streamer bonds which we saw were typical of Ethylene are now spread out and we have two streamers going right round the molecule, above and below the central plane." (See figure 9.5.)

Coulson's picture of benzene, later known as the "doughnut" model, was typical of his approach to applying quantum mechanics to chemical problems. He firmly believed that this application should serve chemistry in a language chemists could understand. To make the molecular orbital theory such a language, he did not hesitate to adopt concepts like hybridization and maximum overlapping. Unlike Hückel, he had no qualms about using the concept of resonance energy, defined as the difference between the total energy of the system and the energy of one of its canonical structures. Years later, he renamed this term "delocalization energy," explaining that the delocalization of electrons was responsible for the increase of binding energy.[37] Coulson was not among the creators of the molecular orbital theory, but he made it a

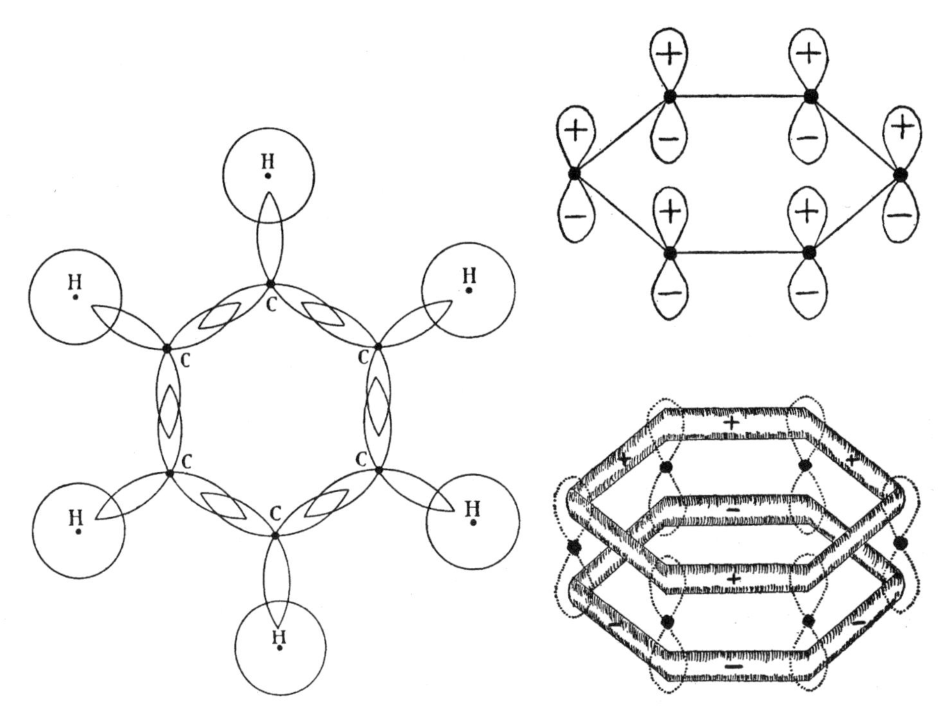


**Figure 9.5**
The formation of chemical bonds in benzene: the construction of the hexagon plane with σ or "sausage" bonds (left); the array of six $2p_z$ orbitals, when they did not interact (upper right), and when they interacted (lower right). Source: Charles A. Coulson, "Quantum theory of the chemical bond," *Proceedings of the Royal Society of Edinburgh* 61 (1941): 115–139, on 128.

chemical language by dexterous translation of the conceptual tools developed in the rival theory.

## The Concept of Hybridization: Coulson versus Pauling

During World War II, Coulson taught many courses at University College, Dundee with such a great passion that his colleagues thought he would burn himself out. Coulson's first book, *Waves* (1941), came out at that time.[38] He was also able to produce dozens of research papers on molecular problems, such as the momentum distribution of electrons, the contour integration method, and two-center integrals occurring in the theory of molecular structure, before he moved to Oxford in 1945 as a lecturer in mathematics. There, Coulson associated himself closely with the chemistry department. He had a room in Cyril Hinshelwood's Physical Chemistry Laboratory.[39]

In 1947, Coulson wrote another important review article on the molecular orbital theory for a new journal, *Quarterly Reviews*. Here again, Coulson stressed the advantage of pictorial interpretations by this theory:

> Like all problems in theoretical chemistry, the detailed working out of any part of the theory requires a certain familiarity with mathematical methods. In this case the essential technique is that of solving the Schrödinger wave equation under certain given conditions; but the fundamental ideas behind the theory are easily understood by experimental workers with little or no mathematical experience. For that reason we have avoided all mathematics, and have insisted everywhere upon the pictorial character of each step in the argument. *It is the visual character of these steps that makes the method of molecular orbitals so fundamentally simple*, but this situation must not blind us to the fact that in practically every case quantitative as well as qualitative predictions have been made.[40]

He also strongly advocated the use of *localized* molecular orbitals over *non-localized* ones whenever possible. "If localized M.O. can be used," he said, "it is better to do so, for they are vastly easier to imagine and handle than non-localized M.O.: and in addition they preserve the idea of a bond connecting two of the atoms in a polyatomic molecule." He even asserted: "If we do not use them for those simple cases where they can be used, we *do violence to the long chemical tradition dating from G. N. Lewis's famous 1916 paper on electron-pair bonds*."[41] Again, he invoked the criterion of maximum overlapping "to give a plausible justification for the use of localized M.O." and the concept of hybridization for the treatment of carbon compounds.

For Coulson, the valence bond theory, the localized molecular orbital theory, and the non-localized molecular orbital theory were three different languages for describing chemical bonds. It was possible to translate some ideas from the valence bond theory to the localized molecular orbital theory, but it was equally possible to transform the representation of localized molecular orbitals into that of non-localized ones. In his 1947 review article Coulson paid particular attention to the compatibility between the two modes of representation in the molecular orbital theory. While he promoted the use of localized molecular orbitals, he added that "if we wish, we may *translate* the whole problem into the *language* of non-localized M.O."[42] The treatment of methane illustrated this point of compatibility: one could use either localized molecular orbitals (following Coulson, who employed the concept of hybridization and the criterion of maximum overlapping) or non-localized molecular orbitals (following Mulliken, who used group theory). What was obviously lost in the latter was the pictorial understanding of chemical bonds. In addition, Coulson made "one important comment": "Group theory tells us what types of M.O. may occur, but it does not tell us the coefficients . . . that occur in them, or the relative order of the energies of various M.O. of different symmetry types. It is possible, however, to estimate these by various empirical rules, so that reliable descriptions of the ground state and the allowed transitions may be obtained qualitatively: a quantitative study is usually quite prohibitively complicated, as some recent work on diphenyl shows." As he acknowledged, the non-localized molecular orbital representation had an unquestionable advantage in dealing with the excited states of molecules, their transition intensities, and selection rules. "This language is essential," he also pointed out, "when we come to aromatic and conjugated compounds." But it is important to remember that even when he was compelled to use non-localized molecular orbitals for benzene, Coulson offered a picture of benzene bonds, drawing on hybridization and maximum overlapping rather than group theory.

Coulson did not stay at Oxford long, because he was appointed in 1947 to a newly created Chair of Theoretical Physics at King's College, London.[43] His research productivity took off during his London years, which saw about seventy papers written alone or jointly with his students. His second textbook, *Electricity* (1951), appeared at that time and was an immediate success. Elected a Fellow of the Royal Society in 1950, Coulson was at the peak of his career.[44]

In 1952 Coulson returned to Oxford University to assume the Rouse Ball Chair of Mathematics at the Mathematical Institute. By that time, he had already written a draft of his third book, *Valence* (1952).[45] This book is indeed a culmination of his efforts to bring quantum-mechanical interpretations of valence to chemists in a comprehensible way and to make a pedagogical push for the choice of molecular orbital theory. The preface clearly epitomizes this point:

> In the last twenty-five years the theory of valence has made enormous progress. To a large extent this has been due to the advent of wave mechanics. The result is that a situation has now been reached in which the education of a chemist is not complete unless he knows at least the main lines along which such progress has been achieved. This does not mean that every chemical student should be able to make his own theoretical calculations—this would be ridiculous, and will probably never happen. But it does mean that he should be sufficiently acquainted with the chief ideas and the essential tools that lie behind the modern theory of valence. The splendid and elegant elucidation of so large a part of chemistry which we have seen in the last two decades should not remain unknown to him. No longer ought he to be content with an electronic theory of valence couched in pre-wave-mechanical terms.[46]

Coulson stressed not only the importance of wave mechanics in understanding chemical bonds but also the possibility of employing quantum-mechanical ideas and tools without too much mathematics. "Practically no mathematics is needed for this purpose," he reiterated, "since almost everything necessary can be put in pictorial terms." Likewise, he believed that the role of the theoretical chemist was to serve chemistry as a chemist. "Contrary to what is sometimes supposed," he said, "the theoretical chemist is not a mathematician, thinking mathematically, but a chemist, thinking chemically.

That is why almost everything in this book should be understandable to a chemist with no mathematical attainments."[47]

Despite many similarities between Coulson's *Valence* and Pauling's *The Nature of the Chemical Bond*, published 13 years earlier, there is a remarkable difference in the two books regarding the parallel development of the molecular orbital and valence bond theories for the interpretation of chemical bonds. Whereas Pauling's book relied only on valence bond theory, excluding molecular orbital theory, Coulson's followed the strategy of treating both theories as equally as possible. In *Valence*, the two theories were described in separate chapters, closely compared in one chapter, and then used almost indiscriminately. Coulson tried to be fair to both of them. In fact, his strategy of taking an impartial stance was an effective way of promoting the molecular orbital theory without downgrading its rival. He paved a way for chemists to convert to the molecular orbital theory without negating what they had believed before.

Indeed, Coulson's review articles and book had just this effect. C. R. Noller of Stanford University said to American chemists in 1950:

> Simultaneously with the development of the valence-bond method of treating the electronic structure of molecules, the molecular orbital approach was being developed by Hund, Mulliken, and others. The mathematics of this method is no more understandable to the average organic chemist than that of the valence-bond concept, and since there was no explanation of the results in terms of the formulas with which he was familiar, the molecular orbital treatment remained unknown to him. Recently, however, a group of younger English chemists, and particularly C. A. Coulson, have interpreted the molecular orbital treatment in terms that appeal to many organic chemists. Although Coulson has written several excellent review articles on the subject, the writer believes that wider acquaintance of American teachers of organic chemistry with the molecular orbital approach is desirable.[48]

"Recently," John Roberts of MIT also noted in his review of *Valence*, "many practicing chemists have become interested in more precise predictions about chemical structures and reactivities than can be gained by the simple qualitative resonance treatment successfully popularized by Pauling and Wheland." Roberts continued: "Coulson has now provided a lucid qualitative account of the mathematics and procedures of quantum theories of valence with applications to organic and inorganic compounds as well as the structure of metals. The valence-bond and molecular-orbital treatments are discussed impartially and the advantages and disadvantages of each emphasized."[49]

In the *Journal of the American Chemical Society*, George Wheland, a strong proponent of the valence bond theory, also praised *Valence*: "When one of the acknowledged leaders in any field writes a book dealing specifically with that field, the reader can usually expect to find an authoritative and stimulating exposition of the subject. In this respect, *Valence* by C. A. Coulson measures up to all reasonable hopes."[50] Wheland valued

Coulson's lucid exposition and wise strategy of keeping the mathematics to a minimum, which allowed the fundamental ideas to be readily grasped by a non-expert in quantum chemistry. In particular, he lauded the side-by-side use of valence bond and molecular orbital theories, which made the treatment "the best rounded and most nearly complete one that is now available." The book was successful in showing that the relation between the two theories was "not conflicting, but complementary." Wheland highly recommended *Valence*, which he thought was more convincing and up-to-date than *The Nature of the Chemical Bond.*

Pauling reviewed *Valence* from a completely different standpoint.[51] Published in *Nature*, his review was more critical than any other. Pauling first questioned whether *Valence* was successful in presenting quantum mechanics to chemists in such a way as to be helpful rather than confusing. To him, the book gave only "sketch accounts" both of the mathematical methods of quantum mechanics and of the facts of structural chemistry. "It is my opinion," he wrote, that "a student [of chemistry] . . . needs to build up a solid and complete framework of one sort or another, without gaps. Although the author states that practically no mathematics is needed for his purpose, the book contains many quantum mechanical equations, some of which are obtained from earlier equations in the book by methods that are not presented to the reader." Pauling called this treatment "descriptive quantum mechanics" as opposed to "theoretical quantum mechanics," fearing that many would be confused by it.

Coulson's tacit endorsement of the molecular orbital theory, concealed under the even-handed treatment of the two theories, did not go unnoticed by Pauling. "The author has allowed his enthusiasm for the molecular orbital method to lead him to make some statements that probably cannot be supported," Pauling maintained. In particular, Pauling could not tolerate the ways in which the concept of hybridization—which he had invented—was used to undermine his valence bond or resonance treatment of some molecules. He used carbon monoxide as an example. For Pauling, this molecule exemplified how quantum mechanics helped solve the long-standing chemical question of whether it had a double bond or a triple bond structure. In *The Nature of the Chemical Bond,* he had explained that neither structure was right, since resonance occurred between them, or possibly among the three structures including the ionized single bond one.[52] These three structures seemed to contribute to the resonance with approximately equal weight, as evidenced by the negligibly small value of the electric dipole moment. It is important to note that Pauling explained the structure of carbon monoxide without using the concept of hybridization.

In contrast, hybridization was the key to Coulson's treatment. For Coulson, the bond in CO could be properly described as a triple bond, similar to that in acetylene: the carbon atom underwent an *sp* hybridization, having two hybrid orbitals ($s \pm p_x$). One of them ($s + p_x$) formed a $\sigma$ bond with the oxygen $2p_x$ orbital (possibly containing a small amount of $2s$); $2p_y$ and $2p_z$ orbitals of carbon formed respectively $\pi_y$ and $\pi_z$ bonds with those of oxygen. All these bonds had the polarity $C^+O^-$, but it was counterbalanced by the polarity of the lone-pair electrons occupying the other hybrid orbital ($s–p_x$). The asymmetry of charge distribution represented by the lone-pair electrons was a distinctive characteristic of the bond in CO, as compared with a normal triple bond. Coulson was convinced of his interpretation: "It is difficult to see how this electronic distribution could have been represented without the use of hybridized orbitals. Certainly early attempts which spoke of resonance . . . are far less satisfactory representations of the molecule than the more recent one reported here."[53]

Pauling's criticism was direct and unmitigated. He charged that "the molecular orbital treatment that [Coulson] presents involves the serious error" of neglecting the *sp* promotion energy. "The consideration of the energy of promotion of an electron from the $2s$ orbital to the $2p$ orbital," he wrote, "shows that, in fact, the unshared pair for each atom occupies a hybrid orbital that is predominantly $2s$ in character, with a small contribution of $2p_x$, whereas the $\sigma$ bond orbital is largely $2p_x$ in character, with a small contribution of $2s$."[54] To him, hybridization was not an important factor in this case, even if it ever occurred; and thus Coulson's molecular orbital interpretation of the CO bond was misleading.

Pauling's pungent book review provoked Coulson to write a letter in response right after its publication.[55] The purpose of the letter was neither to debate which theory was better nor to argue about what amount of mathematics was proper in a book like *Valence*. For Coulson, these were matters of opinion. What worried him most was the criticism about the use of hybridization. Coulson did not retreat from his standpoint that hybridization occurred more readily for the chlorine atom in the case of hydrogen chloride than Pauling had thought. Furthermore, Coulson adamantly defended his treatment of carbon monoxide, relating it to the general question about the possibility of hybridization in oxygen and nitrogen:

> But in the case of oxygen and nitrogen, I am more troubled. Either one of us is wrong, or we are meaning something different by hybridization. (The issue has little or nothing to do with molecular orbitals or valence-bond structures). You say that energy minimisation should give the hybridization ratio. But that is just the grounds on which I made my remarks. [John A.] Pople in the case of water . . . and many others in the case of nitrogen . . . have shown that there is considerable *sp* hybridisation for N and for O. If you care to calculate the electron affinities of atoms in various degrees of hybridisation, as [William] Moffitt did with great care, and as Mulliken has recently been doing, you find that more hybridisation would be expected than you have stated. And surely these calculations do in a very direct manner take account precisely of that excitation state energy which you say was neglected.[56]

Qualitative as it might appear, Coulson's use of hybridization was based on the recent energy calculations. For this reason, he considered the criticism of hybridization to be a matter of scientific truth—whether one was completely right or wrong. Otherwise, he and Pauling were in effect using a different meaning of hybridization.

Pauling wasted no time replying to Coulson's letter.[57] Not only did he uphold his previous arguments, but he took this opportunity to write an extensive criticism of *Valence*, that had not been included in his book review. He gave page-by-page comments on a typographical error, the sentences that might misrepresent his ideas, and the uses of concepts like electronegativity and atomic dipole moment. In addition, he expressed his disagreement with Coulson's treatment of ethylene, casting doubt upon the need to use the trigonal ($sp^2$) hybrid orbitals for the description of the carbon atom in this molecule. Furthermore, Pauling strongly complained that *Valence*, especially the chapter "Hybridization," did not properly mention his 1928 paper in which this concept had been announced for the first time. "It may be that when I speak of hybridization I am using a different meaning for the term than other people, as you say in your paper," he said. But he contended it was he who "discovered (or invented) hybridization, and was the first to publish the major part of the material contained [in that chapter]." He also expressed his pride and confidence in his 1931 paper which gave a detailed discussion of hybridization: "I have felt especially well pleased about this 1931 paper for twenty-one years now. I may say that I have greater confidence in some of my own simple calculations than in some of the very complicated calculations, involving a larger number of arbitrary decisions, that have been published."

Pauling's comment on the simple calculations versus the complicated calculations is worthy of a close examination. These calculations had to do with measuring bond strength, which he considered to be a crucial factor for determining types of hybridization. As a matter of fact, Pauling himself had not gone through the process of minimizing the energy as a function of the hybridization parameter. Instead, he had devised a simple, convenient way of calculating the "maximum bond strength" that orbitals could have. His concept of maximum bond strength was predicated upon the consideration of the angular part of orbitals, with the neglect of their radial part; and it gave only relative values compared to that of the pure $s$ orbital, which was set to have 1. According to this scheme, the $p$ orbital had the maximum bond strength of 1.732, the $sp$ hybrid orbital 1.993, the $sp^2$ hybrid orbital 1.991, and the $sp^3$ hybrid orbital 2.[58] It seemed to explain nicely why methane would prefer the tetrahedral $sp^3$ hybridization to any other kind. However, the theoretical soundness of the idea of maximum bond strength had been questioned almost since its conception, and Coulson insisted in *Valence* that it should be replaced by the overlap integral: "although this concept of orbital strength is undoubtedly a valuable one, it is less satisfactory for explaining detailed variations in bond energy than is the overlap integral described at the beginning of this section."[59] Coulson also spelled out its weaknesses: "In particular it reverses the order of CH bond energy in methane, ethylene, and acetylene, and it predicts the absurd result that the energy of a $\pi$-bond should be zero." In reply, Pauling did not defend his use of maximum bond strength as strongly as he attacked the (mis)appropriation of hybridization for the molecular orbital theory. Instead, he wrote:

> I may say that I feel strongly that the quantity which we have named the bond strength can be used as a means of discussing hybridization of orbitals with great reliability than approximate detailed calculations involving minimization of the energy. In other words, I feel that the overlap integrals or resonance integrals as evaluated with use of approximate wave functions are apt to be the source of larger errors in the conclusions than the simple postulate of proportionality of the energy to the bond strength.[60]

The debate between Pauling and Coulson did not go any further. Despite Pauling's wishes,[61] Coulson did not reply until the summer of 1953 when the steam of contention had already blown off.[62] Only the urgent need to make corrections for the second edition forced him to work carefully through Pauling's criticism. Coulson wrote to Pauling: "Apart from one or two matters, which are, as you say, still matters of opinion, I think that I can agree with most of your comments." He also agreed that "the 1931 paper was one of the best things you have ever done; and this is still true even if we discover, as time goes on, that a good many of the details require a certain amount of alteration." Under this carefully worded compliment, however, Coulson retained his points of disagreement with Pauling, especially on the question of hybridization. He continued to argue for the ease of hybridization when the atom was considered in its molecular environment, rather than in an isolated condition. "I can see," he wrote, "that there will be a stabilising of atomic configurations (within the molecular framework) where electrons are drawn into bonds by hybridisation of *s*, *p*, and *d* orbitals. This again suggests to me that hybridisation may be easier to achieve than we sometimes believe. I shall read your work, when it is published, with great interest, to see how far you have got with these points." No work on this matter came from Pauling afterwards.

Interestingly, Coulson's book made Mulliken feel uneasy, too. Mulliken complained that *Valence* gave the impression that the molecular orbital theory began with Lennard-Jones's 1929 paper.[63] He emphatically asserted that "the general concepts of MOs and of the building-up principle . . . were developed in 1927–8 by Hund and [himself]," and that these ideas were "foreshadowed by Birge, Mecke, Sponer, and [himself] in 1925–26 in studies of the electronic states of diatomic molecules, with pre-quantum mechanical attempts to understand these." For Mulliken, the fundamentals of the theory had already been discovered prior to Lennard-Jones's contribution, which was confined to the development of the LCAO approximation for the treatment of "*valence-shell* MOs," "approximately *calculating* their energies," and thus "leading among other things to a clear understanding of bonding and antibonding MOs." In addition, he pointed out that Coulson failed to mention Hund's early work on the representation of molecular orbitals as localized between atoms. Coulson gratefully admitted that he should have inevitably been tempted to overestimate his mentor's contributions, and particularly appreciated the comment on Hund's papers, which he had read "with enormous profit."[64]

In fact, Mulliken's discomfort with the historical survey revealed an unstated but important aim of *Valence*. Coulson sought to reorient the molecular orbital theory from spectroscopic concerns to chemical applications; from interpretations of band spectra to calculations of bond energies; and from the *aufbau* principle to the Schrödinger equation. In his alternative account of the history of molecular orbital theory, Mulliken ignored Hückel's work on ethylene and benzene, but Coulson took this work as a starting point to study conjugated and aromatic molecules. *Valence* was a landmark for chemical applications of the molecular orbital theory, not only easing the chemist's conversion to it, but also drawing the attention of its users to problems more relevant to chemistry.

### Dewar's Strategy: Polemical Rather Than Neutral

In Oxford, Coulson maintained close ties with chemists, formally teaching courses in the chemistry department, and informally offering his expertise whenever it was needed. However, he was neither a physical chemist nor an organic chemist: he did not work at the laboratory bench, producing thermochemical data or mixing chemicals. His tools were always pencil-and-paper, with a calculator on one side of the desk and books on the other. Perhaps he spent more time reasoning and dreaming with his mind than writing and punching with his hands. In his inaugural lecture, Coulson compared his position to the location of the building in which he worked: "When we awake from our dream we realize how excellent it is that here in Oxford the Mathematical Institute is placed in the Science Area, close, on the one hand, to the Radcliffe Library, where young men see visions and old men dream dreams, and on the other, to the laboratories where solid men and women make solid experiments."[65] In short, his perspective was theoretical, and his aim was to make theory useful for experimental work in chemistry. In this respect, Coulson had much in common with Pauling: he made the molecular orbital theory something useful, convenient, and comprehensible for chemists. However, Coulson did not enjoy blissful plaudits from leaders of the chemical community as

Pauling had received from N. V. Sigwick and C. K. Ingold in Britain and A. A. Noyes and G. N. Lewis in America.[66] While established chemists were lukewarm toward Coulson's campaign for the popularization of molecular orbital theory, some young chemists were enthusiastic.

Except for Coulson's own students, no one was more ardent in advocating and promoting the molecular orbital theory than Michael Dewar (1918–1997). Dewar was an organic chemist by inclination and by training. He considered himself "already sold on organic chemistry" when he began his undergraduate years at Balliol College of Oxford University in 1936, and despite his tutor's efforts to convert him to physical chemistry, he always preferred organic chemistry.[67] The war broke out as he was working on the structure of molecules from natural products in his senior year. Instead of being drafted, he was ordered to stay in Oxford to carry out research on war problems, such as finding better explosives and making new sulfa drugs, the study that eventually earned him a doctoral degree. He joined Robert Robinson's research group as a postdoctoral fellow to work on the penicillin project later in the war years. As part of this project, he studied the structure of stipitatic acid and its parent aromatic system, a seven-membered ring which he termed "tropolone."[68] Because he could draw two equally plausible structures of stipitatic acid, he assumed it would have a resonating structure, a concept from the valence bond theory. This problem, however, led him to investigate more broadly the theoretical ground of organic chemistry. He later recalled: "At that time, organic chemists, including me, knew little or nothing about molecular orbital (MO) theory in general and Hückel's contributions in particular. Indeed, even resonance theory was viewed askance in the D.P. [Dyson Perrins Laboratory]. However, because the war kept me away from experimental work other than my war problems, I became increasingly interested in theoretical organic chemistry as an outlet."[69] This is when Dewar read Coulson's 1941 review article.[70]

After the war, Dewar took a position at Courtaulds, Ltd., a company producing rayon and acetate fibers. While carrying out basic research there, he also wrote a monograph on theoretical organic chemistry for the Oxford University Press. Published in 1949, Dewar's *The Electronic Theory of Organic Chemistry* was not only the first general account of organic chemistry in terms of the molecular orbital theory; it also went one step further than Coulson had as the first outright attack on the valence bond theory or the resonance theory. In the preface Dewar asserted:

Previous writers in this field have approached chemistry from the standpoint of the resonance method and their general interpretation of chemical phenomena is now well known. *This approach is most unsuitable from the organic chemist's point of view since it involves a new symbolism and a novel and uncongenial outlook.* (A cursory examination of the literature shows how difficult it is for chemists to distinguish intuitively between resonance and tautomerism). For these reasons the molecular orbital method has much to offer. It provides a picture of molecular structure closely akin to that of classical organic chemistry; in both, charge migrations, residual affinity, and the like, appear in similar forms. Therefore if the older electronic symbolism is reinterpreted in terms of molecular orbital theory, it can still be used as a valid representation of chemical phenomena. The author has tried to follow this line of approach.[71]

Dewar criticized the valence bond theory, not because it was entirely wrong, but because it was confusing and inconvenient, questioning the very heart of the utility and ease that Pauling had cited as compensations for any theoretical weakness. Dewar saw the confusion between resonance and tautomerism as the main liability of valence bond theory. He pointed out that "in spite of all warnings, many organic chemists will continue to regard the various resonance forms in a hybrid as having some kind of real existence; that is, they regard the hybrid as a kind of tautomeric mixture of forms."[72] He disclosed further problems: "The attempt to correlate stability with the number of classical structures possible was also erroneous and misleading, and, moreover, the resonance theory does not emphasize sufficiently the difference in kind between σ- and π-orbitals, nor the special stereochemical relations of the latter." For Dewar, therefore, the resonance notation was "cumbersome" and tended to "obscure the general features of a molecule behind a fog of formulae." After all, it was a "most unsatisfactory rule-of-thumb substitute for a proper understanding of basic principles."

Dewar did not have Coulson's tact and even-handedness in dealing with controversial issues. Nor did he have Robinson's indisputable authority for handling the vast amount of literature on the study of the reaction mechanism. These flaws did not go unseen by American reviewers. Paul Bartlett classified Dewar's book as one of those written by "young enthusiasts with fresh viewpoints" rather than by "mature scholars bent on accuracy."[73] He found that the molecular orbital account of organic chemistry, though by no means revolutionary, was fresh enough to be stimulating. Yet he was not convinced by Dewar's criticisms of the valence bond theory: "The hope of better things . . . is not fulfilled in the book; for no general operational superiority is demonstrated for the rather nebulous notation of molecular orbitals as an instrument of correlation and prediction, in comparison to the well-developed qualitative resonance scheme. It is healthy, nevertheless, to have this substantial reminder that resonance is not a phenomenon of nature." Similarly, Saul Winstein saw the value of the book in stimulating further research, but also its weakness of "carelessness and even irresponsibility in preparation."[74] He thus could not agree with Robinson, who, in the foreword, enthusiastically recommended the book to students of organic chemistry. Winstein wrote: "The reviewer is somewhat worried as to its effect on the student. Dewar writes with confidence and finality on debatable questions and is inclined to utter generalizations and propose mechanisms for reactions which have not been put to any test." Dewar later recalled:

> The book in fact created quite a sensation. It was late, actually. It took three years to appear, because at that time, after the war, everything was a mess. But it had a whole lot of reviews, each worse than the last, and each time another review appeared the sales went up. It was read by a whole generation of young organic chemists, and it was the thing which really converted organic chemists to molecular orbital theory. It was the first account of organic chemistry in terms of molecular orbital theory.[75]

Dewar thrived on his "unorthodox" approach to contentious subjects.[76] In 1951, he moved to Queen Mary College at the University of London to assume the chair in chemistry. He was also invited to give the Reilly Lectures on the molecular orbital theory at Notre Dame University in the United States for six weeks. In this opportunity to reach an American audience, Dewar not only elaborated on his previous attempt to derive the qualitative relationships between molecular structure and properties, branding the resonance theory as the one based on intuitive arguments rather than rigorous reasoning, but he also explored a very simple way to calculate molecular quantities of chemical importance, such as energy and charge distributions. The source of inspiration was the recent papers by Coulson and H. Christopher Longuet-Higgins, who further advanced Hückel's method in a general way, applicable to hydrocarbons and heterosystems as well as unsaturated organic compounds.[77] Dewar found in their theory what he called "one vital missing link": the consideration of structural changes involving more than alterations in the electronegativities of the constituent atoms, that is, a case like the combination of two aromatic systems to form a larger one.[78] He regarded the formation of the bonds combining aromatic systems as a perturbation, and applied perturbation techniques to obtain corresponding changes in energies and electron distributions. This idea was the backbone of what he later termed the PMO (Perturbation Molecular Orbital) theory of organic chemistry.[79]

It is hard to measure the extent to which Dewar's controversial book and his papers full of mathematical treatments helped organic chemists to convert from the valence bond theory to the molecular orbital theory. But it is certain that, along with Coulson's review articles and *Valence*, they contributed to igniting some interest in this long-neglected theory. Personal as it may be, John Roberts's recollection succinctly captures the situation around 1950: "I enjoyed giving the course [in organic chemistry at MIT], and it loped along until the summer of 1951, when a big change took place. For several years, there was a growing consciousness of molecular orbital theory. Mulliken, Coulson, Wheland [with the resonance theory], and others were publishing what seemed like rather esoteric papers, some of which clearly were relevant to physical organic chemistry." He describes the widespread acceptance of the resonance theory and the reason for it:

Resonance theory was well entrenched, generally highly satisfactory in a qualitative way, even if you could not claim to have the faintest understanding of the quantum mechanics involved and had no real idea of what to make of a statement like "Compound X resonates between structures A and B." To be sure, there were books such as Eyring, Walter, and Kimball's *Quantum Chemistry* or Pauling and Wilson's *Introduction to Quantum Mechanics* that covered matters such as these, but precious little of it was either understood by or useful to organic chemists. Most of us were resigned to explaining resonance to students by some variation on Wheland's rather tortuous qualitative description of pendulums and similar analogs in *The Theory of Resonance* with the hope that, in the repetition, familiarity could mimic understanding.

I was uncomfortable with this approach but, even today, quantum mechanics is bound to be a bit uncomfortable if you require things to be intuitively obvious. Nonetheless, when one gets beyond the description of the general nature of resonance theory, there is little question of the enormous usefulness of resonance theory in correlating, predicting, and providing a kind of understanding of an immense range of organic phenomena. Its use in such connections was very ably documented by Wheland, a consummate scholar, in his 1944 and later books.[80]

In Roberts's opinion, a breakthrough for popularization of the molecular orbital theory came in 1947 with Coulson's article in the *Quarterly Review*, which "made it easier to understand how to formulate atomic orbital models like those of cyclopropane and carbocations." "The winds of change had started to blow," he felt:

In 1950 I got a copy of Michael Dewar's *The Electronic Theory of Organic Chemistry* and was tantalized by the claims, in typical Dewarian style, that resonance was, in effect, old hat and that we should hitch our chariots to molecular orbital theory. . . . I listened to the siren song and, inflamed, told my class in the first lecture of the usual Fall semester that this year was going to be different—resonance would be abandoned and the course would be taught *solely* on the basis of molecular orbital theory.[81]

Roberts became a leading American proponent of the molecular orbital theory, along with Andrew Streitwieser Jr., a chemist from the University of California at Berkeley, who had taken Roberts's lectures at MIT during his postdoctoral period. Both Roberts and Streitwieser used the theory for their research and taught it in class. Like Dewar, they explained chemical phenomena with the new framework of the molecular orbital theory, abandoning the old one of the valence bond theory. In 1961, Roberts wrote a monograph on molecular orbital calculations, which was designed to help organic chemists get over the "activation barrier" for the conversion. Streitwieser's comprehensive textbook, *Molecular Orbital Theory for Organic Chemists*, was published in the same year.[82] Dewar himself contributed to the popularization of molecular orbital theory in America by accepting an offer of a professorship from the University of Chicago in

1959, and, four years later, moving to the University of Texas at Austin as Robert A. Welch Research Professor.[83]

In Britain, H. Christopher Longuet-Higgins, Coulson's protégé, emerged as an important spokesman for the molecular orbital theory. He was also a rising star in theoretical chemistry. He took over the Chair of Theoretical Physics at King's College, London, the position having been vacated by Coulson's move to Oxford. In 1954 he was appointed a Professor of Theoretical Chemistry at Cambridge as a successor to Lennard-Jones. In the prestigious Tilden Lecture he delivered at the Chemical Society, he looked back on the early successes of both the valence bond and molecular orbital theories.[84] He then said: "It was understandable that this [molecular orbital] theory did not have such an immediate appeal as the resonance theory, since the latter used the language of 'structures' whereas the ideal of a molecular orbital has no obvious counterpart in classical chemistry. However, as I hope to show later, there are important questions in aromatic chemistry to which the molecular-orbital theory gives a simple and clear-cut answer whereas the answers provided by the resonance theory may be ambiguous or positively misleading." His point was that the molecular orbital theory was more than a restatement of the resonance theory in unfamiliar language: it occasionally led to different conclusions or to different interpretations of facts. Furthermore, he endeavored to alter a "widespread misconception" that this theory was mathematically demanding and practically useless to the organic chemist who did not have a computing machine. Just like his mentor, Longuet-Higgins reiterated that reasonably reliable results could be obtained only with pencil and paper. "Even a child can use [it]," he assured chemists of the great convenience of the molecular orbital theory.

## Conclusion: Pedagogy in Theory Change

By the 1970s, the molecular orbital theory had not only become an indispensable research tool for chemists, but it had also become firmly entrenched in chemical education. For Pauling, this was a lamentable situation. He commented: "it was a tragedy that the writers of elementary textbooks of chemistry decided to discuss the molecular orbital method, because the introduction of such a discussion in the teaching of elementary chemistry has served to confuse students in such a way that they are often got off to a far worse start than they were back before the modern ideas were developed."[85] He believed that only "one system for treating valence, valence bonds, and molecular structure should be used for the elementary student, in order that he build up a sound picture of molecules and the chemical bond and not be confused. The valence bond treatment is much simpler than the molecular orbital treatment, and it is also more powerful, so far as elementary, non-mathematical discussions are concerned." For this reason, Pauling argued that the molecular orbital theory ought to be "X-rated" until students reached a certain age. Pauling's defensive remark reflects a sea change in the chemical community. It also reveals the extent to which practitioners of quantum chemistry were concerned about pedagogical implications of their theories.

There was no one who single-handedly ushered in the age of the molecular orbital theory. It was a collective effort. Nevertheless, none were more important in making it understandable to chemists than Coulson and Dewar. Coulson gave a balanced account of the two theories and made such conceptual tools as hybridization and maximum overlapping a part of molecular orbital language. His teaching and writing were persuasive and sparkling. Roald Hoffmann, a 1980 Nobel laureate in chemistry, poetically describes the pedagogical work involved: "American and British chemists had secured a place for quantum mechanics in chemistry, through the charismatic exposition of Linus Pauling, the quieter and deep reflections of Robert Mulliken, and the elegant, perceptive teaching of Charles Coulson."[86] In contrast, Dewar undermined the very basis of the resonance approach with his uncompromising criticisms. While Coulson stressed the translatability and compatibility of the two theories, Dewar emphasized the practical convenience of molecular orbital theory. In contrasting yet complementary ways, they helped chemists to consider an alternative to the valence bond theory. The development of effective teaching strategies was essential to theory change in quantum chemistry.

## Notes

The following abbreviations are used: *RMP*: Robert S. Mulliken Papers, University of Chicago. *LPP*: Ava Helen and Linus Pauling Papers, Oregon State University, Corvallis.

1. Thomas Kuhn, "Objectivity, value judgment, and theory choice," in Kuhn, *The Essential Tension: Selected Studies in Scientific Tradition and Change* (University of Chicago Press, 1977), 320–339, on 327.

2. Similarly, John Hedley Brooke has noted Kuhn's double views on textbooks—the place where science is normalized and the place where the processes of scientific change are distorted. Brooke, "Introduction: The study of chemical textbooks," in *Communicating Chemistry: Textbooks and Their Audiences, 1789–1939*, ed. Anders Lundgren and Bernadette Bensaude-Vincent (Science History Publications, 2000), 1–18.

3. Historians of chemistry, among others, have been keen on analyzing the textbook to provide a new perspective on such major events as the disciplinary origins of chemistry and the chemical revolution. See Owen Hannaway, *The Chemists and the Word: The Didactic Origins of Chemistry* (Johns Hopkins University Press, 1975); Bernadette Bensaude-Vincent, "A view of the chemical revolution through contemporary textbooks: Lavoisier, Fourcroy, and Chaptal," *British Journal for*

*the History of Science* 23 (1990): 435–460; Mi Gyung Kim, "The analytical ideal of chemical elements: Robert Boyle and the French didactic tradition of chemistry," *Science in Context* 14 (2001): 361–395. See also the articles in Lundgren and Bensaude-Vincent, *Communicating Chemistry*.

4. John C. Slater, "Cohesion in monovalent metals," *Physical Review* 35 (1930): 509–529; idem, "Note on molecular structure," *Physical Review* 41 (1932): 255–257; John H. Van Vleck and Albert Sherman, "The quantum theory of valence," *Review of Modern Physics* 7 (1935): 167–228. Slater later elaborated his point of view in Slater, "Molecular orbital and Heitler-London methods," *Journal of Chemical Physics* 43 (1965): S11–S17. On the relationship between the two theories in the early years, see D. A. Bantz, "The structure of discovery: Evolution of structural account of chemical bonding," in *Scientific Discovery: Case Studies*, ed. T. Nickles (Reidel, 1983), 291–329; Kostas Gavroglu and Ana Simões, "The Americans, the Germans and the beginnings of quantum chemistry: The confluence of diverging traditions," *Historical Studies in the Physical and Biological Sciences* 25 (1994): 47–110.

5. According to Stephen Brush's counting of papers published in the *Journal of Chemical Physics*, the main outlet for quantum chemistry, the number for the molecular orbital theory far exceeded that for the valence bond theory in the early 1950s, for example, 22 to 7 in 1952 and 41 to 6 in 1953. Brush, "Dynamics of theory change in chemistry, Part 2: Benzene and molecular orbitals, 1945–1980," *Studies in History and Philosophy of Science* 30 (1999): 263–302. Review articles on the subject of quantum chemistry for the *Annual Review of Physical Chemistry* also indicate that there were more papers reviewed about the molecular orbital theory than the valence bond theory in the 1950s.

6. Massimo Simonetta, "Forty years of valence bond theory," in *Structural Chemistry and Molecular Biology*, ed. Alexander Rich and Normal Davison (Freeman, 1968), 769–782. This volume was dedicated to Linus Pauling by his students, colleagues, and friends on his 65th birthday in 1966.

7. On the impact of digital computers on quantum chemistry, see Buhm Soon Park, "The 'hyperbola of quantum chemistry': The changing practice and identity of a scientific discipline in the early years of electronic digital computers, 1945–1965," *Annals of Science* 60 (2003): 219–247. See also Ana Simões and Kostas Gavroglu, "Issues in the history of theoretical and quantum chemistry, 1927–1960," in *Chemical Sciences in the 20th Century: Bridging Boundaries*, ed. Carsten Reinhardt (Wiley-VCH, 2001), 51–74.

8. John R. Platt, "1966 Nobel laureate in chemistry: Robert S. Mulliken," *Science* 154 (1966): 745–747, as quoted in Brush, "Dynamics of theory change," 279.

9. David S. Urch, "Influence of wave mechanics on inorganic chemistry," in *Wave Mechanics: The First Fifty Years*, ed. William C. Price, Seymour S. Chissick, and Tom Ravensdale (Wiley, 1973), 160–173, quotation on 163.

10. Andrew Streitwieser Jr., *Molecular Orbital Theory for Organic Chemists* (Wiley, 1961), preface.

11. Arthur A. Frost and Boris Musulin, "A mnemonic device for molecular orbital energies," *Journal of Chemical Physics* 21 (1953): 572–573.

12. Robert T. Morrison and Robert N. Boyd, *Organic Chemistry*, third edition (Allyn and Bacon, 1973), 925.

13. James E. Huheey, *Inorganic Chemistry: Principles of Structure and Reactivity* (Harper and Row, 1972), 358.

14. Charles A. Coulson, "Recent developments in valence theory," *Pure and Applied Chemistry* 24 (1970): 257–287.

15. Brush, "Dynamics of theory change," 275–276.

16. Andrew D. Liehr, "A comparison of theories: Molecular orbital, valence bond, and ligand field," *Journal of Chemical Education* 39 (1962): 135–139.

17. See Robert Paradowski, The Structural Chemistry of Linus Pauling (PhD dissertation, University of Wisconsin, 1972); Thomas Hager, *Force of Nature: The Life of Linus Pauling* (Simon and Schuster, 1995); Buhm Soon Park, "Chemical translators: Pauling, Wheland and their strategies for teaching the theory of resonance," *British Journal for the History of Science* 32 (1999): 21–46.

18. Kostas Gavroglu and Ana Simões, "One face or many? The role of textbooks in building the new discipline of quantum chemistry," in Lundgren and Bensaude-Vincent, *Communicating Chemistry*, 415–449, quotation on 441. See also Ana Simões and Kostas Gavroglu, "Quantum chemistry qua applied mathematics. The contributions of Charles Alfred Coulson (1910–1974)," *Historical Studies in the Physical and Biological Sciences* 29 (1999): 363–406, esp. 397–401.

19. Brush, "Dynamics of theory change," 271–272.

20. S. L. Altmann and E. J. Bowen, "Charles Alfred Coulson," *Biographical Memoirs of the Fellows of the Royal Society* 20 (1974): 75–134, esp. 77–78. On his educational background and scientific career, see Simões and Gavroglu, "Coulson."

21. Charles A. Coulson, "The electronic structure of $H_3^+$," *Proceedings of the Cambridge Philosophical Society* 31 (1935): 244–259; idem, "The electronic structure of methane," *Transactions of the Faraday Society* 33 (1935): 388–398.

22. Charles A. Coulson, "The energy and screening constants of the hydrogen molecule," *Transactions of the Faraday Society* 33 (1937): 1479–1492. With his Swedish student Inger Fischer, Coulson later demonstrated that when both theories are further refined, their description of $H_2$ is interchangeable: Charles A. Coulson and Inger Fischer, "Notes on the molecular orbital treatment of the hydrogen molecule," *Philosophical Magazine* 40 (1949): 386–393.

23. Buhm Soon Park, "The contexts of simultaneous discovery: Slater, Pauling, and the origins of hybridisation," *Studies in History and Philosophy of Modern Physics* 31 (2000): 451–474.

24. Robert S. Mulliken, "Bonding power of electrons and theory of valence," *Chemical Reviews* 9 (1931): 347–388, quotation on 365.

25. Robert S. Mulliken, "Electronic structures of polyatomic molecules and valence, VI: On the method of molecular orbitals," *Journal of Chemical Physics* 3 (1935): 375–378.

26. John H. Van Vleck, "On the theory of the structure of $CH_4$ and related molecules, Part I," *Journal of Chemical Physics* 1 (1933): 177–182; idem, "On the theory of the structure of $CH_4$ and related molecules, Part II," *Journal of Chemical Physics* 1 (1933): 219–238; idem, "On the theory of the structure of $CH_4$ and related molecules, Part III," *Journal of Chemical Physics* 2 (1934): 20–30.

27. Van Vleck and Sherman, "Quantum theory of valence," 217–222.

28. Coulson justified the use of the criterion within the molecular orbital theory in "A note on the criterion of maximum overlapping of wave functions," *Proceedings of the Cambridge Philosophical Society* 33 (1937): 111–114.

29. Charles A. Coulson, "Quantum theory of the chemical bond," *Proceedings of the Royal Society of Edinburgh* 61 (1941): 115–139.

30. Ibid., 119–121.

31. Ibid., 121.

32. Ibid., 122.

33. Linus Pauling, "The nature of the chemical bond. Application of results obtained from the quantum mechanics and from a theory of paramagnetic susceptibility to the structure of molecules," *Journal of the American Chemical Society* 53 (1931): 1367–1400; John C. Slater, "Directed valence in polyatomic molecules," *Physical Review* 37 (1931): 481–489.

34. Buhm Soon Park, "A principle written in diagrams: The *Aufbau* principle for molecules and its visual representations," in *Tools and Modes of Representation in the Laboratory Sciences*, ed. Ursula Klein (Kluwer, 2001), 179–198; Gavroglu and Simões, "Beginnings of quantum chemistry."

35. Coulson, "Quantum theory of the chemical bond," 123.

36. Ibid., 127–129.

37. Charles A. Coulson, "Representation of simple molecules by molecular orbitals," *Quarterly Reviews* 1 (1947): 144–178, esp. 166.

38. Charles A. Coulson, *Waves: A Mathematical Approach to the Common Types of Wave Motion* (Oliver and Boyd, 1941).

39. For Coulson's scientific work during his years in Dundee and later in Oxford, see Simões and Gavroglu, "Coulson," 376–382; Altmann and Bowen, "Coulson," 80–82.

40. Coulson, "Representation," 144; emphasis added.

41. Ibid., 160–161; emphasis added.

42. For Coulson's comments on the compatibility of the two modes of representation in the molecular orbital theory, see ibid., 160–172.

43. Coulson to J. T. Randall, June 3, 1946, quoted in Altmann and Bowen, "Coulson," 82. Randall had the chair of physics at King's College.

44. For Coulson's London years, see ibid., 82–87; Simões and Gavroglu, "Coulson," 382–387.

45. Charles A. Coulson, *Valence* (Clarendon, 1952).

46. Ibid., preface.

47. Ibid.

48. C. R. Noller, "A physical picture of covalent bonding and resonance in organic chemistry," *Journal of Chemical Education* 27 (1950): 504–510, quotation on 504.

49. John D. Roberts, "Valence," *Chemical and Engineering News* 30 (1952): 5188–5189.

50. George W. Wheland, "Valence," *Journal of the American Chemical Society* 74 (1952): 5810.

51. Linus Pauling, "Valence," *Nature* 170 (1952): 384–385. Pauling's and Wheland's responses to *Valence* are examined in Simões and Gavroglu, "Coulson," 399–401. I detail the debate between Pauling and Coulson to show that they used the concept of hybridization in somewhat different ways.

52. Linus Pauling, *The Nature of the Chemical Bond* (Cornell University Press, 1939), 135–136.

53. Coulson, *Valence*, 211–212.

54. Pauling, "Valence," 385.

55. Coulson to Pauling, September 8, 1952, in *LPP*: 13.14. Pauling's book review was published in the week of September 6.

56. Ibid.

57. Pauling to Coulson, September 25, 1952, in *LPP*: 13.14.

58. Pauling, *Nature of the Chemical Bond*, 81–86.

59. Coulson, *Valence*, 201.

60. Pauling to Coulson, November 4, 1952, in *LPP*: 13.14.

61. Ibid. Pauling urged Coulson to reply to his letter, writing that "I shall look forward to hearing from you about the questions raised in my letter, in particular the question of the extent of hybridization of orbitals."

62. Coulson to Pauling, July 14, 1953, in *LPP*: 13.14.

63. Mulliken to Coulson, November 21, 1955, in *RMP*: 17.8. Mulliken described his complaint as "petty and trivial in any larger perspective," rather than scientifically important.

64. Coulson to Mulliken, December 21, 1955, in *RMP*: 17.8.

65. Coulson, *The Spirit of Applied Mathematics* (Clarendon, 1953), 22.

66. Park, "Chemical translators," 28–36; Kostas Gavroglu and Ana Simões, "Preparing the ground for quantum chemistry in Great Britain: The work of the physicist R. H. Fowler and the chemist N. V. Sidgwick," *British Journal for the History of Science* 35 (2002): 187–212.

67. For Dewar's educational background at Oxford, see Michael J. S. Dewar, *A Semiempirical Life* (American Chemical Society, 1992), 17–50.

68. Michael J. S. Dewar, "Structure of stipitatic acid," *Nature* 155 (1945): 50–51.

69. Dewar, *Semiempirical Life*, 41.

70. See Michael J. S. Dewar, "Mechanism of the benzidine and related rearrangements," *Nature* 156 (1945): 784, footnote 3.

71. Michael J. S. Dewar, *The Electronic Theory of Organic Chemistry* (Clarendon, 1949), ix–x; emphasis added.

72. Ibid., 17.

73. Paul D. Bartlett, "The electronic theory of organic chemistry," *Journal of the American Chemical Society* 71 (1949): 3859–3860.

74. Saul Winstein, "The electronic theory of organic chemistry," *Chemical and Engineering News* 27 (1949): 3440.

75. Michael J. S. Dewar, Oral history interview conducted by James J. Bohning (1991), Chemical Heritage Foundation, Philadelphia.

76. Dewar himself described his whole career as unorthodox. See Dewar, *Semiempirical Life*, 52.

77. Charles A. Coulson and H. Christopher Longuet-Higgins, "The electronic structure of conjugated systems, I: General theory," *Proceedings of the Royal Society of London* A191 (1947): 39–60. There were four more papers in this series: H. Christopher Longuet-Higgins, "Some studies in molecular orbital theory, I: Resonance structures and molecular orbitals in unsaturated hydrocarbons," *Journal of Chemical Physics* 18 (1950): 265–274. This was one of three papers under the same general title.

78. Michael J. S. Dewar, "A molecular orbital theory of organic chemistry, I: General principles," *Journal of the American Chemical Society* 74 (1952): 3341–3345. This was the first of the series of six papers under the same general title, which were published subsequently in the same volume. Despite Dewar's claim of the originality, the referees of his papers, including Mulliken and Wheland, wondered to what extent Dewar's work was new, and to what extent it was a review of the work of Coulson and Longuet-Higgins. See Mulliken to W. A. Noyes Jr. (editor of the *Journal of the American Chemical Society*), November 9, 1951; Wheland to Noyes, undated, in *RMP*: 18.15.

79. For Dewar's continuing enthusiasm for the perturbation molecular orbital theory, see Michael J. S. Dewar, *The Molecular Orbital Theory of Organic Chemistry* (McGraw-Hill, 1969), esp. chapter 6; Michael J. S. Dewar and Ralph C. Dougherty, *The PMO Theory of Organic Chemistry* (Plenum, 1975).

80. John D. Roberts, *The Right Place at the Right Time* (American Chemical Society, 1990), 120–121.

81. Ibid., 122.

82. John D. Roberts, *Notes on Molecular Orbital Calculations* (W. A. Benjamin, 1961); Andrew Streitwieser Jr., *Molecular Orbital Theory for Organic Chemists* (Wiley, 1961).

83. Dewar, *Semiempirical Life*, 95–170.

84. H. C. Longuet-Higgins, "The role of the π-electron in aromatic chemistry," *Proceedings of the Chemical Society* (1957): 157–165.

85. Linus Pauling, "Prospects and retrospects in chemical education," *Journal of Chemical Education* 57 (1980): 38–40.

86. Roald Hoffmann, "Kenichi Fukui (1918–98)," *Nature* 391 (1998): 750.

# IV Generational Reproduction

# 10 The Foundations of a Canon: Kohlrausch's *Practical Physics*

**Kathryn M. Olesko**

## A Canon among Physics Manuals

No manual of practical physics has had a longer publishing record than Friedrich Kohlrausch's *Leitfaden der praktischen Physik*.[1] First printed in 1870 as an instructional manual of experimental techniques encased in a mere 121 pages, the book was an immediate success. Within ten years it went through four editions; by the eve of World War I the eleventh edition appeared, bringing the total number of copies in circulation to 42,000. The twenty-fourth edition (1996), titled *Praktische Physik: Zum Gebrauch für Unterricht, Forschung und Technik*, spanned three volumes and 2,300 pages. Still bearing Kohlrausch's name even though compiled by teams of specialists, *Praktische Physik* has proven remarkably adaptive to different national systems of instruction with widely varying student clienteles, to conceptual and technical advances in physics, and even to the changing industrial applications of physics, including micro- and nanomanufacturing.

Although physics instructors at all levels had been creating their own written and practical exercises for most of the nineteenth century (if not earlier), Kohlrausch's *Praktische Physik* was without a comparable *published* competitor until the end of the century. Nonetheless, Kohlrausch's manual outlived its nineteenth-century competitors, including Adolph Wüllner's *Lehrbuch der Experimentalphysik* (six editions, 1862–1907) and Eilhard Wiedemann's *Physikalisches Praktikum* (six editions, 1890–1924). To date it has even had wider circulation than other popular physics manuals of the twentieth century, including Wilhelm H. Westphal's *Physikalisches Praktikum: Eine Sammlung von Übungsaufgaben mit einer Einführung in die Grundlagen des physikalischen Messens* (13 editions, 1938–1971) or Waldemar Ilberg's *Physikalisches Praktikum* (12 editions, 1967–2001). Originally conceived for a local audience—beginning students in the mathematico-physical seminar at the University of Göttingen—Kohlrausch's manual became a part of the canon of texts used in physics instruction nationally and then worldwide, a pedagogical aid used by undergraduates, graduates, and professionals alike.[2]

Practical problems in physics—written, experimental, and oral—were not without precedent in the German states when Kohlrausch's manual first appeared at the dawn of Germany's industrializing Second Empire. Eighteenth-century manuals and textbooks in mathematics and applied fields like business arithmetic, architecture, military mapping, geometrical drawing, water management, and civil engineering often included written and practical exercises for training in basic and advanced mathematical and manual skills.[3] A state examination system inaugurated in the eighteenth century for civil servants—secondary school teachers, architects, surveyors, foresters, civil engineers, and other bureaucratic administrators—used written and oral practical problems as a means to guarantee quality control in the execution of state responsibilities, especially in matters concerning state finance, natural resources, public safety, and education at all levels.[4] Despite the strong theoretical orientation of the Prussian-led university reform at the beginning of the nineteenth century embodied in the ideals of *Wissenschaft* (pure scholarship) and *Bildung* (self-cultivation), a tradition of practical learning based on practice rather than theory persisted in university seminars, exercise sessions, nascent university laboratories, and even state certification examinations for university-trained teaching candidates in several, but not all, major disciplines.

By the 1830s written and manual exercises in physics instruction had taken shape in seminars and private professorial or university laboratories nearly everywhere in the German states, notably at Göttingen under Wilhelm Weber, at Königsberg under Franz Neumann, and at Berlin under Gustav Magnus, and then later at Heidelberg under Gustav Kirchhoff and at Munich under Wilhelm Beetz, to name a few. Characterized by local variations at first, practical exercises in physics eventually converged around common models as the century closed. Kohlrausch's *Praktische Physik* played a decisive role in that process of standardization. In this sense Kohlrausch's manual can be regarded as part of a "literary canon" in physics. It has continued successfully to list and provide widely accepted, community-endorsed, and reliable instructions for techniques that physicists (as well as other scientists) should know or may wish to know. A roster of other post-World War II canonical works in physics might also include Jackson's *Classical Electrodynamics*, Goldstein's *Classical Mechanics*, or Merzbacher's *Quantum Mechanics*, as well as other standard texts in the field.[5]

How can we understand the historical process of canon creation in physics? How does a textbook or manual become part of a canon? Is the historical analysis of canonical texts merely a matter of studying the *written material* of a text, or is something more involved? In identifying Kohlrausch's *Praktische Physik* as part of the literary canon in physics, I wish to stress two characteristics of what *becoming and being a canon* means. First, when we go back to the root meaning of "canon" in classical Greek, we find that canon included both the idea of a *list* as well as the connotation of *rule*.[6] Hence the word "canon" applies to foundational texts for determining truth (as it did in ancient philosophy), but it also signifies norms of actions as well as measures for assessment (as it does in the New Testament, Gal. 6: 16). A canon is thus constituted of words, but it also invokes standards of behavior, whether mental or social.

How, though, can we understand the "canonical" status of a text beyond the words it contains? To address this question I propose a second characteristic of a canon: embedded in it are traces of the circumstances under which it took shape. Although we may instinctively be drawn to analyze the *dissemination* of a text to explain its canonical status, it seems necessary first to consider its *foundation* in actual practices. Words on a page are but the remains of the activities, mental and social, that led to them being there.

Accordingly this essay begins by analyzing the foundations of Kohlrausch's practical exercises before they were written up in manual format. From official seminar and institute reports, Kohlrausch's personal laboratory and reading notebooks, and most importantly, from his meticulous teaching records of the practical exercises he designed before the appearance of the first edition of *Praktische Physik*, we can reconstruct in detail the foundations of his manual. This essay then proceeds to extract from these results three elements of the local culture of physics at Göttingen and to consider these in conjunction with the student-assisted investigations that were a part of the process of creating practical exercises. These three elements are: cognitive and practical preferences, epistemological guidelines, and social norms. As the "rules" that accompany the "list," these elements are equally a part of the manual's foundations as are the actual exercises included. Although a text eventually only becomes canonical by virtue of its reception, the choices made in assembling it—viewed from the practices that preceded it—help us to recapture the cognitive, epistemological and social dimensions it was intended to disseminate, and so to understand more fully what others were likely to have seen in the manual's early editions.

## Student Learning and Pedagogical Innovation at Göttingen

Kohlrausch's first exposure to rigorous practical exercises in physics occurred while a student under Wilhelm Weber (1861–1863) and later as Weber's assistant in the Göttingen physical institute (1866–1870).[7] By the time Kohlrausch arrived at Göttingen, Weber had standardized the practical exercises that supplemented his teaching in the university's mathematico-physical seminar, established in 1850 ostensibly on the model of Franz Neumann's at the University of Königsberg. The three outstanding features of Weber's exercises were his focus on research techniques, especially in determining the reliability and sensitivity of an instrument; his concentration on the methods for measuring the earth's magnetism, derived from an international project he began with Carl Friedrich Gauss in 1833[8]; and his reliance on advanced students for pedagogical innovation, especially in handling introductory students. Although interested initially in a more inclusive approach to physical exercises that would accommodate students from all levels and a variety of disciplines,[9] after 1850 Weber settled into a pattern of relying on advanced students to support his own research and his teaching responsibilities in the seminar and practical exercises.[10] He required of older students that they teach the sessions on advanced issues in terrestrial magnetic measurements, while Weber himself led the practical exercises on bifilar systems, the determination of the moment of inertia of a vibrating magnetic bar, theories of instruments, determination of longitude and latitude, measuring absolute intensity and magnetic inclination and declination, and correct experimental results for constant and random errors (although advanced students eventually took over teaching many of these methods and techniques). Weber evaluated and separated students according to their abilities in terrestrial magnetic measurements, and even assigned advanced students the task of reducing the data from the Magnetic Association he and Gauss had founded—although finer data reduction was always secondary to more thorough instrument analysis in his exercises.[11]

The prominent role of advanced students in Weber's seminar and exercises not only prepared them for independent research projects, but also placed extraordinary responsibility on them for the proper teaching of beginners, and hence for pedagogical innovation. The results were mixed. Weber's own track record in shepherding student projects into print was restricted to a few notices on collaborative work done with postdoctoral students. A rare published report from 1858 reveals much about his pedagogical goals.[12] On the theme of the polarity of diamagnetic force, the investigation took as its point of departure what Weber considered to be John Tyndall's lack of understanding of his instrument in measuring diamagnetism; two postdoctoral students recalibrated the instrument's accuracy and reliability and from there tested the law of the proportionality of diamagnetic forces. In the report Weber also announced that one of the postdocs also undertook precise measurements of the temperature-dependency of the resistance of metals with results purportedly accurate to the ninth decimal place. While these postdoctoral students managed to find their name in print alongside Weber's, the results were *his,* not theirs.

So despite the impressive string of students between 1850 and 1870 who later became physicists, astronomers, or mathematicians—in addition to Kohlrausch, Bernhard Riemann, Eduard Riecke, Ernst Abbe, Ernst Julius Schering, Friedrich Eisenlohr, Wilhelm von Bezold, Karl Vondermühll, and Wilhelm August Nippoldt all took his seminar and practical exercises—Weber's research mentoring was based on subordination and the hierarchical delegation of authority, on a narrow set of issues in terrestrial magnetism, and on the assumption that technical specialization would result in the transfer of skills to other areas in physics. For some select few students his pedagogy worked, but in the end, his ideology was that of an older generation that ignored beginning students and that believed in the best rising to the top through innate "inclination and talent"—a position that Franz Neumann at Königsberg had explicitly rejected in the construction of his own well-known practical exercises and, more importantly, in the design of his introductory course on theoretical physics (an idiosyncratic representation of issues in classical mechanics) specifically for introductory students. The difference in pedagogical styles at the two locations extends further. Neumann gave students his own data to use freely; Weber took data from students to use in his own projects. Yet whereas Neumann had placed the entire burden of the certification of results on the individual, and endowed the ability to do so with virtue, Weber distributed the responsibility around, asking students to evaluate each other's data, creating a collaborative and cooperative atmosphere in which the certification of knowledge could only be viewed as a community effort based on discussion of results, whether or not those results were produced locally or arrived from elsewhere in published form.

To the degree, then, that beginning students in the Göttingen seminar and practical exercises were able to move to advanced levels at all, they were reliant upon the skills of advanced students to craft appropriate introductory exercises that could guide the passage from fundamental manipulations to research techniques. For the most part we can only guess how they fashioned their pedagogical styles. We know that Schering and Riecke taught the method of least squares, although the method never became as prominent in the practical exercises at Göttingen as it was at Königsberg, where the computation of random errors was regarded not only as a measure of the precision of results, but also of the skill and integrity of the investigator.[13] We also know that the task of turning recent literature in physics into topics for the practical exercises, which the new seminar statutes of 1861 identified as the *director's* responsibility,[14] was passed off to advanced students or assistants in charge of conducting elementary exercises.[15] The accommodation of beginners at Göttingen was thus dependent upon the pedagogical creativity of advanced students under Weber's supervision. Indeed, this was the only route through which pedagogical innovation occurred.

## Expanding the List of Practical Exercises

Yet there is much to suggest that Göttingen's introductory practical exercises were neither for beginners nor very systematic when Kohlrausch became Weber's assistant in 1866.[16] Although trained by advanced students, Kohlrausch devoted an extraordinary effort to trying to understand what type of problem *could be* molded into a suitable practical exercise. When Kohlrausch's responsibilities for the practical exercises expanded to include a separate elementary practicum for a service clientele—chemists, pharmacists, physiologists, medical students, and teaching candidates in mathematics and physics—the pedagogical challenge facing him intensified.

To understand how Kohlrausch accommodated beginners and a service clientele, we need to go back to his earliest extant notebook: a combined personal reading, seminar, and laboratory notebook from 1861–62, when he was an advanced student in the seminar but also directed the elementary astronomical observations next to Schering's introductory seminar exercises.[17] More an intellectual diary of occasional thoughts than a systematic recording of data, Kohlrausch's notebook shows him exploring a wide variety of issues but also intensely focused on a few; organized and systematic but also spontaneous and unsettled; critical of published scientific claims, but willing to turn even questionable conclusions into potential practical exercises. Only a large section on terrestrial magnetic data, at the beginning of the notebook, is systematically organized. Nearly every day he was at the laboratory, recording data, commenting on his readings, and from time to time mentioning the weather ("18 January 1862. Saturday. Weather: clear, cold!"), criticizing his own data ("Again much too low!"), or checking himself ("Measure: does the needle hang straight down? Yes."). Issues that could become practical exercises started as questions inspired by recent literature, comments by Weber in the seminar, and probably also by Schering's practices with younger students. They included the following:

Can free water contain electricity?

If an elastic body changes its shape, does its volume also change? Experiment with glass filled with a liquid.

How does light pass in media of various refractive indices?

What is the influence of an electric current on a diamagnetic body?

What is the effect of heat on electromotive force?

What is the duration of charge in a Leyden jar?

What is the dependency of elasticity constants on temperature?

About a hundred suggested problems are found in the notebook; some were complicated enough to become the foundation of original investigations. The overwhelming majority, however, constituted criticisms or assessments of claims in the published literature. Sometimes after asking a question, Kohlrausch designed an experiment (or suggested that the original protocol be repeated), cited the relevant literature and the state of the problem, and recorded and evaluated his data. (He omitted his protocol.) Not all questions required numerical answers, but Kohlrausch seemed especially drawn to those that could be settled quantitatively.

One of the most pressing historical issues in the pedagogical history of any science is understanding how research results become exemplars for the classroom. Kohlrausch honed his own pedagogical skills on two recent tomes on electricity: Gustav Wiedemann's on galvanism and electromagnetism, and Peter Theophil Riess's on frictional electricity.[18] Both works were theoretical but were based on the compilation and assessment of a half century of experiments carefully chosen for the soundness of their results. Riess in particular found the selection of examples difficult because of the "extraordinary amount of material" and the "lack of a comprehensive theory by which experiments could be put in order."[19] To aid the sifting process, Wiedemann repeated critical experiments in galvanism, omitting from the volume "observations and theoretical considerations . . . which are recognized as incorrect . . . or which . . . have yielded only special results of little use."[20] From Wiedemann, Kohlrausch noted different views on the electrolysis of water (1: 405), electricity in salt solutions (1: 98), and electromotive force (1: 245–246). From Riess, Kohlrausch selected sections on electricity and specific heat (1: 147), the measurement of amounts of electricity with a test body (1: 117, 119, 133), transfer of electricity between two charged bodies (1: 209), the electrophor (1: 293), experimental proof of some of Riess's tables (1: 512), and changes in the torsional resting position of a hanging body (1: 84).

Take, for instance, Kohlrausch's criticism of Riess's claim that the distance **S** between two charged electric spheres was proportional to their density **D** (Riess, 1: 329, 377). Kohlrausch combined data from two different sections of Riess's analysis and showed that the verbal claim of proportionality did not follow from the actual numerical ratios if the distance of the charged spheres was anything but infinity, when the ratio was one. (See table 10.1.)

**Table 10.1**

Kohlrausch's computation of the ratio of the distance **S** and the density **D** of two charged electric spheres, using data from Riess's book on frictional electricity.

| S/D | 1 | 1.02 | 1.006 | 0.917 | 0.812 | 0.607 |
|---|---|---|---|---|---|---|

Kohlrausch explained:

> I believe the proportionality is probably true, but it doesn't follow from Riess's trials. . . . He himself then says: the deviation—is explainable, in part—through the restricted accuracy [of the measurement of the distance]—mainly through the influence of the moving sphere on the density of the first one. But with such unknowns in the calculation, which have an influence of 35%, one cannot claim to have discovered a law. Page 377 follows again with a series of trials with somewhat regular deviations from the calculated values, whereby the correctness of the law is supposed to be confirmed. Rather than "density" it might be better to set "differences in the density" of the two spheres as the variable in the law. The latter must, however, be determined in individual trials. This is the law that should be controlled through these measurements.[21]

A notebook from later in 1862 shows him following the same pattern of developing workable problems from criticisms of data and claims in the literature.[22]

Kohlrausch's penchant for precise numerical calculation being well known,[23] it comes as no surprise that his pedagogical proclivities should also move in this direction. Weber's terrestrial magnetic exercises suggested as much. Yet Kohlrausch's notebook shows that he was decidedly not following the example of Weber's terrestrial magnetic exercises, but was rather assuming responsibility to keep up on the literature, critique it, and from that, extract usable problems for students. In this first attempt to create problems, he focused primarily on electricity. By the time he returned to Göttingen as assistant in the physical institute in 1866, in charge of introductory exercises, a more complicated task awaited him: what were those exercises to be, and what purpose were they to serve?

### Adjusting to Student Clienteles

Both Kohlrausch's student clientele and the introductory exercises that he developed for 1867–68 indicate, though, that he did not significantly differentiate at first between training a service clientele and a more specialized one in mathematics and physics. With only 13 students enrolled in his physical practicum in that year, Kohlrausch developed a set of instrument-based exercises that included calibrations, comparisons of instruments, and determinations of the errors, reliability, and sensitivity of instruments. Specific exercises included measuring specific gravity; comparing the Daniell and Regnault hygrometers; calibrating micrometers and thermometers; determining electrical resistance units by different methods, including Siemens's; learning how to use the spectroscope and determining Fraunhofer lines; as well as gaining skill with the theodolite, calorimeter, galvanometer and Wheatstone's bridge. But in contrast to Neumann's exercises,[24] which demanded immersion in one issue or at most a few for the entire semester, examining every possible error, contingency, and uncertainty,

Kohlrausch's exercises gave students the opportunity to become versatile in a wide variety of standard basic operations using a small set of customary instruments. There was no pretense here of performing research or even of striving to end up with something original. For most exercises he assigned to his students, Kohlrausch first did them himself and meticulously recorded his data—to which he later referred when evaluating the results reported by his students. He did the same for instrument calibrations, especially with thermometers.

What is not clear is whether or not in 1867–68, his second in the assistant's position, he had a plan for what sequence of operations any single student should work through. Instruments were limited, so each student, each week, had to work on a different exercise, although over the course of the semester nearly all students completed nearly all the same (or similar) exercises. Kohlrausch kept meticulous records of what each student had done. In the table reproduced here as table 10.2, exercises actually completed are underlined. These exercises are telling of how Kohlrausch designed a sequence of exercises for advanced students. Kohlrausch's notes indicate he had specific recommendations for how each of these exercises should be carried out, for his notebook ended with references to articles in Poggendorff's *Annalen der Physik und Chemie* that treated instruments, calibrations, the determination of constants, specific heats, and other topics contained in his exercises.[25]

In the following semester, the summer semester of 1868, Kohlrausch began to keep more detailed records of his students' assignments, as well as to prepare them more thoroughly for the exercises, even though he still had not instituted a fixed format for the sequence of the exercises. He now began to offer lectures on standardized protocols,

**Table 10.2**
Comparison of Aurel Voss's and Carl Pape's exercises in Kohlrausch's practicum, 1867–68.

| Voss | Pape |
|---|---|
| Which crystal angle? | Calibration of glass tubes? |
| Nichelson's aerometer. | Types of rotation? |
| Scale-aerometer. Freezing point. | Testing glass? |
| Microscope = magnification. ~~Resistance.~~ | ~~Specific gravity.~~ |
| Focal length. ~~Comparison of barometers.~~ | Scale-aerometer. ~~Focal point.~~ |
| Sextant. Calibration. Tubes. | Nicolson resistance |
| | Testing of a scale. Theorems of weighing. |
| | Sensitivity. Microscope = magnification |
| | Resistance. ~~Regnault's hygrometer.~~ |
| | Specific gravity in calibrated tubes. Glass. |
| | Volume = measurement. ~~Sextant.~~ |

and he also asked the more advanced students to make presentations for him. One such set of lectures and presentations included the following:

25 April 1868 Calibration of tube #1 [thermometer]; range of measurements
6 May 1868 Comparison of Dove's barometer with Weber's
18 May 1868 Moment of inertia of the magnetometer
21 May 1868 Wöhler's elements of the tangent magnetic compass
29 May 1868 (Saturday)
Afternoon: Magnetization of the six magnets; specific gravity of the magnet
*Eggers:* Coefficient of refraction of Fraunhofer lines
16 June 1868 *Albert, Hoogewerff*: on the volt meter
23 June 1868 *Nippoldt* on the bifilar magnetometer [includes notes from Nippoldt's protocol][26]

For the 13 students taking the practical exercises, Kohlrausch seemed to decide on their assignments as he went along. Not all students chose to participate to the same degree. Although he offered them the opportunity to use the laboratory twice a week for two hours each time, only about half did; most completed about 16 sessions over the course of the semester, or only about one two-hour session per week.

In 1868 exercise topics still varied, but small clusters of related ones were now apparent. One student spent two weeks on magnetic inclination, followed by one on the intensity of terrestrial magnetism. Even the determination of the length of the simple seconds pendulum took one or at most two weeks; at Königsberg the same problem would have taken the entire semester due to its pivotal role in Neumann's course on theoretical physics, where it served as a model for all measuring exercises. Indeed, the brevity of time spent on the seconds pendulum suggests that not only had Kohlrausch designed do-able exercises, but that he had either stripped them down to their most essential elements, or packaged them in such a way that some elements were already completed and only a few remained to be performed. The norm at Königsberg, spending several weeks on a single issue, was at Göttingen a sign either of an inability to master the fundamentals or of readiness to undertake an independent research project, as was the case with Eggers, who spent eight weeks on topics in magnetism. More common was a sequence like Carl Pape's, from the summer semester of 1868:

April 27 Testing of the Mendelsohn-Fortin theorem on weighing
May 2 Elasticity coefficient of the cathetometer
May 4 (same)
May 9 (same); microscope
May 11 conclusion; density of brass
May 16 Transport of the magnetometer

| | |
|---|---|
| May 23 | Moment of inertia |
| May 25 | Transport. of the magnetometer |
| June 13 | Torsion coefficient |
| June 20 | Spectral apparatus |
| June 22 | Absolute intensity of terrestrial magnetism |
| June 27 | Length of the seconds pendulum |
| July 4 | Indices of refraction |
| July 6 | Density of steam, begin |
| July 18 | (same) |
| July 25 | Moment of inertia |
| July 29 | Indices of refraction |
| August 1 | Refraction indices |

Kohlrausch produced similar lists for all students in attendance that semester. His attention to each student's progress was noted in the margins of his calendar of their seminar exercises as additions or deletions to their list of exercises, along with comments on their sicknesses, their absences, and their need to attend the practical exercises more often.[27]

Holding a common set of lectures on protocols for all students in the winter semester of 1868–69, Kohlrausch for the first time engaged them in a common project, the calibration of thermometers, in which all students' results were shared and evaluated among themselves, and then compared to his own results. But when student enrollment increased from twelve to thirty in the spring of 1869, Kohlrausch returned to his earlier method of assigning different sequences of exercises to each student. Six two-hour laboratory practica with one, five, or eight students punctuated his week, including two sessions on Saturday, in addition to attending the seminar, held on Thursday evenings for two hours. Once again it is unclear if the assignments were made on the basis of aptitude, gaps in knowledge, or the availability of equipment, although the latter was certainly a factor in scheduling. As in the previous year, not all students elected to attend two sessions per week; most went to only one. Next to the list of exercises actually completed, Kohlrausch noted possible future exercises, as well as suggestions and evaluations, usually asking for more effort on the part of the practicant: "Needs to come more often," "Attend seminar," "Give a lecture," "Determine results," "Check results," "Incorrect results," and "Provide numbers." Results independently achieved were preferable, but Kohlrausch sometimes assisted in the taking of measurements, even for advanced students.[28]

Students most often completed exercises not too different from Pape's of the summer semester 1868, with the notable exception of laboratory sessions devoted exclusively to special instrumental concerns. One type was devoted to the assessment of the "steadiness" and "sensitivity" of the instrument, especially when measuring weight. The other was in the construction of instruments, such as scales and multiplicators. Students' results were not merely signs of what they had learned and how well they had learned it, though. Near the end of his notebook, Kohlrausch gathered together student measurements (e.g., the horizontal component of terrestrial magnetic intensity, elasticity coefficients, or barometric measurements). He compared their results to one another to see if he could achieve consistency among them, as he had done in his reading notebook when evaluating published numerical results. In that sense he made students active critics of and contributors to the determination of fundamental physical constants. As his clientele, and with it, the exercises, expanded, Kohlrausch noted the need for more and different instruments (e.g., a small air pump, jars with sprayers attached, metal pyrometers whose resistance could be measured).[28]

In the Göttingen mathematico-physical seminar students learned some advanced mathematical methods in physics, including number theory, but the practicum seemed to draw very little upon them.[30] By contrast, Königsberg student exercise notebooks amply demonstrate that students themselves derived the formulas they needed, even though many could have been adopted ready-made as they were at Göttingen. Similarly, at Königsberg, students computed logarithmic tables themselves (or sometimes Neumann's second wife lent a helping hand), while at Göttingen students did not engage in the tedious process of calculating logarithms and other numbers they needed. Instead from the literature Kohlrausch assembled tables of fundamental constants which were added to the technical apparatus of the seminar in the summer semester of 1868. The product of experiment, worked over by combining results of different investigators and interpolating missing values, Kohlrausch's tables were a testament to the convergence of results that improvements in precision measuring techniques had achieved by the 1860s. Relieving students of the responsibility to compute physical constants each time for themselves, these tables became essential reference tools in practical exercises. By the end of the next decade, Kohlrausch had assembled over three dozen tables of fundamental constants for his manual, including tables of: the density of matter, specific gravity, the expansion of water from 0° to 100°, the reduction of a volume of gas to constant pressure and temperature, the wavelengths of the most important spectral lines of chemical elements according to the Bunsen-Kirchhoff schema, and values for terrestrial magnetic inclination, declination, and horizontal intensity. As for strictly mathematical tables, he included only three: squares, square roots, and their inverses; trigonometric numbers; and logarithms to four decimal places.[31]

Phenomenologists of daily life like Erving Goffman and E. T. Hall argue that in every profession, including teaching, there is a "behind the scenes" behavior that reveals what a person "really thinks" and makes visible professional discourse and action look scripted and artificial.[32] Yet Kohlrausch in private appeared very much like Kohlrausch in public. His own private notebook on the practicum in the three years from 1867 to 1869—a notebook that was at least one level removed from the ones in which he kept a diary of student assignments—logs where he directed his systematic effort over that period, as well as what he valued in the construction of the exercises.[33] It reveals another dimension of his pedagogy-under-construction. The notebook, which begins on October 25, 1867, logs the thinking behind the doing in the practical exercises. For some five dozen problems, the notebook sketches out, not always completely, his protocols, calculations, and remarks to himself about how to translate a published or customary protocol into a better practical exercise. All of the nascent exercises concerned instruments, protocols, calibrations, or standards determinations; all minimized the use of mathematical formulas but accentuated numerical calculations. That protocols produced usable numbers for the original investigators was insufficient for Kohlrausch; they had to work in the practicum. When he could say "the formula works for all parts of the operation"—which meant that students could actually produce numbers—then the protocol was on its way to becoming a practical exercise.[34] He compared students' numerical results with his own, and even based some exercises on student-designed protocols, as he did with those of Wilhelm August Nippoldt, who received his doctorate in physics at Göttingen two years later, in 1869.[35]

## The Rules behind the List

As much as Kohlrausch's notebooks help us to understand the origins of his practical exercises, they are limited in their ability to tell us broader details about the foundations of his canon, details found in the local culture of physics in Göttingen. We need to go beyond the *list* of exercises to the *rules* that accompanied them—in short, to the norms that guided their execution. To uncover those rules, we turn to publications either by advanced students, or about the work they completed in conjunction with Göttingen's practical exercises. Between 1868 and 1870 over a dozen investigations or reports of data appeared. Most were authored by Kohlrausch but nearly all were based in part on work completed by advanced students or postdoctoral assistants who had themselves come through the ranks of the exercise sessions and were now placing into practice what they had learned. Kohlrausch even remarked on students' performance in them. This published literature reveals the cognitive and practical preferences, the epistemological guidelines, and the social norms that constituted the environment in which Kohlrausch's exercises took shape.

### Cognitive and Practical Preferences

The intellectual center of gravity around which all practical exercises were related was the original Gauss-Weber project on terrestrial magnetism. Kohlrausch's first published mention of original work completed by students occurred in the context of a more general report on terrestrial magnetic data from 1867.[36] The project continued in part due to technical improvements, in part out of respect for Gauss and Weber, but mostly due to the project's suitability to accommodate vast numbers of students of varying ability in the collection and reduction of data. Unlike at Königsberg where the only larger-scale project was a failed one on geothermal measurements, at Göttingen the magnetic project thrived as a local site of data collection, practical exercises, and scientific socialization, even though as an international project it had long ago foundered. What Kohlrausch brought to the project were better instruments and a greater efficiency of operation, including ways of preserving the accuracy of the data while decreasing the frequency of the observations. The students Eggers, Nippoldt, Sleumer, and Voss were especially involved in taking measurements in 1867; Kohlrausch praised them for the quality of their measurements which were "an important proof of the accuracy and the exact interlocking of the instruments and the methods of measuring."[37] Refined measurements of the temperature dependency of terrestrial magnetism led to Eggers's first publication.[38]

The pliability of the methods for measuring terrestrial magnetism was especially pertinent to the determination of electrical standards of resistance, a project of utmost international significance with the expansion of electrotechnologies.[39] Weber, who had developed a standard of resistance based on his system of absolute units of length, mass, and time, was a major player in the debate, so interest in electrical standards determination at Göttingen was high. But Weber's standard, expressed as a set of metal coils, was actually the British favorite, while German scientists and engineers preferred Werner Siemens's electrical unit based on a column of mercury. Kohlrausch brought the literature from the debate into the seminar and practical exercises and enlisted advanced students to work with him on a comparison of the two units. Although Kohlrausch's stated preference was for Siemens's unit, he nonetheless analyzed Weber's unit with students in order to establish a "control" for the British trials.[40] Compelled to determine "with the greatest possible accuracy the ratio of Siemens's to Weber's absolute unit, which by accident (as everyone knows) is equal to 1 earth-quadrant/second," Kohlrausch had Eduard Riecke work on a correction that would eliminate the induced current in the magnetic needle, thereby achieving (at least for this perturbing effect) a more precise analysis of the measurements taken thus far.[41]

The debate over the measure of electrical resistance not only meshed well with experimental techniques and practical exercises at Göttingen, it is also illustrative of a cluster of cognitive preferences that were evident in other investigations in which students took part: an intense interest in the quality and accuracy of standards;[42] an overriding concern for the determination of fundamental constants;[43] the extension of known results to new physical cases;[44] and the determination of the limits of reliability of existing laws.[45] Kohlrausch in particular felt an obligation to correct measurements and conclusions based on them found in textbooks, the sources his students would first consult.[46] Many of these, it could be argued, were also a part of Weber's cognitive preferences, intellectual agenda, and practical style, but Weber did not integrate them into his teaching to the degree that Kohlrausch did. Weber certainly did not make the correction of textbook entries a part of his pedagogic agenda, nor did he engage students in investigations to the degree Kohlrausch did. Kohlrausch was also more open to new and different kinds of problems, as when he introduced experiments on the mechanical theory of heat into his exercises in 1869.[47]

Two additional cognitive preferences are revealed by the investigations based on techniques learned in practical exercises. The first is a preference for mathematical expressions amenable to certain kinds of decisive experimental trials. Kohlrausch appears not to have sharply separated theory and experiment, but to have fostered a deep understanding of the mathematics used in theoretical expressions, even though pure theory was not the focus of either his practical exercises or his research. And yet it was not Weber's mathematical formulations of physical theory that were the starting point of practical investigations, but rather Franz Neumann's. Eduard Riecke's electromagnetic investigations can serve as an example. In 1870 Riecke drew upon the work of Franz Neumann to understand how galvanic electricity could be replaced by a system of magnetic masses.[48] He then turned to Neumann's expression for the distribution of magnetism on a rotating ellipsoid, not only because an experimental proof of the law "was not yet available," but also because the law hitherto could only be used as an "approximation formula"[49] that did not lend itself to consistent physical measurements. Riecke's data demonstrated enough variability to render the law inadequate for the situation it claimed to describe.

A second, corollary cognitive preference further reinforces the conclusion that the concrete was preferred over the abstract in Göttingen exercises. Kohlrausch's "art" of physical manipulation depended upon visualization. This predilection had its strongest origins in the visual representation of terrestrial magnetic data in maps under Gauss and Weber.[50] It was reinforced in later years by the modeling of other investigations on terrestrial magnetic measuring, and became a natural counterpart to mathematical expression.[51] So when Riecke, for instance, wanted to understand further Neumann's rotation ellipsoid, he asked the mechanic Wilhelm Apel to draw seven ellipsoids of different sizes, and then to manufacture them out of wrought iron.[52] The reliance upon visualization helps to explain, too, Kohlrausch's positive view of interpolation when observational data demonstrated regularities—meaning, in part, that one could "see" the data graphically.[53] With regard to both visualization and interpolation, by contrast, Neumann and his students showed little evidence of the former and expressed great doubt, if not complete uncertainty, about the latter. Inductive thinking thus played a larger role at Göttingen than at Königsberg.

Complementing these cognitive preferences were practical ones related to the material conditions of exercises or experiments. Reliability and sensitivity were two key characteristics of instruments that were much sought after. With a reliable instrument the number of analytical corrections was reduced; at Königsberg, by contrast, analytical corrections were the norm.[54] Instruments that demonstrated those traits were preferred objects of exercises, such as Weber's bifilardynamometer, Siemens's resistance scale, and Weber's magnetometer,[55] to name a few. Sensitive instruments that were either finicky or that only the most skilled could deploy were inappropriate for practical exercises. Kohlrausch often found himself holding up the ideal of the precision in measurement achieved by Gauss while acknowledging that important, and sometimes compromising, decisions had to be made in trying to achieve them. The way around the dilemma, he reasoned, was to choose the right technique. Gauss's methods for determining the intensity and declination of terrestrial magnetism, he pointed out, were "distinguished by a classical simplicity of instruments," but required "the highest care of the practiced observer to attain the accuracy achieved by Gauss." So Kohlrausch developed new methods for taking measurements simply, quickly, and simultaneously.[56] Yet some techniques, regardless of the skill required to execute them, held near-sacred status within the repertoire of physical manipulations due to their reliability in producing consistently precise measurements. For instance, in re-determining the influence of temperature on the elasticity coefficient of metals, Kohlrausch abandoned the customary technique of measuring the lengths of metal bars at high temperatures and instead used weights hanging from different wires whose torsion elasticity coefficient could be measured by observing the periods of vibration of the wire. "Period of vibration measurements," he explained, "belong, according to Gauss's instructions, to the finest of all possible measurements in physics. Small changes in the coefficient [of elasticity] can be attained with great sharpness."[57] Finally, Kohlrausch, like Weber before him, valued transportable instruments, for in addition to the seminar, physical institute, the magnetic and astronomical observatories, and the Göttingen Academy of Sciences, both viewed the city and surrounding area as potential sites of physical investigation.[58]

## Epistemological Guidelines

As Hugh Gusterson's essay in this volume emphasizes, there is perhaps no more important quality a science student could acquire than the ability to make sound judgments. In his teaching and mentoring, Kohlrausch provided guidelines for evaluating the epistemological foundation of scientific claims. Students were not merely "solving a problem" when they worked through his exercises; they presented their results according to how they believed scientific claims *should be* presented. The same guidelines served as standards of scientific criticism. Trials, for instance, should provide enough data to go beyond scientific approximations.[59] Kohlrausch made it clear that one judged data on the basis of its transparency, its reduction by means of error analysis, and its aesthetic qualities. Transparency and openness in scientific reporting were especially important. Kohlrausch undertook a redetermination of the specific heat of air at constant volume because "the details of the observations were not published in earlier investigations [which were by Laplace]."[60] A good part of German criticism of the British determination of the standard of electrical resistance was directed at the relative paucity of data and error reporting. Kohlrausch explained that because the British reported "only average values" there was "nothing to go by in the assessment of the irregularities which are to be ascribed" to various instrumental disturbances, especially the magnetism affecting the galvanometer needle.[61] So Kohlrausch completed new trials, meticulously accounting for deviations in average values by means of error analysis.

The type of error analysis he presented in these student-assisted investigations, as well as in his practical exercises, was more streamlined than what one would find at Königsberg. To be sure, like other German physicists, he made good use of the method of least squares, but he placed greater weight on the analysis of the constant or instrumental errors of the experiment. Hence Kohlrausch's exercises did not engender endless trials, as did Neumann's; more data could simply be produced by assembling the results of different investigators. To be "precise" was thus not the outcome of seemingly endless data analysis, especially when some errors could be neatly incorporated into reference tables.[62] The analysis, proper construction, and handling of instruments and experimental conditions were far more important.

Transparency combined with a simpler but thorough reporting of errors can be regarded as aesthetic features of a scientific investigation. In print Kohlrausch praised the "great circumspection" of certain trials, as well as the "completeness and elegance" of error analysis.[63] Aesthetic preferences also guided the construction of exercises.

Above all, exercises had to admit of a "simple" expression, and the simpler the better.[64] Simplicity was also apparent in the regular deployment of approximation formulas which eased otherwise burdensome numerical calculations. When Kohlrausch entered a string of exclamation marks ("!!!!") next to a student's list of exercises, it was a sign of consternation and call to seek a simpler solution.

## Social Norms

Kohlrausch and others at Göttingen taught the practical methods of physics in settings varied in their social organization: lecture, seminar, physical institute, magnetic observatory, and astronomical observatory. Yet no matter where the learning or execution of those methods took place, Kohlrausch's teaching inculcated social norms that shaped the student's professional development (in physics, in teaching, or in another profession) as well as his attitude toward the larger community of which the physics profession was a part. He (as well as Weber before him) empowered students, made them partners in research, and established guidelines for interacting with experienced and supporting members of the community.

By creating the means for students to take part in their own education, as well as in educating others, Kohlrausch empowered them. For the most part their relationship to one another was characterized by receptivity to criticism and a cooperative spirit. But they also learned to respect one another through the implementation of student-led innovation in the practical exercises. The roles and responsibilities of teacher and student were shared as the presentations rotated among them. Making a presentation on recent literature or a new protocol was an "as if" experience: students learned to present themselves and their topic *as if* they were professionals in the field. The importance of an "as if" experience for professional development—whether as a physicist, secondary school teacher, or in another profession—was, I believe, deep and long-lasting. Being chosen an assistant, either as an advanced student or postdoc, was clearly a predictor of future professional success.

Second, in making presentations, advanced students were participating not merely in the transmission of knowledge, but also in its creation and organization. In the German tradition of *Wissenschaft* dating back to Alexander von Humboldt and Friedrich Schleiermacher, lecturing that incorporated new results was a research act, a form of original thinking.[65] Physicists like Franz Neumann and Wilhelm Weber were practiced at this art, and Kohlrausch, in extending the introduction of originality to the creation of practical exercises, was not breaking new ground, but joining the very few who did the same. Among all the pedagogical tasks undertaken by students, none, I would argue, was more important than instructing beginners. Here students themselves were the agents of social inclusiveness in practical instruction. Through their example, they disseminated the idea that the acquisition, practice, transmission, and application of knowledge were in principle open to *anyone*. And surely the pattern set up at Göttingen, where generation after generation of advanced students taught beginning ones, helped to create within physics teaching, locally and then nationally through Kohlrausch's manual, a persistent obligation for and commitment to elementary science training, not only by book and words, but through deeds and actions in the practical exercises. The pattern of practical teaching at Göttingen, especially under Kohlrausch, thus helped to create the means whereby physics as a discipline regenerated itself from the bottom up.

Among the social relations in physics instruction at Göttingen, none was more important than the student-teacher relationship. At the most basic level, Kohlrausch guided professional development through his example and his evaluations. Of Nippoldt's trials on the extension of Ohm's Law to electrolytes, Kohlrausch remarked that Nippoldt had "great knack" in executing them and achieved "completely satisfying results."[66] Yet sometimes students could not perform with the skill of the experienced, so in addition to recommending persistent practice to achieve manual control, Kohlrausch developed compensating techniques that would allow students to remain a part of the investigation.[67] Their roles in the investigation were as partners in research; to a certain degree there was a temporary fusion of their identities as researchers, as Kohlrausch's reports indicate (while the "voice" of reporting was Kohlrausch's, the performance was completed in tandem). The student thus had the opportunity to act as a researcher before he actually was one; as in teaching, acting "as if" in research was an important step toward *becoming* a researcher.[68] Kohlrausch's relationship with his advanced students was thus one filled with mental stimulation (which motivated students), with creativity (rather than conformity or uniformity), and with trust (for Kohlrausch relied on their data). About this intense form of communication that developed between teachers and students in the nineteenth century, Stichweh has aptly concluded that "science itself becomes the medium of the disciplinary formation of men."[69]

A third and final social norm cultivated in the seminar was the relationship between students and the larger community of physics practitioners and those who made their work possible. Students in the practical exercises at Göttingen had an intellectual and moral obligation to review and critique results from elsewhere (British measurements of the standard of electrical resistance, Franz Neumann's electrodynamics, Laplace's computation of the specific heat of air at constant volume, Alfred Clebsch's extensions of Neumann's work, to name a few). Moreover, at Göttingen, there was no separation of the academic and the artisan when it came to physical exercises. From the late eighteenth century onward, mechanics and instrument-makers had grown in cooperation and conjunction with Göttingen's university culture. Nearly all instrument-makers and clockmakers in Göttingen worked in tandem with university professors before 1850 rather than establishing independent shops for marketing instruments.[70]

By the 1860s, when Kohlrausch took over the practical exercises, Göttingen had a total of about a dozen small and large firms specializing in *Feinmechanik* that employed a workforce of 50 masters and 270 apprentices and assistants. Two firms dominated university instrument culture: those of Wilhelm Apel and Moritz Meyerstein. Each focused on a different clientele. Cognizant of the market that had opened up as a result of the construction of physical laboratories in classical gymnasiums, the growth in the number of *Realschulen,* as well as the growth in the popularity of the *Realwissenschaften* after 1848, Apel turned the family firm into one of Germany's first suppliers of mass-produced school apparatus. At Göttingen, Weber, Benedict Listing, Friedrich Wöhler, and later Kohlrausch used Apel's instruments for practical exercises, as did Robert Bunsen and Gustav Kirchhoff at Heidelberg and Justus Liebig at Giessen. Meyerstein's instruments, by contrast, were custom-made, and so were preferred over Apel's for both teaching and research at Göttingen, although Apel's were most likely used for the introductory exercises. Meyerstein's instruments gave the older students in the practical exercises the opportunity to specialize in research methods as well as to achieve a degree of precision above that of mass-produced instruments.

Instrument culture permeated the practical exercises at Göttingen, and Kohlrausch, like Weber before him, capitalized on the rich resources available to them. Both physicists referred to all instruments by their maker or designer: Steinheil's heliometer, Mendelsohn's scale, Riemann's scale, Dove's barometer, Reichenbach's meridian circle, Weber's magnetometer, R. Kohlrausch's electrometer, M. Jacobi's etalon for determining the electrical resistance standard, Apel's thermometer, Nicolson's scale, and others.[71] Proper names attached to instruments reminded the students that the instrument was linked to a specific protocol that was reproduced, criticized, and improved upon in practical exercises. As he had commented on student performance in his reports and published investigations, Kohlrausch lavished praise on Meyerstein's construction of instruments, and always noted that mechanics *constructed* an instrument, but physicists set, to a greater or lesser degree, the *specifications*.[72] The message was clear: instrument-makers were an integral part of the community of physics practitioners, and if one wanted to become a physicist, one had to work with them. Even for the non-specialist student, he delivered an important social message: the construction of knowledge was not an elite enterprise, but rather one that relied on the fast-growing elements of industrial culture. To reinforce this message, Kohlrausch in his exercises recommended that his students visit local instrument-makers.[73]

### The Manual and Its Audience: The Making of a Canon

In the late 1860s, after Kohlrausch had been offering his practical exercises for only a few years, Weber and Wilhelm Beetz gathered together Kohlrausch's laboratory assignments and distributed them for local use. Kohlrausch reported on this and other developments in his teaching in 1870.[74] The need for laboratory training such as that he had designed, he emphasized, "was recognized everywhere but has not been satisfied anywhere in a suitable fashion."[75] He viewed the success of the practicum in its appeal to a broad-based clientele of beginners. In the first six semesters of its operation, he noted, there were 104 practicants from the following areas of study[76]:

| | |
|---|---|
| Mathematics and physics | 37 |
| Chemistry | 24 |
| Medicine | 2 |
| Philosophy | 2 |
| Pharmacy | 36 |
| Telegraphy | 3 |

His practicum, he explained, concentrated *"overwhelmingly on elementary physical exercises*, primarily in measurement, which can be applied to other natural science disciplines or which have already attained prominent significance in practice."[77] For this purpose he used, in addition to the physical laboratory, the magnetic observatory, located far from the disturbances of street traffic and protected from local magnetic influences, both of deleterious effect on precise measurements. Despite the practical difficulty of accommodating a large number of students at these two locations—a condition that had to be put aside for the moment but that he hoped would be improved in the future with the addition of more instruments—teaching the practicum was made considerably easier, he noted, by the publication of the *Leitfaden der Praktischen Physik* (first edition, 1870), which "placed in the hands of every participant the rules and numbers for the reduction of the observations."[78] Yet the manual alone was insufficient in at least one respect for teaching beginners. No matter what their field of study, practicants had to "be able to carry through the calculations" of the exercises, but most did not possess the mathematical skill to do so. In an adaptation of one of the social norms that had characterized Kohlrausch's teaching style over the decade before, Nippoldt, Riecke, and another recent doctorate in physics then assumed responsibility for providing the mathematical instruction the beginning students needed.[79]

An important segment of Kohlrausch's audience was future secondary school teachers, for whom he adapted his introductory exercises to teach them how to deploy instructional apparatus.[80] Here, too, though, he emphasized precise measurement for its ability to prevent the execution of trials in the classroom from becoming a game. He explicitly rejected the idea of creating a special course on instructional experiments, believing instead that anyone who had learned how to set up quantitative trials could master demonstration experiments without difficulty.[81] In an era when the number of physical laboratories to German secondary schools grew rapidly, Kohlrausch's targeted training of secondary school teachers enhanced the impact his manual would have on the larger process of regenerating the ranks of the discipline.[82]

Out of the hundreds of problems that Kohlrausch had worked out in his notebooks, 41 appeared in the first edition of 1870; ten years later that number had grown to 92, with an additional 21 on the system of absolute units in electricity and magnetism that Weber developed at Göttingen. He grouped his exercises under ten different categories—weighing and the determination of density, air pressure, heat, elasticity, light, auxiliary observations for electricity and magnetism, magnetism, galvanism, electrostatics, and, in a gesture to the movement to establish a universal standard time, the determination of time and geographical location—giving his exercises a thematic arrangement and structure they had lacked in his notebooks.[83] Some themes, like the determination of time and the measurement of terrestrial magnetism for naval use, proved to be a passing fashion. By the turn of the century his manual began to serve, in addition to students, scientists required to keep up with the latest methods of measurement in research and industrial production.[84] Although in citing predecessors to his exercises in the eleventh edition (1910) of his manual, he remarked that "probably only Neumann at Königsberg and Magnus at Berlin, and a bit later Kirchhoff at Heidelberg, offered *systematic practical instruction*," Kohlrausch in fact systematized his exercises only to a point.[85]

Despite the didactic and terse nature of his notebook's style, Kohlrausch did not intend that the exercises become rigidly formulaic. The cognitive and practical preferences and epistemological guidelines that had shaped his early teaching were embedded in his manual. There was always an interplay, he admitted, between procedural steps and the desired degree of precision and accuracy: the greater the latter, the more detailed the former. His own preference was that there would always be "room to play" because for instructional purposes it was "superficial" to carry error analysis and the determination of correction formulas to their limits. So for his exercises he tried to keep those constant errors he did not address to within the range of the unintentional errors of observation of the average observer using customary instruments—although he admitted that he would be unable to please everyone because some would want a more thorough treatment of the errors, while others would regard even what he had done as pedantry.[86] His principal guideline for error analysis, like protocols, was "simplicity," and so he advocated what Königsberg-trained students for the most part rejected: interpolation between measured values.[87] Even in his *Kleiner Leitfaden* for beginning students—for whom one might have expected "canned" exercises—Kohlrausch emphasized that the "main goal of the exercise practicum" was

> not in carrying out the problem according to a schema, but in the intellectual property which the practicant acquires in its execution. This goal will be all the more restricted the more the exercise is schematized. For that reason the presentation of the methods and instruments used should not be pushed further than is required for the necessary order and simplicity of the operation of the laboratory.[88]

Likewise for students with a professional interest in physics, he believed that instruction could teach research only in the restricted sense, by acquainting students with the "scientific tools of the trade," for there would always be more than enough particulars that had to be created independently in an investigation.[89] So his exercises were codified only to a degree. Their flexibility was sufficient enough for him to cultivate a high degree of originality through them, as demonstrated in his student-assisted investigations leading up to the first edition of 1870. In his report on the practicum from 1870, he mentioned no less than twelve undertaken in six semesters.[90] This success he attributed to a special extension of his exercises, similar to what he had done for secondary school teachers, so as to foster "a school for scientific physical research."[91]

A factor in understanding why Kohlrausch's *Praktische Physik* became a canon certainly has to do with the composition of his audience, as well as the audiences that found later editions to their liking. Yet the composite nature of his clientele—beginners, a service clientele, future secondary school teachers, and future physicists—had not been a novel feature of physics instruction for about fifty years. All four groups were found in physics courses nearly everywhere throughout the German states. The research ethos that had transformed universities into training grounds for original investigators dated from the 1820s; it played a critical role in shifting the emphasis in instruction from recapitulation and repetition to active engagement not only in the production of knowledge, but also in its representation. Using disciplinary material, rather than pedagogical theory, for teaching secondary school teachers also had been done since early in the century thanks to the structure of state examination systems; indeed an original investigation of sorts was required for the state teaching certification in Prussia and elsewhere.

Kohlrausch's accommodation of beginners—surely one of the most outstanding features of his manual, since his was the first aimed specifically at beginners—also continued practices of a half century earlier. The need to establish an introductory conceptual and methodological foundation for learning in the disciplines, rather than to rely exclusively on what secondary schools might teach as preparation for university learning, was in part a consequence of the introduction of research methods into university instruction, as Neumann's seminar at Königsberg amply demonstrates. Neumann's introductory course on theoretical physics, a course on mechanics, was designed to provide the intellectual groundwork for more advanced courses in physics and practical methods.[92] Through the teaching of Neumann's students, the content, style, and demographic focus of his physics instruction spread elsewhere, to both universities and secondary schools.[93] By no means, though, was there a pan-German acceptance of his pedagogy, although there was widespread international admiration and appreciation for the *systematic* nature of both his courses and the practical exercises in the seminar based on them. In this regard Kohlrausch's pedagogical achievement was certainly greater than Neumann's; for *Praktische Physik* reached a national audience within a decade of its first edition, precisely during the period when the number of teaching laboratories began to expand. Certainly another novel feature of Kohlrausch's exercises was their appropriateness for a service clientele—students from a wide variety of disciplines who needed physical methods for their major subject—but here, too, Kohlrausch built upon earlier practices rather than having inaugurated something entirely new.

Nonetheless, we need to consider more closely Kohlrausch's initial accommodation of both beginners and a service clientele from two social perspectives. Both were central, I believe, in contributing to the canon formation of his manual by expanding the social base that supported it. First, Kohlrausch's social inclusiveness may well have been the result of changing student demographics at Göttingen during the years of his teaching. Seminar attendance doubled during the 1860s when Kohlrausch took over the exercises, a development that worried Hannoverian educational officials because the students would not find positions as teachers in Hannover's limited number of secondary schools.[94] Over the course of the decade, too, Weber and Kohlrausch bent the rules of admission to the seminar, and allowed students from polytechnical schools to attend. In addition, more students from the middle class beyond the *Bildungsbürgertum* (the educated middle class) appeared on the rosters of the seminars and the practical exercises, including the sons of instrument makers.[95] Overall the social shift at Göttingen (as well as Königsberg and elsewhere) was downward toward the lower levels of the middle class, outward toward students from the "real" schools interested in applied subjects, and away from the neohumanist non-utilitarian conception of instruction toward the manual skills Weber taught the talented few and Kohlrausch opened up to everyone. Second, Kohlrausch did not make any reference to the highly charged terms that had been a part of pedagogical debates over the course of the century: "inclination and talent" as well as the related concept of "genius." His predecessors did. In his science pedagogy, Weber had tacitly accepted "inclination and talent," while Neumann explicitly rejected it. In the German states, discussions of all three were linked to the vast social transformation of the nineteenth century affecting self-definition and self-identity, and especially to the growing social inclusiveness of educational systems. One problem was the sense of "natural privilege" that all three terms suggested; another, the misfit between, on the one hand, inclination, talent and genius, and the other, the expansion of disciplinary learning.[96]

The curricular implications of inclination, talent, and genius are worth noting. A genius needed no teachers. (The philosopher and psychologist J. F. Herbart, active at both Königsberg and Göttingen and known to both Neumann and Weber, described the genius as someone who did not know the rules by which he thought.[97]) A curriculum that upheld inclination and talent (such as during the Renaissance) adjusted to individual differences rather than to create and enforce a more standardized curriculum that accommodated all students. To argue, then, that *anyone* could in principle participate in disciplinary training—as Kohlrausch clearly did in the construction of his manual, especially in the all-important accommodation of beginners—was a rejection of inclination and talent and an expression of a willingness to work with those who might not have initially demonstrated any visible signs of giftedness but who nevertheless expressed an interest in disciplinary study. In this regard, Kohlrausch's manual is more important than the sets of practical exercises that preceded it (Weber's, Neumann's, Magnus's, and Kirchoff's, for example) because of the limited reach of their pedagogical innovations. Kohlrausch was the first to have made the social inclusiveness of disciplinary study in the practical exercises of physics a widespread reality through his accommodation of introductory students, and to have done so without preference for those demonstrating inclination and talent.

## Conclusion: Reading Canonical Texts

In an essay that broke new ground in anthropology, Claude Lévi-Strauss pointed out that texts customarily are read front to back, left to right (or right to left) and top to bottom. But there are some texts, such as musical scores, for which this diachronic reading is insufficient. To fully appreciate them one had to read the score synchronically in order to see other patterns, such as harmony.[98] In a similar vein, the phenomenologist of daily life Edward Hall argued that there are texts (again musical scores are invoked) that are impossible to describe entirely in words, that had to be learned by imitation because the written document did not verbalize all that it was about, but whose potential could be exploited only when the text (or score) was actually written.[99] I have attempted here to "read" Kohlrausch's *Praktische Physik* in a somewhat similar fashion, by emphasizing the preferences, guidelines, and norms embedded in it but not explicitly verbalized, and by arguing that imitative practices, found in student-assisted publications, offer clues to what is "contained" in the text but not expressed in words. In the manual's formative stages, those unwritten components included: cognitive and practical preferences, epistemological guidelines, and social norms. These preferences, guidelines, and norms are the *rules* that accompany the *list* in canon formation.

This essay has been concerned primarily with the formative years of Kohlrausch's manual, with some reference to editions published before 1910 when Kohlrausch died at the age of 70. After then there is a wider set of issues to consider in canon formation, such as how a canon helps to define a community, represents authority, and persists over time. To maintain its status, for instance, a canon has to be open to interpretations over generations, and be flexible enough to respond to changing social conditions and intellectual developments. A particularly important dimension of the continuing success of Kohlrausch's manual is its persistent ability to accommodate introductory students in the wake of the international expansion of secondary and tertiary education in the twentieth century. The analysis of the text's canonical status should continue to take into account not only its content, also its unverbalized components expressed in documentary evidence drawn from actual practices, including student-assisted or student-directed investigations. But there are also new and different elements of the text's canonical status that need to be examined from a historical perspective when the manual is considered over the long term, such as the question of authorship (Kohlrausch remains "author," although the manual is written by teams of specialists), relevance (the persistence of interpretive openness in changing intellectual and social conditions), and social inclusiveness (not only the accommodation of beginners and a service clientele, but adaptation to the changing social constitution of the discipline, including issues of race and gender). Eugene Ulrich has pointed out in his own work on biblical canon formation that "intellectual argument cannot safely proceed without a clear definition of terms."[100] The purpose of this essay has been to clarify what "canon" means when applied to the historical study of texts in science pedagogy.

## Acknowledgments

My thanks to David Cahan, Klaus Hentschel, Robert Kohler, and an anonymous referee for their comments on earlier versions of this essay. As editor, David Kaiser offered advice, inspiration, and leadership.

## Notes

1. Friedrich Kohlrausch, *Leitfaden der praktischen Physik* (B. G. Teubner, 1870; ninth edition, 1901); idem, *Lehrbuch der praktischen Physik* (B. G. Teubner, tenth edition, 1905; seventeenth edition, 1935); idem, *Praktische Physik: Zum Gebrauch für Unterricht, Forschung und Technik* (B.G. Teubner, eighteenth edition, 1943; 24th edition, 1996). For beginning students Kohlrausch extracted the *Kleiner Leitfaden der praktischen Physik* (B. G. Teubner, 1900; second edition, 1907), but in later editions elementary exercises were a part of the *Lehrbuch* or *Praktische Physik*.

2. David Cahan discusses Kohlrausch's popularity in "The institutional revolution in German physics, 1865–1914," *Historical Studies in the Physical Sciences* 15 (1985): 1–65, on 48–50, esp. note 128 on 50. See also David Cahan, *An Institute for an Empire: The Physikalisch-Technische Reichsanstalt, 1917–1914* (Cambridge University Press, 1989), 19–20; Christa Jungnickel and Russell McCormmach, *The Intellectual Mastery of Nature: Theoretical Physics from Ohm to Einstein* (University of Chicago Press, 1986), volume 1, 107; Kathryn M. Olesko, *Physics as a Calling: Discipline and Practice in the Königsberg Seminar for Physics* (Cornell University Press, 1991), 410–411.

3. Pertinent to the development of physical exercises at Göttingen, for instance, is Abraham Gotthelf Kästner's *Anfangsgründe der Arithmetik, Geometrie, ebenen und sphärischen Trigonometrie und Perspective* (Vandenhoeck and Ruprecht, 1758), which in 1800 was in its sixth edition and was still used at the university. Problems in the sixth edition included, besides strictly mathematical ones, exercises in perspective and surveying. A textbook specifically dedicated to problems in applied mathematics for beginning civil engineers was Johann Albert Eytelwein's *Aufgaben, größentheils aus der angewandten Mathematik zur Uebung der Analysis: Für angehende Feldmesser, Ingenieurs und Baumeister* (Friedrich Maurer, 1793).

4. State examinations were not only a route to social advancement for poor students, but also the only way for anyone to enter the bureaucracy.

5. Herbert Goldstein et al., *Classical Mechanics*, third edition (Addison-Wesley, 2002), which is used to teach the Lagrangian formulations necessary to do problems in quantum mechanics; John David Jackson, *Classical Electrodynamics*, third edition (Wiley, 1999); Eugen Merzbacher, *Quantum Mechanics*, third edition (Wiley, 1998).

6. This twin association of "canon" is evident even in the earlier Sumerian root of the word: it meant a reed or measuring stick, and from there acquired the connotation of standard, norm, or ideal. The etymology of "canon" as well as a discussion of other issues raised in this paragraph are

found in Eugene Ulrich, "The notion and definition of a canon," in *The Canon Debate*, ed. Lee M. McDonald and James A. Sanders (Hendrickson, 2002), 21–35.

7. Kohlrausch's father, Rudolf, was also a physicist and friend of Weber, suggesting that Kohlrausch the younger may very well have first been exposed to practical exercises through his father. Presently, however, we have no strong evidence to back up the claim. On Kohlrausch's family background and education as well as his years at Göttingen to 1870, see David Cahan, "Kohlrausch and electrolytic conductivity: Instruments, institutes, and scientific innovation," in *Science in Germany: The Intersection of Institutional and Intellectual Issues*, ed. Kathryn M. Olesko (*Osiris* 5, 1989: 167–185, esp. 168–173).

8. Carl Friedrich Gauss and Wilhelm Weber, *Resultate aus den Beobachtungen des magnetischen Vereins, 1836–1841* (Dieterich, 1837–1839; Weidemann, 1840–1843); *Atlas des Erdmagnetismus nach den Elementen der Theorie entworfen: Supplement zu den Resultaten aus den Beobachtungen des magnetischen Vereins. Unter Mitwirkung von C. W. S. Goldschmidt*, ed. Carl Friedrich Gauss and Wilhelm Weber (Wiedemann, 1840).

9. Weber to unknown, February 14, 1848, Wilhelm Weber Briefe, Sig. F2e 1833, Darmstaedter Sammlung, Staatsbibliothek zu Berlin-Preußischer Kulturbesitz [SBPK], Handschriftenabteilung. Also see Wilhelm Weber to Edward Sabine, February 20, 1845, rpt. in *Wilhelm Weber's Werke*, 6 Bde., ed. Königl. Gesellschaft der Wissenschaften zu Göttingen (J. Springer, 1892–1894), volume 2, 274–276.

10. H30 Mathematisch-physikalisches Seminar, Sommersemester 1850—Sommersemester 1856. Historische Sammlung, I. Physikalisches Institut, Universität Göttingen [PI-UG]. Weber kept the log from 1850 to 1852–53, while Ernst Julius Schering did so from 1853 to 1856, with supervision from Weber. I thank Dr. Gustav Beuermann for permission to see the historical material in the I. Physikalisches Institut.

11. For the voluminous primary data that lay at the disposal of students, see Nachlass Magnetischer Verein, Abteilung für Handschriften und Seltene Drucke, Niedersächsische Staats- und Universitätsbibliothek, Göttingen. The teams of observers conducting and reducing measurements included advanced students and recent doctorates from Göttingen and elsewhere.

12. Wilhelm Weber, "Bericht über einige im physikalischen Institut in Göttingen gemachte Versuche," *Nachrichten von der Georg-August-Universität und der Königl. Gesellschaft der Wissenschaften zu Göttingen* [hereafter *Nachrichten*] Nr. 6 (1858): 68–76.

13. Schering lectured on the method in 1862–63 (Königliches-Universitäts-Curatorium zu Göttingen. Universität Göttingen. Akten betr. die Einrichtung eines mathematisch-physikalischen Seminars [1850–1883], Universitätsarchiv Göttingen [UAG] 4/Vh/20, Bl. 233r). Riecke left notes, but no date for his course. H26 Eduard Riecke, "Grundzüge der Wahrscheinlichkeitsrechnung." Historische Sammlung, PI-UG. On the method of least squares at Königsberg, see Olesko, *Physics as a Calling*.

14. "Statuten des mathematisch-physikalischen Seminars zu Göttingen," June 1861, 4/Vh/20, UAG, Bl. 206–207.

15. Paid at first from the seminar's premium budget, the position of directing younger students attained higher status in 1859 when Ernst Schering was appointed associate professor in charge of elementary seminar exercises. Older students still assisted, but received no more than 20 taler for their efforts.

16. Weber designed the position so as to remove from his responsibility onerous tasks associated with conducting the institute; the position included much more than directing the practical exercises or "introducing students-in-training in how to set up observations and trials." Weber to the University of Göttingen Curator, October 18, 1866, 4/Vh/21 Assistentenstelle Physikalisches Institut, UAG. See also 4/Vb/156 Personalakte Friedrich Kohlrausch, UAG and Weber to unknown, February 19, 1869, Wilhelm Weber Briefe, Sig. F2e 1833, Darmstaedter Sammlung, SBPK, Handschriftenabteilung.

17. Nachlass Friedrich Kohlrausch [NFK] 2498 / Tagesbuch I, 1861.1862./ 5. Physik. Protocolle aus dem Phys. Cabinet 1861. 1862. / Rückstand 1861. 1862./Notizen. Archiv des Deutsches Museum [ADM] (incompletely paginated).

18. Gustav Wiedemann, *Die Lehre vom Galvanismus und Elektromagnetismus nebst ihren technischen Anwendungen,* 2 Bde. Erster Band. *Galvanismus. Die Lehre vom Galvanismus.* Zweiter Band. *Elektrodynamik, Elektromagnetismus, Diamagnetismus, Induktion und Schlusskapitel. Die Lehre von den Wirkungen des galvanischen Stromes in die Ferne* (Vieweg, 1861); Peter Theophil Riess, *Die Lehre von der Reibungselektricität* (Hirschwald, 1853).

19. Riess, *Reibungselektricität,* volume 1, iii.

20. Wiedemann, *Galvanismus und Elektromagnetismus,* volume 1, vi–vii.

21. NFK 2498.

22. NFK 2499/3. 1862. Naturw[issenschaften]. ADM.

23. On Kohlrausch's leitmotif of precision measurement, see Cahan, "Kohlrausch and electrolytic conductivity," and idem, *An Institute for an Empire,* 128–136, 147–150, 156–159, 168–175.

24. Student notebooks of the written physical exercises from Neumann's seminar are found in Abt. Va, Rep. 11, Planck, 1836/26 to 1836/33 inclusive, Archiv zur Geschichte der Max-Planck-Gesellschaft, Berlin. I am grateful to Dr. Karl von Meÿenn for drawing my attention to these notebooks in 1995.

25. NFK 2500/3. 1867. 1868. Prak[tikum]. ADM. Student names and suggested exercises are on 82–84; Voss and Pape, on 84.

26. NFK 2501/4. 1868. $H_2SO_4$ $ZSO_4$. ADM. The names of the advanced students who made presentations are italicized.

27. NFK 2501.

28. NFK 2502/5. 1868. 1869. [Protocolle. 1868. 1869.]

29. Ibid., 156r.

30. In research published by advanced students, though, more abstract mathematically oriented problems were to be found. See below on Riecke's publications.

31. Kohlrausch, *Leitfaden,* fourth edition (1880), 280–307. 311–351.

32. Erving Goffmann, *The Presentation of the Self in Everyday Life* (Doubleday, 1959); Edward T. Hall, *The Silent Language* (Doubleday, 1959).

33. NFK 2601. Varia I. 1867. 68. 69. Enthält großentheils die ersten Uebungsaufgaben aus dem Praktikum in Göttingen, ADM.[Only the odd pages are numbered.]

34. Ibid., 3.

35. Ibid., [40], [60].

36. Friedrich Kohlrausch, "Resultate aus den magnetischen Beobachtungen im Observatorium zu Göttingen vom Jahre 1867," *Nachrichten* (1868): 160–163.

37. Ibid., 163.

38. H. Eggers, "Ueber den täglichen Gang der horizontalen Intensität des Erdmagnetismus in Göttingen (Mitgetheilt von F. Kohlrausch)," *Nachrichten* (1869): 162–165.

39. British and German scientists and engineers were the most active in the determination of this standard. On the German contributions, see Kathryn M. Olesko, "Precision, tolerance, consensus: Local cultures in German and British resistance standards," *Archimedes* 1 (1996): 117–156. On the British contributions, see Bruce Hunt, "The ohm is where the art is: British telegraph engineers and the development of electrical standards," *Osiris* 9 (1994): 48–63; Simon Schaffer, "Late Victorian metrology and its instrumentation: A manufactory of ohms," in *Invisible Connections: Instruments, Institutions, and Science,* ed. Robert Bud and Susan E. Cozzens (SPIE Optical Engineering Press, 1992), 23–56; Hunt, "'Practice vs. theory': The British electrical debate, 1888–1891," *Isis* 74 (1983): 341–355; Schaffer, "'Accurate measurement is an English science,'" in *The Values of Precision,* ed. M. Norton Wise (Princeton University Press, 1995), 137–172; Graeme Gooday, "Precision measurement and the genesis of physics teaching laboratories in Victorian Britain," *British Journal for the History of Science* 23 (1990): 25–51.

40. Friedrich Kohlrausch, "Resultate der Beobachtungen im magnetischen Observatorium zu Göttingen vom Jahre 1868; insbesondere eine Bestimmung der absoluten Intensität des Erdmagnetismus auf galvanischem Wege," *Nachrichten* (1869): 35–42.

41. Friedrich Kohlrausch, "Beobachtungen im magnetischen Observatorium aus dem Jahre 1869, insbesondere Bestimmung der Siemens'schen Widerstandseinheit nach absolutem Maasse," *Nachrichten* (1870): 513–524; quotation on 514. See also idem, "Ueber einige hydro- und thermoelektromotorische Kräfte, zurückgeführt auf Siemens'sches Widerstandsmaass und Weber'sches Strommaass," *Nachrichten* (1870): 400–404.

42. Kohlrausch, "Resultate aus den magnetischen Beobachtungen . . . vom Jahre 1867."

43. Friedrich Kohlrausch, "Eine Bestimmung der specifischen Wärme der Luft bei constantem Volumen," *Nachrichten* (1869): 160–162.

44. Friedrich Kohlrausch, "Ueber die Gültigkeit der Ohm'schen Gesetze für Elektrolyte," *Nachrichten* (1869): 14–16.

45. Eduard Riecke, "Ueber die Ersetzung eines auf einer Oberfläche befindlichen Systems galvanischer Ströme durch eine Vertheilung magnetischer Massen (Vorgelegt von F. Kohlrausch)," *Nachrichten* (1870): 103–108.

46. Friedrich Kohlrausch, "Ueber den Einfluss der Temperatur auf die Elasticitätscoefficienten einiger Metalle," *Nachrichten* (1870): 257–262, on 258.

47. Kohlrausch, "Eine Bestimmung der specifischen Wärme."

48. Riecke, "Ueber die Ersetzung eines auf einer Oberfläche befindlichen Systems galvanischer Ströme."

49. Friedrich Kohlrausch, "Mittheilung einer von Herrn E. Riecke im physikalischen Institut ausgeführten experimentellen Prüfung des Neumann'schen Gesetzes über den Magnetismus der Rotationsellipsoide," *Nachrichten* (1870): 396–399. Riecke continued this line of investigation with his "Ueber das von Helmholtz vorgeschlagene Gesetz der elektrodynamischen Wechselwirkungen (Vorgelegt von Wilhelm Weber)," *Nachrichten* (1872): 394–404.

50. Gauss and Weber, *Atlas des Erdmagnetismus.*

51. Visual thinking was also reinforced by Benedict Listing's seminar exercises at Göttingen, not discussed here. On Listing's talents in visualization and drawing, see Klaus Hentschel, *Mapping the Spectrum* (Oxford University Press, 2002), 124, 429. Problems in visualization also fell within the scope of Kohlrausch's research; see his "Ueber eine durch die verschiedene Brechbarkeit des Lichtes hervorgebrachte stereoskopische Wirkung," *Nachrichten* (1870): 415–416.

52. Kohlrausch, "Mittheilung einer von Herrn Riecke . . . ausgeführten experimentellen Prüfung des Neumann'schen Gesetzes," 397.

53. Friedrich Kohlrausch, "Ueber galvanische Widerstandsbestimmung flüssiger Leiter, insbesondere über die von Herrn A. Nippoldt im hiesigen physikalischen Institute ausgeführte Messung des Widerstandes der verdünnten Schwefelsäure," *Nachrichten* (1868): 415–420, on 419.

54. Kathryn M. Olesko, "When instruments disappear from view," in *Abriendo las Cajas Negras. Colección de instrumentos cientificos de la Universitat de València*, ed. José Ramón Bertomeu Sánchez and Antonio García Belmar (Universitat de València, 2002), 409–414.

55. Friedrich Kohlrausch, "Das Weber'sche compensierte Magnetometer zur Bestimmung der erdmagnetischen Intensität," *Nachrichten* (1871): 50–57.

56. Kohlrausch, "Resultate der Beobachtungen im magnetischen Observatorium zu Göttingen vom Jahre 1868," 36.

57. Kohlrausch, "Ueber den Einfluss der Temperatur auf die Elasticitätscoefficienten einiger Metalle," 259.

58. Hence Weber's magnetometer was valued because it was transportable. Kohlrausch, "Das Weber'sche compensierte Magnetometer," 51.

59. Kohlrausch, "Ueber galvanische Wiederstandsbestimmung," 416.

60. Kohlrausch, "Eine Bestimmung der specifischen Wärme der Luft bei constantem Volumen," 160.

61. Kohlrausch, "Beobachtungen im magnetischen Observatorium aus dem Jahre 1869," 517.

62. As evident in the tables he prepared for his practical exercises. NFK 2601, 3, 35, 49, 54, 91, 99.

63. Kohlrausch, "Beobachtungen im magnetischen Observatorium aus dem Jahre 1869," 518.

64. NFK 2601, 1, 3.

65. Rudolf Stichweh, "The unity of teaching and research," in *Romanticism in Science: Science in Europe, 1790–1840*, ed. Stefano Poggi and Maurizio Bossi (Kluwer, 1994), 189–202, on 194.

66. Kohlrausch, "Ueber galvanische Widerstandsbestimmung flüssiger Leiter," 417.

67. Ibid., 418.

68. In this regard, Sharon Traweek's invocation of Roland Barthes bears repeating: "The origin of work is not in the first influence, it is in the first posture: one copies a role, then by metonomy, an art; I begin by reproducing the person I wish to be." *Beamtimes and Lifetimes: The World of High Energy Physicists* (Harvard University Press, 1988), quoted on 82–83. The thrust of Barthes's statement, Traweek's argument, as well as the example discussed here runs counter to postmodern assessments of identity formation in science, such as Jean-François Lyotard's remark that "a person [in science] does not have to know *how to be* what knowledge says he is." On the contrary. Jean-François Lyotard, *The Postmodern Condition: A Report on Knowledge* (University of Minnesota Press, 1979), 26.

69. Stichweh, "The unity of teaching and research," 192.

70. On the *Feinmechanik* industry in and around Göttingen before 1900, see Hermann Wellenreuther, "Vom Handwerkerstädtchen zur Universitätsstadt," *Göttinger Jahrbuch* 49 (2001): 21–37; O. Behrendsen, *Die mechanischen Werkstätten der Stadt Göttingen: Ihre Geschichte und ihre gegenwärtige Einrichtung* (Kiepert, 1900); O. Behrendsen, "Zur Geschichte der Entwicklung der mechanischen Kunst: Neue Beiträge zur Geschichte der Mechaniker Göttingens im 18. und in der ersten Hälfte des 19. Jahrhunderts," *Deutsche Mechaniker-Zeitung (Beiblatt zur Zeitschrift für Instrumentenkunde und Organ für die gesamte Glasinstrumenten-Industrie. Vereinsblatt der Deutschen Gesellschaft für Mechanik und Optik)* Heft 10 (1907): 93–96, 101–107, 115–121, 129–137; Klaus Hentschel, *Gauss's unsichtbare Hand: Der Universitäts-Mechanicus und Maschinen-Inspector Moritz Meyerstein—Ein Instrumentenbauer im 19. Jahrhundert* (ms., version August 14, 2002). I thank Dr. Hentschel for sharing his manscript with me, and for exchanging views on the role of the instrument maker Meyerstein in the seminar, practical exercises, research, and state projects (notably weights and measures determination) in Göttingen.

71. H30 Mathematisch-physikalisches Seminar, [Sommersemester] 1850—Sommersemester 1856. Historische Sammlung, PI-UG; 4/Vh/20, UAG, Bl. 76r, 122v, 198r-298v; NFK 2502.

72. For example, Kohlrausch, "Beobachtungen im magnetischen Observatorium aus dem Jahre 1869," 58.

73. NFK 2502.

74. Friedrich Kohlrausch, "Bericht über das Physikalische Institut, Abtheilung für Experimentalphysik, aus den Jahren 1866 bis 1870," *Nachrichten* (1870): 417–420. Reprinted in *Gesammelte Abhandlungen von Friedrich Kohlrausch* , 2 Bde. (Barth, 1910–11), volume 1, 1006–1008.

75. Kohlrausch, "Bericht . . . aus den Jahren 1866 bis 1870," 417.

76. Ibid., 418.

77. Ibid., 417 (emphasis added).

78. Ibid., 419.

79. Ibid., 417, 418.

80. Ibid., 418.

81. Kohlrausch, *Leitfaden der praktischen Physik,* first edition (1870), iv.

82. Kathryn M. Olesko, "Physics instruction in Prussian secondary schools before 1859," in Olesko, *Science in Germany*, 92–118.

83. Kohlrausch, *Leitfaden der praktischen Physik,* fourth edition (1880).

84. Ibid., ninth edition (1901); tenth edition (1905).

85. Notably, Kirchhoff was trained at Königsberg, reducing the cited loci of systematic instruction to two, and Weber was not mentioned, undoubtedly due to the narrow focus of his practical exercises. Kohlrausch, *Lehrbuch der praktischen Physik,* eleventh edition (B. G. Teubner, 1910), vii (emphasis added). Cahan also points this out in "Institutional revolution," 48, where he uses a slightly different translation. The foreword is reprinted as "Vorwort zur 11. Auflage des *Lehrbuchs der praktischen Physik," Gesammelte Abhandlungen von Friedrich Kohlrausch,* ed. Wilhelm Hallwachs, Adolf Heydweiller, Karl Streker, and Otto Wiener (J. A. Barth, 1910), 1084–1088.

86. Kohlrausch, *Leitfaden der praktischen Physik,* second edition (1872), iv–v, 17.

87. Ibid., fourth edition (1880), 21.

88. Kohlrausch, *Kleiner Leitfaden,* "Aus dem Vorwort zur ersten Auflage [1900]," fourth edition (1907), v.

89. Kohlrausch, *Leitfaden der praktischen Physik,* second edition (1872), iv.

90. Kohlrausch, "Bericht . . . aus den Jahren 1866 bis 1870," 419–420.

91. Ibid., 418.

92. For an analysis of his course on mechanics as theoretical physics, see Olesko, *Physics as a Calling*, 172–204.

93. For a regional assessment of the adaptation of Neumann's pedagogy, see ibid., 317–365 (in secondary schools), and 366–450 (in universities).

94. UAG 4/Vh/20, Bl. 86r-87r, 204r, 368r. Weber correctly pointed out that most students obtained positions elsewhere.

95. For example, the son of Wilhelm Apel, A. Apel, whose notebooks provide a rare look at Weber's lectures. H346 Physik. Weber. Wintersemester 1858–59 & Wintersemester 1865–66 [A. Apel], and H14 Experimentalphysik. Sommersemester 1860 [A. Apel], Historische Sammlung, PI-UG. For a later period, 1875–1885, medical students were the fastest growing element of Kohlrausch's clientele. Cahan, "Institutional revolution," 45–46.

96. A study of inclination and talent in the nineteenth century must begin with their early modern meanings. See Richard M. Douglas, "Talent and vocation in humanist Protestant thought," in *Action and Conviction in Early Modern Europe: Essays in Memory of E. H. Harbison*, ed. Theodore K. Rabb and Jerrold E. Siegel (Princeton University Press, 1969), 261–298; Anthony LaVopa, *Grace, Talent, and Merit: Poor Students, Clerical Careers, and Merit in Eighteenth-Century Germany* (Cambridge University Press, 1988). Contemporary criticisms in the popular press included: Anon., "Ueber den Kantischen Begriff vom Genie," *Philosophisches Archiv* 2.2 (1794): 13–25; Salomon Maimon, "Das Genie und der methodische Erfinder," *Berlinische Monatsschrift* 26 (1795): 362–384; Anon., "Genie und Verstand," *Neues Hannöverisches Magazin* 12 (1802): 745–748.

97. Johann Friedrich Herbart, *Ueber die Möglichkeit und Nothwendigkeit, Mathematik auf Psychologie anzuwenden* (Bornträger, 1822), 21–23. By the middle of the century, "genius" had become an object of satire. See Bogumil Goltz, "Der deutsche Genius und seine Bedeutung für die Welt," in *Exacte Menschen-Kenntniß in Studien und Stereoscopen.* Erste Abtheilung*: Characteristik und Natur-Geschichte der Frauen.* Zweite Abtheilung*: Physiognomie und Characteristik des Volkes.* Dritte Abtheilung*: Die Deutschen: Ethnographische Studie* (O. Janke, 1859–60), volume 3, 1–19.

98. Claude Lévi-Strauss, "The structural study of myth," *American Journal of Folklore* 68 (1955): 428–444.

99. Hall, *Silent Language*, x.

100. Ulrich, "The notion and definition of a canon," 21. The reference is to Plato's plea for clarification.

# 11 Generating High-Energy Physics in Japan: Moral Imperatives of a Future Pluperfect

Sharon Traweek

## Introduction

In this essay I am interested in how and why particle physicists simultaneously engage in three projects: generating new knowledge makers, generating new knowledge, and generating new sites for their work. Specifically, I want to explore the pedagogical aspects of this process. I use three Japanese sites: KEK, now called the Japanese National Accelerator Research Laboratory; Sokendai, a new university for interdisciplinary graduate education; and the GLC, the Global Linear Collider that Japanese particle physicists want to see built in Japan. KEK has been in use since 1976; major new facilities were added in 1986 and 1999; Sokendai opened in 1988. The GLC is to be built by 2006, either in Japan or the United States. Each site is an incubator for different kinds of new physicists.

Physicists have promoted new projects by promising that they will generate new knowledge in theory, experimental design, data analysis, detector construction, and accelerator design. They want new projects in order to maintain skills at a site, to generate new skills at that stable site, and to transfer skills to new sites. Who will make all that new knowledge? They need new projects to create the best possible next generation of physicists, against the anxiety that left unchallenged, the next generation will not be able to make new knowledge. Lurking in the background is the fear that the best and brightest of the next generation might have already been attracted to the biological and computing sciences. To make new knowledge in particle physics, then, new projects must be created to lure the prodigal best and brightest of the next generation. I will argue that the community can only learn and teach with its prostheses; it is a cyborg.[1]

### Oral Traditions

Knowledge is acquired orally in the international high-energy physics community.[2] High-energy physicists doing experimental research at the best labs around the world do not read frequently; they rarely write, and crucial details of their knowledge are never written. No one could fully understand an experiment or an accelerator from reading the published articles. Drafts of articles, called "preprints," circulate internationally, both on paper and electronically; people scan through these casually, formulating their questions, and then contact the authors to talk. Their American English shop talk is in the sort of vernacular spoken by men in working-class and lower-middle-class American work groups. More recently, a less gendered and less American form of International English with a more middle-class vocabulary and grammar, devoid of American slang, has become widely spoken among physicists currently working in Japan, as it is among the growing traffic of experts circulating around the world. The accent remains American.[3]

Almost all the articles published about experiments in the two major journals (*Physical Review* and *Physical Review Letters*) list all the members of the experimental research group alphabetically as authors, often more than a hundred people. The authors' separate institutional affiliations are given, but their titles are not. No outsider knows who among the authors are technicians, graduate students, postdoctoral research associates, or professors (of whatever rank), and no one can discern who is responsible for what part of the experimental equipment, research design, research equipment operation, data collection, or data analysis. Just because an outsider cannot determine this from a list of authors does not mean that insiders do not care about these issues.

Insiders must know whom to contact in the group in order to learn important details about their experiments, and to know whom to believe. All that knowledge is transmitted orally.[4] These physicists talk, continuously it would sometimes seem, about who knows what, whom to credit, whom to fault, and the quality of everyone's current reputation. They also talk a lot about the past, telling stories. They often disparage the histories of their field written by professional historians, in part because they are based only upon published articles, laboratory records, formal interviews, and funding documents.

The basic knowledge of how to conduct an experiment in the "state-of-the-art" laboratories of the world is not written. This tacit knowledge, these "rules of thumb," this common sense can only be learned inside the community, inside a lab, inside a research group. High-energy physics is held in the minds and bodies of the practitioners and in their talk; it is transmitted through participating in an experiment with people who are knowledgeable. One of the ways to find out who is not knowledgeable is to see who is reading: usually these are graduate students. If students are learning only from articles in journals, they do not have access to the basic knowledge of the community. If students are not at one of the five state-of-the-art labs in the world or in one of the fifty or so top university physics departments in the world it is difficult to learn this oral knowledge.[5]

At two major high-energy physics laboratories in the United States, established physicists gossip endlessly about the differences between the two labs' leaders, equipment design, operation of experiments, and decision-making processes. At KEK, the laboratory where I have done most of my research in Japan, the gossip and talk are on approximately the same issues: apprenticeship, leaders, equipment, beamtime, decision making, and funding. Access to this gossip is crucial for a full apprenticeship; participating in it is a necessary sign of membership in the community.[6]

### Infrastructure: Library to Laboratory

If labs or departments are trying to improve their status, they must import some of the knowledgeable ones and send some of their own for sojourns in the top places; they must also build and sustain the infrastructure necessary for basic research. That infrastructure includes (1) sustained funding for education, research, and laboratory equipment at all levels, from elementary schools to national laboratories; (2) routine allocation of a certain proportion of the country's gross national product for scientific work; (3) involvement of a certain proportion of the country's population in scientific work; and (4) consistent international engagement in high-level information exchange about and documentation of their work. Much more significant is the unqualified acceptance of a country's researchers' observations and analysis by the core scientists in the international research community.[7] The physics community in the United States achieved this infrastructure during the 1940s and the early 1950s. The Japanese have been building these resources during the last several decades; in the last 15 years they have made it into the world of state-of-the-art labs.

Basic research in experimental science—particularly in so-called big science, with its stunningly expensive research equipment—can only be conducted in the very richest countries. At the other end of the spectrum, many universities around the world are not able to afford subscriptions to the major research journals, much less easy access to the Internet. Only during the past generation has Japan been able to support the kinds of basic research that require state-of-the-art research equipment, state-of-the-art communications technologies, and the frequent international circulation of many scientists.

In 1976 the Ministry of Education was questioning the postage physicists needed to send their articles abroad, claiming this was personal correspondence, and the physicists knew they were not receiving all the preprints circulating among the top laboratories. Most now have easy access to preprints, electronic mail, fax, and telephones, and

Japanese physicists are always involved in planning the workshops and conferences held nearly monthly around the world, and occasionally those gatherings are held in Japan. Nonetheless, not all Japanese participate easily in the most important aspect of those meetings, the "corridor" talk that occurs before, during, and after the meetings; nor are there many who are comfortable being "on the circuit," giving informal talks and visiting colleagues at their home institutions, as Americans are groomed to do while they are still apprentices.

Before the 1980s many Japanese graduate students in physics apparently wrote dissertations based upon analytic summaries of research journal articles, a clear sign that those candidates and their departments were still only at the edge of the international research community. By the mid 1980s the physicists at KEK were trying to get permission from the Ministry of Education both to bring Japanese graduate students to KEK and to grant degrees to these students, based upon their day-to-day work collaborating on KEK experiments. (Many of the returnee Japanese physicists described this apprenticeship style of research education as American or European.[8]) KEK, along with several Japanese labs in other fields built in the 1970s, eventually received this permission and by 1993 there were many such graduate students at KEK, including several foreigners, primarily from Asia and Latin America.[9] Situated learning in big science is also a luxury only the richest countries can afford.

## Personae and Pedagogy

Since most of our discourse on the learners and teachers is rooted in so-called Western assumptions about "individuals" and "selves," I want to explore those assumptions.[10] Our discourse, our analysis, and our subject of "the self" replicate each other. "The self" is just one of a fascinating list of singular generics that permeate a certain sort of discourse, usually academic, namely self, justice, beauty, man, truth, logic, identity, culture, reason, woman, individual, and so on. Changing any text's singular generics into plurals (or into gerunds) very quickly reveals important analytic slippages. For this project on pedagogy it is important for us to remember that the persons who are learning and teaching physics are not singular generics and they are not in oppositional categories.

My goal is to write about physicists engaged in learning as an everyday process with multiple, accumulated experiential strategies for knowing, often in the company of others using different ways of knowing. It is necessary for us to remember that the persons who are learning and teaching, the community that is learning and teaching in various settings all may be culturally different from the persons, groups, and settings in other cultural sites, even if those sites are just across the room at some international laboratory. Consider how persona are presented in various social situations. In the Japanese *bunraku* puppet theater three people dressed in black manipulate the colorful puppets and all are visible to the audience. One operates the head and right arm, a second operates the torso and left arm, and the third moves the legs; the second and third puppeteers are masked. Very quickly an audience member can learn to "watch" either the puppets or the puppeteers, but not both simultaneously.[11] Similarly, we can see the person or we can see the person embedded in a web of social relations.

During the mid 1980s a senior position at a middle-rank Japanese university was vacated by a retirement and one group, outside that university, wanted a junior, very talented physicist to be given the job, rather than the person who simply would have been promoted into it by the traditional way of managing these transitions. Negotiation nationally focused not on the young man's undisputedly superior work in physics, but on how the department would function if a young person were to have high status, and, most importantly, on why a young person ought to have high status. The visiting American physicists found this confusing: truly believing in meritocracy, they thought status ought to be the outward sign of an interior state, namely talent in physics. The Japanese physicists thought high status ought to correlate with another kind of talent: capacity to guide and nurture others, rather like an exceptional parent. Certainly there ought to be jobs and recognition for exceptional physicists, but high position was reserved for those who could recognize and nurture the exceptional ones. The Americans were startled by this notion that high status should be conferred for generativity. Some of the Japanese physicists were trying to insert "foreign" criteria in the promotion process.

At the same time the Japanese physicists also were actively discussing who ought to succeed the director of the major physics laboratory in Japan upon his retirement in a few years. Everyone seemed to believe that their opinions ought to be and would be heeded. A real leader of the lab, one said, ought to lead by *hara-gei*, literally by the "art of the stomach."[12] This means that by the time a decision is made or an action is to be taken a leader has so responded to and shaped the views of others that the leader's will is known and the others will act upon it without the leader needing to speak. Others said that the next director should be someone in whom everyone could trust that he would make decisions in terms of what was best for physics, not in terms of personal ambition.

This disjunction was particularly confusing for the Americans who thought that whatever was good for a talented physicist was good for physics. As one Japanese physicist at the laboratory began to receive important positions on government committees, adjunct university professorships, laboratory committees, and so on, many other Japanese remarked upon his sacrifice: obviously he would prefer to think only of physics, but he was taking on these grand positions in order to advance the whole field of physics in Japan. He was choosing to be generative.

Thinking about learning, teaching, meritorious thinking, and leading thinkers is not always organized in the way that we conventionally talk about these subjects. The personae dramatis are not located in insolated actors, but are realigned, rather like a stage filled with those *bunraku* puppets, each animated by three visible black suited artists, with the puppets' voices all chanted by another three artists, where no one chanter speaks only for one puppet. Site, action, dramatic situation, motivation, and intention are all present, along with pathos, heroism, comedy, and tragedy, but they are reconfigured.[13]

Our conventional American webs of signification about selves emphasize that people own their attributes, enact their decisions, acquire reputations, have thoughts, and possess emotions, maintain privacy—in short, are good capitalists. Encountering people in another culture our local notions about selves are challenged by theirs. Dorinne Kondo wrote that as a fieldworker in Japan she was propelled toward a "collapse of identity" which "was carefully negotiated by the ethnographer and her informants within particular configurations of power."[14] Different generations of American physicists of differing ethnicities and gender, working in different parts of the world have different ways of being physicists, just as Japanese physicists of different generations and gender from different regions of Japan have different ways of putting together "Japaneseness" and physicist. Encountering each other they must learn the meanings of their differences.

## Apprenticeship Learning in High-Energy Physics: Variation in High-Energy Physics

Situated learning in high-energy physics, of course, lasts an entire career, but there are discrete stages of a career in each country and specific qualities that one is expected to learn to display during each of them.[15] American undergraduates learn analogical thinking from formulaic textbooks, tricky problem sets, and choreographed experiments; the dominant pedagogic style is authoritarian, whether the teacher is lecturing, demonstrating, or engaging in so-called Socratic questioning. Graduate students are eventually given a mundane task, usually rebuilding a piece of malfunctioning research equipment, a task which requires dedication and meticulous attention to detail; they are also given enormous time pressure. The students learn, by trial and error, to find shortcuts, and to fear, above all else, losing time; they also learn to ask laboratory technicians most of their questions. Their supervisors are postdoctoral research associates who only hope the graduate students will not kill themselves with exploding equipment; they tell many stories about near-misses.

The postdoctoral research associates learn to differentiate themselves from the graduate students, in part by always discussing their own ideas with brash confidence and challenging the ideas of all others harshly. But like the graduate students, they too are given a very time-consuming and tedious task: postdocs are expected to build, usually from scratch, some complex but mundane software or hardware the group very much needs for its experiment. Gradually, the postdoc must learn, with no instruction on this crucial point, that to merely complete this necessary task, like a good grad student, is inadequate. Without instruction, again, they must quietly anticipate and inform themselves about some bit of knowledge or equipment the group might "unexpectedly" need to complete the experiment well. If their predictions are correct, their expertise eventually will be absorbed quickly by the group, with appreciation, of course, but often without comment.

The postdocs I met at Fermi National Accelerator Laboratory (Fermilab) in the 1970s and the 1990s all said that they saw themselves as "independent." They knew that most of the senior experimentalists doing research at Fermilab were university-based and actually would not be at the lab on a regular basis; these postdocs wanted to join the other younger physicists whom they saw as "really running the experiment." Being "independent" from the senior physicists did not mean that they meant to be independent from the other young physicists. The postdocs at the Stanford Linear Accelerator Center (SLAC) said that they thought all the groups were very "tight-knit" and that communication in the groups would be good. They thought that they could learn a lot from being among experimentalists who had so much experience working together. They were also eager to learn from the senior experimentalists whom they saw as being in the lab all the time. I asked the postdocs how they learned what they needed to know as postdocs. Almost all the Fermilab and SLAC postdocs indicated that they tried to learn by doing and watching, rather than by asking any questions. The period of doing menial tasks seemed longer at SLAC than at Fermilab.

Most of the postdocs at SLAC, closely watching physicists at various stages of their careers, concluded, without having been told explicitly, that it was important to learn three sets of skills: experimental design, data analysis, and how to "run experiments." They pointed out to me that anyone who learned only one of these kinds of skills would not become a research group leader. Fermilab postdocs' comments about how to become a group leader were less specific, referring only to "getting results" and "having good ideas," their same criteria for being a good postdoc. Both Fermilab and SLAC postdocs told sobering stories about inappropriate ways of calling attention to one's work and the consequences. The SLAC postdocs had a much more detailed sense of how to present their ideas and work within their research group, perhaps from having attended weekly group meetings with the full array of middle-rank and senior physicists. All the postdocs told stories of camaraderie among the postdocs and the necessity of being at the lab all the time.

Postdocs have learned to be quite anxious about the future. Holding their positions for 3–6 years, they are nearing the end of their lengthy apprenticeship. They know that about three-quarters of those who have finished their 11–18 years of training in high-energy physics do *not* remain in the community.[16] Those successful few who are incorporated into the community soon learn to feel that time is a most desirable, but scarce commodity; they also learn very quickly to fear the impending obsolescence of their once successful ideas and skills.

The personal qualities expected in the successful initiate are not attractive, engaging, or compelling to most foreign students in the American high-energy physics laboratories; if they do learn to display these qualities, they later remark on how difficult it was, upon their return home, to shed these new features of their personalities which their compatriots found so unappealing. Some, of course, enjoy performing this style and very much miss it after they leave the United States. Many Americans, including many minorities and women, are also quite unwilling to acquire these traits novices must perform so adroitly; consequently, they choose not to pursue a career in one of these fields, even though they are quite interested intellectually in the subject matter.[17]

Elsewhere I have argued that this American apprenticeship learning in high-energy physics is profoundly engendered.[18] I have also argued that it is profoundly American. There are several features of this apprenticeship that, taken together, seemed to me distinctive to contemporary American education in the professions:

1. erecting barriers to the successful completion of tasks by the newcomers
2. tacitly teaching novices that to succeed they must simultaneously obey dutifully and yet learn to break their leaders' rules, covertly and cleverly
3. expecting newcomers to speak brashly and harshly to everyone as an outward sign of their commitment to truth, instead of to social niceties
4. the years of spending almost all one's waking hours in the lab
5. believing that acquiring knowledge outside their field shows a lack of commitment to the subject and is a waste of time better spent on their research
6. taking pride among the novices in never asking questions and the feeling among teachers that providing detailed explanations is inappropriate
7. harboring complex anxieties about time

8. the feeling that men are better suited to these tasks. (Folk theories about the relationship between sex hormones, gender orientation, rationality, and creativity are widespread.)

Several popular books rehearse the formulaic trials of this American novitiate.[19] Only a few academics have studied this process: exemplary are the psychiatrist Benson Snyder's study of MIT undergraduates and the study of the training of surgeons by the sociologists Charles Bosk and Stephen Hilgartner.[20]

### Explaining Variation in Apprenticeship Learning

All but one of these eight qualities have been used to describe Japanese apprenticeship.[21] I have two possible explanations. The first is that this novitiate in the professions in the United States is not distinctively American and that it strongly resembles apprenticeship around the world. It is characteristic of apprenticeship learning worldwide that, invited into the site where real work is being done, the apprentice at first is not allowed to do significant work and is expected to stay out of the way of people who are. Beginning apprentices are usually assigned to do some necessary, but tedious task. During that time, which could last many months, if not years, they are to learn diligence, commitment, and determination, but they might not realize they are being evaluated for these qualities. If successful, the apprentices gradually begin to be assigned less menial tasks. They learn, not because they are given instruction, but because they take it.

Many of the American senior high-energy physicists came from rural, lower-middle-class, or working-class families and were often the first in their families to attend college; these men may have unconsciously replicated the world of learning a trade as they formed a way of learning and teaching expertise in research. Nonetheless, training for the professions and arts in other countries appears to be more strongly marked by the manners and expectations of higher social classes.

Another argument is that this novitiate is similar to the preparation for priesthood in some monasteries and that all novitiates in the professions and the arts might be patterned on these prior models from religious institutions.[22] The historian of science David Noble has asserted that the modern international scientific community bears many striking resemblances to the international religious communities in medieval Europe.[23] One difficulty with this argument is that the European high-energy physicists, physicians, and lawyers do not seem to have endured this kind of an apprenticeship. What of the Japanese high-energy physicists and their models of learning and teaching: does their apprenticeship resemble a novitiate or an arts training by an *iemoto*?

There is another important issue. Each generation learns, matures, and tries to teach the next in specific historical, social, material, intellectual, economic, and political ecologies. When there are sudden and pervasive changes in those ecologies, it may make the generational transfer of craft knowledge impossible. This problem saturates the novels of Kawabata, in particular *The Master of Go*.[24] A master of the board game of *go* could not find the old kind of student; the eager students of the current generation of Japanese high-energy physics masters are also not like their teachers. In fact, some are not even Japanese. Finding the right students is a problem for them, just as it was for Kawabata's master; finding the right teacher (and the right practice) is a problem for the new kinds of students, just as it was for the provocative younger player in Kawabata's story.

Since 1945 the Japanese high-energy physics community has experienced multiple ruptures in these ecologies, ruptures so serious that teaching the next generation the older ways of making experiments may be inappropriate. Nonetheless, it may be quite useful to pass on their accumulating knowledge of various ways to do physics in fluctuating ecologies. My story here is about the multiple ways of learning and teaching high-energy physics in multiple kinds of institutions while coping with ruptures and strategically building continuities.

## Sixty Years of Generational and Class Differences among Japanese Physicists in the Present

Since the Meiji Restoration of 1868, establishing Japan's modern nation-state and, subsequently, its modern university system, there have been at least four modes of teaching physics in Japan, and there are now four currently active generations of physicists there.[25] Retired, senior, mid-career, and junior Japanese physicists have been educated in physics quite differently and at very different historical moments in Japan's political economy. These generations differ in several ways: social class, physics language, physics education, career paths, gender, and leadership and negotiating styles.

### 1870–1950

The first mode, practiced from the 1870s through the late 1940s, was based on German university models, including German-language textbooks.[26] Physics concepts were described with German words in an otherwise Japanese discourse. Physicists born up to the mid 1930s (now in their late sixties or older) received this sort of education. Opportunities for experimental research were quite limited, but some good research equipment was available at Osaka University and at a few research institutes. Most students concentrated on theoretical issues, requiring only paper and pens, access to the journals, and conversation with colleagues. The very best students might have hoped to visit a German university for a year or two and then return to Japan to teach. Hideki Yukawa did his Nobel Prize-winning work in this sort of setting.[27] From Yukawa's autobiography and the reminiscences of others in this generation, I gather that there were a few very collegial working groups of physicists, but there was not much interaction among them.[28]

It appears that most of the Japanese physicists of this generation were from academic families; their German teachers and peers had similar class backgrounds.[29] Visiting a laboratory of a physicist from this generation I was struck by how cluttered and even dirty the lab, the offices, and the hallways were. I was told that the group leader had decided to use the salary of a janitor to support a graduate student; the whole group was supposed the clean the rooms regularly, but they rarely resolved to do this. I was witnessing a reenactment of the Spartan ethos of the old prewar Japanese elite secondary schools.[30] Physicists of this generation received the resources for their research entirely from the Ministry of Education, staffed by career bureaucrats who would have attended the same sort of primary, secondary, and undergraduate schools as the physicists; they would understand each other.

### 1950–1975

During the Allied occupation of Japan physics began to be taught from American textbooks, written, of course, in American English. The physicists of this generation tend to use English-origin words for basic physics concepts, although some may write them in *katakana*, a Japanese phonetic script used for foreign words. Many of them were able to take advantage of the newly instituted Fulbright program to attend graduate school in the United States and some stayed for postdoctoral research positions. Most speak an American English of the sort spoken in American university research labs rather comfortably, but they do not read much in English, aside from scientific reports. Some have told me that they learned English as children from American movies and from GIs.

This generation of Japanese physicists was born between about 1935 and 1955; that is, they are now in their late forties to late sixties. The elder ones in this group were young during World War II; most experienced widespread hunger. Some witnessed the Allied fire bombings of the cities, and a few were in Hiroshima or Nagasaki during the atomic bomb blasts. Many say that as children they became fascinated and horrified by the physics of the war and resolved to study how physics could be used peacefully. At every annual meeting of the Japan Physical Society (*Nihon Butsuri Gakkai*) there is a notice on the front cover of the program announcing the group's commitment to the peaceful uses of physics.

Many of the men in this generation who are physicists have been raised in rural areas or in working-class parts of the cities; they often are the first in their families to have received a university education. They tell poignant stories of finding some confusing book about physics in their middle school libraries and struggling alone to understand what they now regard as very elementary ideas. Several have told me that they would never have had access to post-secondary schooling if it had not been for the land and educational reforms of the Allied occupation policies. Some have complained that their Japanese college physics teachers did not teach: that is, as students they were assigned problems and when, without help, they reached a solution, they were simply given another problem. A few are proud of the fact that they learned how to teach themselves physics from the textbooks they found in libraries. Some have confided to me that they were decidedly uncomfortable with the higher social class style of their teachers; they realize that their more rustic or rougher ways may have annoyed their teachers.[31] American men I have studied in this generation came either from rural areas or from working-class urban families, especially the children of Jews who immigrated from rural Eastern Europe and Russia about the time of the Russo-Japanese War. Consequently, the Japanese Fulbright students often found their American teachers and peers to have rural or working-class backgrounds similar to their own.

After Hideki Yukawa won the Nobel Prize in 1949 he was able to establish a Research Institute for Fundamental Physics at Kyoto University, primarily for theorists.[32] Those who have spent time at Yukawa's institute said it was an extremely important part of their education. I was told that Yukawa had thought it very important that the various groups of physicists in Japan communicate more freely and so he arranged for his institute to host many young Japanese postdoctoral physicists for five years. In Japan prestigious departments admit to their graduate programs very few students educated elsewhere and hire almost no one educated elsewhere to their faculty, but they were willing to hire the Yukawa postdocs. My interlocutors said that they thought the institute's openness had been tolerated because of the enormous prestige Yukawa enjoyed throughout Japan. When KEK was being planned, it was decided that in staffing the new lab they should try to follow Yukawa's model, so that it too would be a place bringing together good physicists from all over Japan and building strong ties throughout the community in Japan.

Several physicists of this generation speak with great fondness of a few Japanese laboratories, located in Kansai, Kanto, and Tohoku, where they did experiments in the 1950s and the 1960s. They describe the rudimentary equipment and sparse resources with a certain pleasure, pointing out that they had to be very clever to get any physics done at all. They enjoyed the fact that everyone needed to work on every part of the experiment and they all learned a great deal. They speak in especially glowing terms about the men who gathered the resources to start these labs and argue that such labs are incomparable facilities for training young physicists. Those educated in the southwestern part of Japan (loosely speaking, from Nagoya to Hiroshima and on to Kyushu) emphasize that physics discussions there were especially open because the groups were run "democratically," that is, with very little attention to hierarchies of age or position.[33] Some were shocked when they first traveled to Europe and found the labs there extremely hierarchical, even more than they would expect in the northeast of Japan; they find labs in the United States run in a manner that would register about midway on this scale.

Like Americans and Europeans, Japanese of this generation often discuss the leadership styles of laboratory directors. Leading by hierarchical authority or by force of personality are thought to be less desirable than leading by *hara-gei*. As I described earlier, this is an art, power, and knowledge thought to emanate from the lower stomach area.[34] When I suggested to physicists that leadership by *hara-gei* also demanded a corresponding sensitivity on the part of the followers, several agreed and pointed out that the younger generation is much less equipped with that skill.

Another part of the anatomy was invoked in the 1980s by a Japanese physicist when discussing the leader of his lab: he said that the shoulders of his director were very large and strong. Confused, I asked what he meant; it was explained to me that everyone could take their problems, problems of any kind, and the director would carry all of them on his powerful shoulders. The physicist agreed that he was invoking *amae*, a mature leader's capacity to encourage interdependence in the community and a capacity among the members for appreciating this interdependence.[35]

One day at the cafeteria we were discussing the leadership styles of the men who might succeed the then retiring director of KEK; it was agreed that one could lead by *hara-gei*. Later an American physicist joined Japanese physicists in discussing the same topic; when asked to describe the special qualities of the same potential director, a Japanese physicist said, "I would give him a blank check." The closest American equivalent I can find for *hara-gei* is the capacity to recognize and act on one's "gut reactions," but in the United States that expression suggests a sensitivity to one's own desires rather than to other people's. The process of responding to *hara-gei* is not unlike that sometimes attributed to women when using the expression "women's intuition," but that usage does not invoke a notion of leading others by force of one's intuition.

One physicist contrasted leading by *hara-gei* to Americans who lead from the mouth, by talking all the time. When I asked if they thought any American physicists led by *hara-gei*, a few said that they thought they knew at least one who was learning. It is intriguing to watch the interactions between the eldest and the next generation of Japanese physicists and the European and North American physicists who come to Japan seeking support for international collaborations based elsewhere: those who lead from the mouth are speaking to those who are sending other kinds of signals. The gap is not unlike the one existing in the United States between the stereotypic communication styles of men and women.

During the mid 1980s the Ministry of Education had begun to require funding proposals be written without foreign words or *katakana* (phonetic script).[36] I found many then mid-career and senior Japanese physicists struggling to create *kanji* (ideographic characters) for ideas that they knew only in English and German; sometimes they would ask me for the etymology of the English and German words so they could try to parallel the ideas behind the words as they designed their *kanji*. In each case the Ministry was causing the physicists to find some way to represent their work and their goals as distinctively Japanese. Many of the men discussing these issues with me had spent their childhoods in villages; they found it ironic that they had to demonstrate the Japaneseness of their work to government bureaucrats who had probably been raised in big cities.

My research in 1986–87 was funded by the Fulbright Association; from their excellent staff I received extensive orientation materials about doing research in Japan. I was embarrassed to read that my behavior during my first visit to Japan had been quite inappropriate and I resolved to do better. When I greeted the same people I had met 10 years earlier I carried a suitable gift and I described my research and my goals more politely. My interlocutors patiently watched my awkward performance, and then asked why I was acting so. When I apologized profusely for not having known how to act properly during my last visit, they then asked where I had learned about such manners. I showed them my written instructions from the Fulbright people in Tokyo. They read them avidly and then laughed, saying that they now understood people in Tokyo a little better and that might come in handy when they next visited the Ministry of Education. Everywhere I went physicists said they had heard I had some interesting instructions from Fulbright and asked for a copy.

### 1975–1990

Even the most ambitious Japanese of this generation can expect to spend their entire careers in Japan, if they wish. They know that Japan has state-of-the-art research equipment in physics, that international conferences are frequently held in Japan, and that foreigners come to Japan to do research and collaborate with Japanese physicists. They are less aware that the so-called infrastructure for basic research in science is not yet firmly in place, even though Japan has been a rich country their entire adult lives.

With a few notable exceptions, the physicists who are now in their late thirties and early forties appear over the last 10 years to be much less comfortable with foreigners than either their elders or the younger group. They seem to see the foreigners at the lab as having useful kinds of expertise, but they also seem to regard the lab as essentially a Japanese space managed by Japanese for goals defined by Japanese. They want all the activities at the lab conducted only in Japanese, including all announcements. There is a place for foreigners in this plan, but not as full colleagues. They see this as the way American and European labs were when they were visiting graduate students and postdocs; they do not seem aware that the American and European labs and universities have become much more cosmopolitan over the last few decades. They are incredulous when I mention the foreign-born leaders of American physics laboratories and when I estimate that about 20 percent of the faculty at leading research universities in the United States are foreign-born. They tell me they cannot imagine anytime in the future when foreign-born faculty at Japanese universities would reach those numbers or that the director of their laboratory or any of its division leaders could be foreign-born.

They have learned most of their physics in Japanese from Japanese textbooks. The youngest have even learned *kanji* for basic physics concepts. Unlike their elders, they do not speak as easily with foreign colleagues. These physicists tend to socialize with the foreigners only in activities arranged for the entire research group or the whole lab. The foreigners in this age group, inclined to see Japan as fully the equal of their home countries, some of whom have spent several years at KEK, lament that they have no good Japanese friends at KEK, although they do have close Japanese friends who work outside the lab.

I should add here that the lab is like a village: there is a dormitory at KEK that has rooms for about 200 physicists, there are about 40 apartments, and there are a cafeteria, a restaurant, a convenience store, a bookstore, an ATM, a large gym, a fitness course, several tennis courts, a soccer field, jogging trails, and a baseball diamond. A barbershop is open two afternoons a week; a travel agent and a medical doctor each visit the lab once a week. It is quite possible to live entirely within the laboratory indefinitely. Three bus lines have a stop directly in front of the lab; one connects the lab to Tsukuba Center, 10 kilometers away; from there buses go to Narita and Haneda Airports. Another bus travels between KEK and the nearest train station; the line connects to north Tokyo in 60 minutes. An express bus links KEK directly to central Tokyo in 90 to 120 minutes. In 2005 there will be a train line connecting Tsukuba Center to Tokyo.

### 1990–Present

Since the mid 1980s many graduate seminars in physics have been taught in English! Understanding the oral transmission of knowledge, some older Japanese faculty have explained to me that they want their students to be able to participate in the face-to-face free exchange of ideas that occurs "in the corridors" at the frequent international conferences, so they teach in English. Their students tend to have urban middle-class backgrounds; relatively few have traveled outside Japan for graduate school or postdoctoral appointments. Most speak a heavily accented, American English; the grad students explain that they have learned from American rap and rock music.

Since the 1990s I have noticed changes in the Japanese graduate students. They tend to live in private apartments in Tsukuba, rather than at the lab, and they have their own cars; this could not be said of any of the earlier generations of physics students I have seen at KEK. They have dressed similarly to their worldwide counterparts (jeans, polo-style knit shirts or short-sleeved woven shirts, and running shoes). This generation includes too the first noticeable cluster of women scientists; many of them are married to colleagues. The graduate students at KEK in the 1990s knew that there were not many positions available in the major universities in Japan. As they learned to do experiments at KEK and interact with people from the private companies building equipment for KEK, these graduate students became interested in taking positions at some companies; several have done so. They shared all these qualities with the physicists of their age group at labs in the United States and Western Europe.

The Japanese graduate students and postdocs in the 1990s were very curious about the young foreign postdocs and graduate students working in the international collaborations based at KEK; they were friendly and socialized with them a bit, but not nearly as much as one would have found in the same age groups at SLAC, Fermilab, or CERN. This may have been due to the fact that the foreigners in their twenties at KEK then had very little money, due to the then rapidly fluctuating foreign currency exchange rates, and consequently could not afford to leave the lab very often; the Japanese students were relatively more prosperous.

During one of my stays at KEK in the early 1990s I had a desk in an office with several Japanese graduate students; they asked me many questions about the conduct of American and Japanese academic life, including the shape of physicists' careers. When I asked why they were asking me these kinds of questions, they replied that I answered more fully than anyone they had ever met, including their own teachers. This reminded me that their elders had complained about the lack of communication with their teachers.

I asked physicists of the elder three generations if there were any *shinjinrui* at the lab. *Shinjinrui*, a term invented by journalists in the 1980s, is roughly translated as the "new breed"; it is used derogatively by older generations to refer to young people who were raised entirely during the years of prosperity. They all answered yes, emphatically. The elders, senior scientists, and the mid-career physicists all found the *shinjinrui* physicists altogether strange. Clearly, every living generation of Japanese physicists sees this youngest generation as significantly different from all their predecessors.

## Generational Continuities and Differences in Iconographic Devices

Elsewhere I have written extensively about generational differences in some strategies for making knowledge in high-energy physics.[37] For example, in my last book I described how Japanese research equipment of the 1970s, like that of earlier decades, had been designed to last for the lifetime of the research group, unlike American detectors where the group was designed to last the lifetime of the detector.[38] Because funding in Japan for research equipment traditionally was allocated to *koza* (rather like chairs in the German academic system) only at the time when the *koza* was being established, research equipment was designed to last at least 25 years, technologically and intellectually. Equipment was designed to be responsive to coping with a very large array of questions about a range of subjects that had already been established as significant. Such equipment was designed conservatively, in part, if not entirely, because of massive funding constraints. By definition such equipment cannot be "state-of-the-art" intellectually or technologically.

The most interesting European and American high-energy physics research equipment built during this same period, by contrast, was designed to answer a specific set of questions on problematic subjects. Some of these detectors were completely dismantled after 5–10 years of use; many were radically revised during that interval. Obviously, the latter procedure requires not only a great deal more money, but also a different set of skills. Furthermore, when KEK got its initial funding the government inserted the constraint that no equipment would be built at KEK and very little would be bought outside Japan. This required the Japanese physicists at KEK to learn to work closely with engineers from the Japanese firms building the equipment designed by the physicists. During this period Americans had at their laboratories large fabrication facilities with large groups of master technicians—really more like artisans—interested in working at such labs because they were always building prototypes and often asked to build the impossible.

By the late 1970s and the early 1980s Japanese physicists at KEK were receiving radically new levels of funding. This, coupled with the radical changes in research design worldwide due to computing, meant that the conservative way they had been designing research equipment for decades, utterly appropriate to the constraints of their then current ecology, might no longer be useful. Those Japanese experimentalists who entered graduate school before 1975 or so, those now over 45, had skills that might be obsolete.

Certainly, computers unite younger physicists in both Japan, the United States, and elsewhere, and separate them profoundly from their elders. Computing had been used initially to run the particle accelerators and collect huge data sets from the research equipment called detectors. Gradually, computers were also used to monitor the state of the detectors during experiments and then to correct any problems in the operation of the equipment. In the past physicists had used the huge data sets to confidently generate analyses of large-scale phenomena. However, increasingly sophisticated computer monitoring of the research process eventually led them to be confident that even anomalous data were real signals and not epiphenomena of the experimental process itself. They began to design experiments to explore those anomalies. Furthermore, by the late 1970s they could do preliminary on-line analysis of the collected data and eventually detectors were designed to be modified in mid-experiment in order to reconfigure the experiment in response to new data.

Just as the physicists turned toward anomalies in the design of their experiments and their equipment, they also began to turn toward simulations in their data analyses. By the mid 1980s younger physicists, in particular, routinely built scenarios of possible particle paths through all the parts of their detector. These plot lines are written in their computers; sub-plots are circulated not just among the group, but also among friends, worldwide, through electronic mail. The physicists over about 45 years of age tend to find this free exchange of simulations rather indiscriminate and indiscreet, and potentially embarrassing for the group. To them it is as if the group freely discussed how they calculated the "error bars" in their own data; in their papers and talks error bars graphically acknowledge noise and uncertainty, but do not announce the details.

There was a profound shift away from a knowledge culture in which people justified their interpretations with talk of "log books," "physical intuition," "physics judgment," and intimate mechanical knowledge of their detector. The craft knowledge of physicists who had finished their postdocs before 1975 seemed to be obsolete. However, it is just those physicists who were becoming the research group leaders and would eventually become laboratory directors in the 1980s and the 1990s. Some of the young physicists in their laboratories today have been using computers to think since they were in elementary school.[39] By now there is a very large research literature on how "computer assisted instruction" and various interactive technologies have been altering pedagogies and learning strategies over the last generation. We know much less about how generations using different knowledge-making practices are working together in the same environment.

I have argued that a practice linking these generations is the making of graphic images for learning, teaching, thinking, proposing, and organizing knowledges.[40] Of course, imaging has a history as old as humans. Scientists have been using illustrations and charts strategically for centuries; they have also served as mnemonic devices for a complex set of conceptual relationships. Consider the mutations in the Period Table of the Elements, Feynman diagrams, and the Standard Model of Fundamental Particles and Interactions.[41] The charts, tables, and graphs used in high-energy physics are pedagogic devices and tools for making new knowledge: "A lengthy education in high-energy physics enables the adept to follow the iconography of these images much better than the novice. [. . . Furthermore, a good] chart should enable adepts to use it to predict new elements of the system by noting the characteristics on the coordinates pointing to any blank boxes on the chart."[42]

Graphic presentations of data also serve iconographically as signatures for the research teams/devices that generated them. The J and psi particles were named for the shape of the data on the computer-generated images from specific detectors' graphic data analyses. Similarly, over the last three decades far less known graphic data displays from many groups are immediately recognized by adepts as both signatures and claims of ownership by those groups.

Simulations have emerged in the context of this long and powerful history of using graphic images strategically for making knowledge, asserting ownership of ideas, and seeking support for truth claims. It would appear from this history that simulations would be immediately accepted as another tool among many similar devices. However, they are far more contested for three reasons. First, older physicists firmly believe that simulations are generated, examined, and defended by the young with no regard to what is physically possible. They say that this shows that many of the younger generation have "no physics judgment."

Secondly, they criticize the young for freely circulating their simulated analyses of the group's data among their friends outside the group. To the elders this practice shows no commitment or loyalty to the group and it undermines the right of the group to claim ownership for the data and analyses they have generated. The ownership of data and analysis are volatile subjects. The elders know too that there are many arguing that all "raw data" be made available to anyone as soon as it is collected. The very idea of sharing data elicits intense conflict within the community. The elders realize that the casual circulation over the last two decades of simulated analyses of current data have undermined the traditional claims of research groups to control and own their equipment, their data, and any interpretations of that data.

Finally, the older generation worries that simulations will be used to replace their accelerators and their detectors. In high-energy physics simulations have been used for about 20 years to assist in the design of experiments, data collection, and detectors, as well as data analyses. These physicists are well aware that in some other research fields simulations have come to replace the mechanical, physical experimental process itself. In a field with equipment as expensive as high-energy physics the practices of these other fields have emerged as a significant challenge. To even raise the subject triggers intense and angry responses among almost all the elders, especially the leaders.

## Gender, Generation, Generativity, and Reproduction

Elsewhere I have analyzed the generational differences among women physicists in Japan and the United States and their various strategies for making knowledge.[43] In some ways their generational differences and strategies are the same as the men's. However, there are interesting differences. The proportion of women physicists in the two countries is about the same (circa 5 percent) for those over 50 years old; that number began to rise in the United States during the 1980s and in Japan during the 1990s. Nonetheless, women in high-energy physics remain a small minority, especially in comparison to the major changes made of the last generation in many other fields, including the sciences, especially in biology.[44]

With a few very notable exceptions, elder women in both countries have been excluded from significant roles and resources throughout their careers. They and their younger counterparts have benefited from changing laws and cultural attitudes about women, especially in the decades after the enactment of the U.S. civil rights legislation (which included gender) in 1964 and the Japanese version in 1986. In both countries women scientists now at mid-career have begun to gain greater access to important resources. For the last couple decades some have been admitted to the major graduate programs and invited to the major research laboratories as postdoctoral researchers, and a few have been hired as scientists at these laboratories. There are almost no senior women in those laboratories; those few are concentrated in computer-based work. Women high-energy physicists are at the center of the action, but primarily in marginal (albeit crucial) roles, at least so far. Again, this is in sharp contrast to the situation in biology.

Because women, as a group, have been so underrepresented in the field and have had so little access to significant resources they have not been in a position to allocate or bequeath resources. Certainly, individual women have been able to do this. In addition, individual women have developed innovative strategies for surviving and succeeding in this hostile working environment; some have also created original ways to teach and mentor students with considerable success. However, as a group, in my opinion, women in high-energy physics have not been in a position to influence pedagogy in physics significantly. Again, this stands in sharp contrast to the situation for biology in the United States.[45]

However, in both countries women have forged important coalitions and organizations. For example, both the American Physical Society and the Japanese Physical Society have sections for women scientists and each society has had commissions to study the conditions for women in the field and to recommend policy changes. The International Union of Pure and Applied Physics held an important and widely publicized International Conference on Women in Physics at Paris in March 2002; most nations' physics societies had prepared data on the condition of women physicists in their countries and then sent men and women delegates to participate. In many cases those delegates were internationally prominent scientists.

National organizations, including American Women in Science and the Society of Japanese Women Scientists, have created coalitions among women across scientific fields. These coalitions have been able to lobby governments, professional societies, and universities to alter policies hostile to women scientists and to create resources for women scientists. Some of these organizations have also created pedagogic programs to encourage girls studying science and to mentor young women scientists.

## Gendered Global Political Economies

There are important demographic changes globally in who are scientists, engineers, and physicians.[46] This is especially noticeable in the United States where there is a decline in the size of the social cohort from which U.S. researchers usually have been drawn, namely white men from rural areas, the urban working class, and the lower middle class. Some university faculty have said that researchers can compensate for this decline simply by recruiting more foreign students; over the past 20 years the proportion of foreign graduate students and postdoctoral research assistants has dramatically increased. For example, the highest status U.S. physics departments now usually have more than

75 percent foreign students. Nevertheless, there has been pressure for other kinds of changes in the recruitment of scientists, engineers, and physicians, and this pressure has come from some interesting sources.

For example, the military wants scientists, physicians, and engineers with U.S. citizenship. States with a large part of their economies based on the production of knowledge, such as California and Massachusetts, do not want a decline in their prosperity over the next 25 years simply because they have a decline in their population of white men in rural areas and the working class and lower middle class. They want to replenish their knowledge workers from other parts of their population. Both the military and states like California are now sponsoring programs to increase the recruitment and retention of women and minorities in science, medicine, and engineering.

It seems plausible to me that similar changes will come to Japan. As Japan shifts in the next decades from a manufacturing-based economy to a knowledge-based economy, it will need even more scientists, engineers, and physicians. It has probably already recruited 10 percent of its population of ethnic Japanese men for this kind of work. The Japanese government is already sponsoring the temporary recruitment of Chinese computer scientists. Unless Japan is willing to import large numbers of foreign knowledge workers, the society will need to turn to other parts of its own population, namely women who are ethnically Japanese and both men and women from minorities and other ethnic groups in Japan, especially Korean-Japanese, Chinese-Japanese, and Burakumin. I suspect that women who are ethnic Japanese are likely to benefit first as the government recognizes the need for many more knowledge producers.

Such women are already working in the Japanese scientific, technical, and medical communities, often in the margins and usually without power. Because of these conditions they are sometimes closer to the world of international science, medicine, and engineering than their male colleagues. These women scientists, physicians, and engineers know how to translate between the international discourse and the national research discourse. I would argue that in those countries in which women are accepted as translators, carrying the words of men to other men, softening the negotiation of power, women also would be able to find work as liaisons in the world of science and technology. These women occupy a doubly negative social category: as scientists who are not men and as scientists who are not European or North American. In this space of double jeopardy they have created, through bricolage, through tinkering with the two systems that exclude them, new strategies for making science and for making scientists.

All the Japanese physicists at KEK, like Japanese scientists at the other new important national laboratories in Japan, work at the edge of two empires, the very well established Japanese national scientific community, based in universities and established over a century ago, and the immensely powerful multinational scientific enterprise based in North America and Europe, launched during World War II. Mid-career Japanese women physicists usually work in those established Japanese universities, but at the edge of three worlds: at the edge of the established national scientific community, at the edge of the communities of newly enriched Japanese physicists, biologists, and astronomers who are building "world-class" laboratories in Japan, and at the edge of the multinational scientific community based in Europe and North America.

## Generating the Future: Designing the Future of High-Energy Physics

Since the end of World War II high-energy experimental physicists internationally have continuously discussed what accelerators and new detectors ought to be built to help them answer the new questions they want to address. They have successfully amassed the resources to do their experiments and build their labs in the context of fluctuating global political economies and they have passed these skills to each new generation. At the beginning of this period those who were participating in these discussions were primarily European trained scientists, whether they were based in Europe or the United States. By 2000 the geography of knowledge making in high-energy physics was a far larger world: there were doctoral students and postdoctoral researchers in north America, east and north Asia, western Europe, and Russia; they were doing frontier research at state-of-the-art facilities in east Asia, northwestern Europe, and the United States. The 500 or so most prominent members of the community now include significant contingents of people educated outside western Europe and north America. The largest group is from Asia (China, India, Israel, Japan, Korea, Russia, and Taiwan, in particular). Australia, Brazil, New Zealand, and South Africa are probably next. Even among the Europeans there has been an interesting change over the last 50 years: there are now many influential high-energy physicists from the "periphery" of Europe (Scandinavia, Scotland, Ireland, Spain, and Greece). In 1945 almost all would have come from Denmark, England, France, Germany, Hungary, Italy, the Netherlands, and Poland.

In 1960 there were about twenty state-of-the-art laboratories in high-energy physics; they were located in Europe (England, France, Germany, Italy, and Switzerland), the former Soviet Union, or in the United States (Brookhaven, Berkeley, Los Alamos). By 1990 that number had been reduced to four or five: one in Japan, one or two in Europe, and two in the United States. In 2010 there might be only one. Over the last 10 years, the high-energy physics community worldwide has reached a carefully crafted consensus on the questions they want to address next, the order of their importance, and the devices they would need to do that body of research. Top priority has been given to the construction of an accelerator called a linear collider. If that machine is built, most physicists assume that it will be located in the country that pays the most for its construction; if so, it would probably be in Japan or the United States because only those countries could afford even half the projected price of several billion U.S. dollars.[47]

If there is to be only one such facility in the world, high-energy physicists have come to agree that the laboratory must be organized "globally," rather than "regionally" (like CERN, the European lab) or "nationally" (like DESY, Fermilab, KEK, and SLAC). Even this classification system is new; until recently, the physicists were proud of the fact that each of these labs welcomed researchers from around the world, independently of how the labs were funded. They liked to call the labs "international" even if the funding and management were not. The prospect of one state-of-the-art lab in the world has prompted them to foreground what they have always known, but not liked to discuss: there is a correlation between a lab's funding source and the nationality of researchers using the lab.

Though the world community of high-energy physicists has become much more multi-national, the sites where they can do their research is increasingly restricted to the small number of richest countries in the world that can afford to build their laboratories. The same process has occurred in other "big sciences": astrophysics, plasma physics, and physical oceanography. Big sciences are defined as those requiring extremely expensive research devices—now costing more than a billion U.S. dollars—with restricted access.[48] In each of these fields the number of "state-of-the-art" research facilities in the world is dwindling to one. (By contrast, scientists in other fields like computing and genetics may use equipment costing up to $100 million each; there can still be more than a dozen such facilities around the world.) Scientists in every big science field are trying to learn how to design a global laboratory and how to convince their governments to support these projects. All these facilities are being justified, in part, by arguments on why they are crucial for the development of the next generation of research scientists.

### The Future Pluperfect: The Global Linear Collider (GLC)

As physicists learn that they have never been global, they also realize that they must design a global laboratory; several models are emerging. Physicists in Europe, Japan, and the United States are generating organization plans for going global. The goal is twofold: convince their foreign colleagues that theirs is the best way to converge their resources and yet minimize the problems of having all those resources concentrated in another country. The second goal is to convince the politicians or bureaucrats in their national capitals that supporting 50 percent of the laboratory's cost is a good investment in their national future.

Meanwhile, the international UNESCO (United Nations Educational, Scientific, and Cultural Organization), based in France, the regional OECD (Organization for Economic Cooperation and Development), based in Belgium, and the G8 (Group of Eight Industrialized Nations), a set of annual conferences held for presidents and prime ministers, finance ministers, and science policy advisers (from Canada, France, Germany, Great Britain, Italy, Japan, Russia, and the United States), have all convened panels to explore how such global scientific laboratories might be financed and managed. At the end of World War II a large set of international organizations were designed and inaugurated, primarily by diplomats. At that time some Europeans imagined a united Europe with a parliament, a common market, and a common currency. They decided to begin by designing Europe-wide research facilities; CERN was one of the first.[49] There is some anxiety among European diplomats and scientists that this first generation of global organizations will be located outside Europe; a compensatory strategy might be to locate the management of these distributed global facilities in Europe.

The shift from "national" to "international" and then to "global" corresponds to a geographical shift in the "critical mass" of research scientists and the infrastructure for their work. In 1900 that mass was distributed along a corridor stretching between universities in Great Britain to those in Italy, as it had far before 1900. By 2000 that mass was distributed in the northern hemisphere in two clusters, one centered in the north Atlantic and another in the north Pacific.[50] By the time the GLC finishes its most productive cycle of experiments (circa 2025) that mass most likely will be distributed in the northern hemisphere in a dense band above the thirtieth parallel. There is also likely to be another critical mass of research scientists and infrastructure emerging along a band in the southern hemisphere, linking Australia, Brazil, South Africa, and Indonesia, at least.

Which model of the GLC and the other new global laboratories will prevail during the next few years as these labs get established? Will these new infrastructures represent rear-guard efforts to maintain the status quo in the face of massive demographic and economic changes in doing physics research? Will these labs be designed for the next generation of scientists, far more global than their teachers, who will be spending their research careers at the GLC? If the new global labs are operating by 2010, then the first postdoctoral research associates there are now undergraduates; those who will be postdocs in 2025 started elementary school this year. In what kind of physics pedagogy should they participate now?

### Sokendai

During the 1960s the Japanese government decided to develop an array of national research institutes. KEK was the first; there are now nearly 20, and more are planned. In the mid 1980s some of these institutes began to pressure the government to allow them to have graduate students. Eventually, the Graduate University for Advanced Study (Sokendai) was established for that purpose; the students work both at the various institutes and at the campus in Hayama City, Kanagawa Prefecture. Sokendai now actively encourages students to do work which is at the interface of at least two institutes. (The humanities are not represented, as yet; and only a few social science projects are visible.) I know of no comparable institution in any other country, but Sokendai is not an isolated invention: in the 1990s many Japanese universities restructured their departments into interdisciplinary programs, creating opportunities for new kinds of teaching and research collaborations. A quick glance at the Ministry of Education's website, exploring the structure of universities in Japan, reveals many interdisciplinary programs and fewer traditional programs.[51]

These new institutes and university programs are not under the control of the traditionally powerful universities in Japan, although several are managed by the same governmental ministry.[52] (The conventional hierarchy begins with the former imperial universities of Tokyo, Kyoto, Osaka, Nagoya, Tohoku, and Hiroshima; resources have been distributed along that hierarchy for over 100 years.) During the planning stages it appeared that faculty from those powerful schools would keep the institutes bound to the hierarchy, but that has not happened. In several cases the international reputations of researchers now at institutes certainly exceeds that of the faculty at any university; this represents a fundamental revision of the geography of knowledge in Japan in one generation.

When Sokendai was first established they required that students already have received a master's degree before admission to the graduate program at Sokendai. This year they are developing programs for students who will enter Sokendai directly after their undergraduate education. They will receive an interdisciplinary education focused on apprenticeship learning of research practices at the various national research institutes. The president of the university, an astronomer, has asked two high-energy physicists to help design the curriculum. One is organizing a set of STS (science, technology, and society) courses for these apprentice scientists doing interdisciplinary research.

The president of Sokendai is eager to attract many foreign students to the university. Some courses will probably be taught in English. For the last few decades many graduate seminars in engineering and the sciences have been taught in English in Japan. The goal is to prepare students for full participation in the global research community, and English is the language of that realm, at least for the time being.

## Going Global from Japan

Many scientists around the world are arguing that the new global laboratories should not be built in their countries. They prefer to maintain the status quo of laboratories that are financed and managed nationally or regionally, with a significant majority of the scientific personnel coming from that country or the immediate region. If the high-energy physicists in each of these areas leave to work elsewhere in the world, local resources will be released for other purposes.

In Japan two different groups support this argument. Some high-energy physicists would like to see the primary KEK accelerator upgraded into a "Super B Factory" in order to continue the current, highly successful research program. They also use a sentimental argument: Japan first achieved international prominence in the field of high-energy physics with theories of B mesons articulated during the middle decades of the last century. A Super B Factory might enable them to provide the elusive experimental evidence that would corroborate the latest implications of those theories. Another group, not incompatible with the first, wants to keep Japanese resources concentrated in Japan on projects operated primarily by and for Japanese scientists. It is only during the last several years that Japan has had world-class research facilities; many would like to have such resources remain under Japanese control for as long as Europeans and Americans had nearly exclusive control of state-of-the-art facilities.

Yet another group in Japan would like the high-energy physicists to leave KEK for a GLC located elsewhere; then nuclear and cosmic ray physicists would control the existing resources at KEK. In 1997 (26 years after its founding), KEK, the Japanese national High Energy Physics Laboratory (*Ko Enerugie Butsurigaku Kenkyusho*), changed its name to High Energy Accelerator Research Organization (*Ko Enerugie Kasokuki Kenkyu Kikoh*). The name change reflected various structural changes, including the fact that it had incorporated two other laboratories: the former Institute for Nuclear Studies (INS), affiliated with the University of Tokyo (Todai) for 43 years, and the Meson Science Laboratory, affiliated with Todai for 19 years. The acronym KEK has been retained on all the signs, stationery, and publications. Most have seen this consolidation of Japanese particle physics resources at KEK as emblematic of the dominance of high-energy physicists among all other particle physicists.

The previous director of the Institute for Cosmic Ray Research (ICRR), also affiliated with Todai, became the director of KEK on April 1, 2003. Almost all the members of the committee that selected him had strong ties to the University of Tokyo. Some speculate that it is only a matter of time until the ICRR joins KEK. Many see this as part of a decades-long process of Todai regaining control of the state-of-the-art physics resources in Japan. Does this also mean that nuclear and cosmic ray physics would regain their position as the patrons of high-energy physics, rather than being patronized?

Could the status quo of 1960 be reestablished by occupying the territory of their prodigal sons who had become nouveaux riches? Should younger readers find this incredulous, I would like to remind them that there are similar efforts going on now, in universities and laboratories around the world, to reassert the academic ecology in which the currently retiring faculty were educated. The agents in this process, of course, are not the retirees; they are the protégés who had expected to inherit the resources first amassed by those retiring faculty. They are finding their patrimonies drastically eroded and want to restore them to the grandeur they thought they witnessed from behind the chairs of their mentors.[53] Each of the generations I have described has also produced its clones as well as spawning prodigals who moved on to build new labs in the new ecologies. The clones and the prodigals want different futures.

### Conclusion: The Future Pluperfect

Grammatically, the "future perfect" in English expresses an action that will be completed by a specified time. For example, we might say that by 2010 physicists will have finished their first experiments at the GLC. However, no one is certain that will happen. The "future pluperfect" tense is more apt: it speaks of an indefinite time yet to come, an action that has not yet occurred, states not yet in existence, a world that might be. We can say that important knowledge in particle physics will have been discovered by 2010 if the GLC were to be built. Using the future pluperfect in our statements about the GLC sounds and looks decisive, even though the decision has not yet been made. English speakers have come to be aware of this "booster" grammar from advertisements, politicians, and salespeople.

In Japanese there are additional, very interesting variations in verb forms. Our grammar can signal that we are discussing the intentions of the speaker, the intentions of subject actors, or the indeterminacy of the outcome. We could discuss the GLC while letting our interlocutors and our readers know that we are ourselves committed to building the GLC, that we know that certain people intend to build the GLC, or that building the GLC is as yet uncertain. It is important to specify the state of mind of the various actors; the verb forms and sentence structure could display that judgment. Conveying specificity in one's commitments is very important, as is expressing sincerity in those commitments. Of course, Japanese have become astute readers of the ways these forms are used by advertisers, politicians, and salespeople.

Just as there are multiple ways of figuring the past, there are many strategies for building the future. Any cultural anthropologist would assert that making the past and the future are tightly interwoven in any community, and the variations in their narratives represent different factions within the community.[54] I would add that differences in pedagogy correspond to different versions of the past and future. Since 1945 the high-energy physics community has been imagining, designing, and campaigning for the next machine that they insist is needed to solve most of the questions they had just then begun to pose. Over the last half century those very expensive machines have usually been built and the best physicists have found some way to gain access to them. The withdrawal of funding for the Superconducting Super Collider (SSC) in east Texas while it was under construction was a huge shock to the community. Many lament that they "lost a generation" of physicists that day. Even more grieve for the futures they never had; the trajectories of their careers suddenly shifted into another shape, never to be regained. Few in that generation have reproduced.

In that context the diverse designs and campaigns for the Global Linear Collider are competing plans for regeneration of the community. They do not discuss the possibility that the GLC will not be built; privately, senior physicists say that they cannot imagine how the community will survive without this next machine because without it there will not be a next generation of high-energy physicists. Pedagogy requires its devices; without those tools they believe that the community cannot reproduce itself and it cannot make new knowledge. As I claimed in the introduction, this community can only learn and teach with its prostheses; it is a cyborg.

Even if the GLC is built, will the entire community gather at the new site? How can the national communities maintain their resources at home if they leave them for others to use in their absence? One solution is to operate the GLC remotely, just as astronomers often do with their observatories nowadays. Some say this is impossible with an accelerator, and only occasionally do-able with a detector, but plausible for data analysis. Others notice that accelerators and detectors are already operated at distances of 20–100 meters for weeks at a time and that maintenance and modifications are done only during "downtime" when there are no accelerator particle beams. This kind of future is not acceptable to most of those over the age of 50. If it comes to be, the younger generations will need to invent a new set of pedagogies for that strange new world. Some of them, along with a few renegade elders, are already doing that.

The benefit of a remotely operated, "virtual laboratory" is that the community could be dispersed far more widely than it is now. Anyone with a good computing system and access to a satellite dish could join high-energy physics experiments. Access to big science would no longer be so restricted. Big science laboratories have been industrial, in that they require a sophisticated manufacturing-based infrastructure and connection to an industrial power grid. A new kind of high-energy physics laboratory could be sited in such a political economy, but operation and manipulation of that prosthesis could be distributed globally. Most of those who currently control access to the state-of-the-art laboratory sites are invested in that restricted access and the hierarchies that generates. If access to laboratories were to become far more dispersed, the geography of knowledge making might become far less hierarchically structured. So might the pedagogies.

## Acknowledgments

I would like to thank David Kaiser for the invitation to participate in the workshop that generated this volume. Many physicists have graciously and generously answered my queries and allowed me to witness their activities over many years and I remain indebted to them. I also appreciate the generous support for my research provided by the Ministry of Education, Culture, Sports, Science, and Technology (MEXT) of the Japanese government.

## Notes

1. On the concept of cyborg in science studies, see Donna Haraway, *Simians, Cyborgs, and Women: The Reinvention of Nature* (Routledge, 1991); Gary Downey and Joseph Dumit, eds., *Cyborgs and Citadels: Anthropological Interventions in Emerging Sciences and Technologies* (School of American Research Press, 1997); Chris Hables Grey, Heidi Figueroa-Sarriera, and Steven Mentor, eds., *The Cyborg Handbook* (Routledge, 1995).

2. For a fuller discussion of oral communication in this community, see Sharon Traweek, *Beamtimes and Lifetimes: The World of High Energy Physicists* (Harvard University Press, 1988), esp. 75–78 and 114–123.

3. For a study of the talk between a faculty member and graduate students in an American university, see Elinor Ochs, Sally Jacoby, and Patrick Gonzales, "Interpretive journeys: How physicists talk and travel through graphic space," *Configurations* 2 (1994), winter: 151–171.

4. Occasionally it emerges in letters of recommendations, although those usually are written in a highly formulaic mode. It is important to remember that those formulaic modes differ from country to country.

5. On learning from practice, see the other essays of this book. See also Jean Lave, *Cognition in Practice: Mind, Mathematics, and Culture in Everyday Life* (Cambridge University Press, 1988); Jean Lave and Etienne Wenger, *Situated Learning: Legitimate Peripheral Participation* (Cambridge University Press, 1992).

6. Part of the preceding paragraph is taken from Traweek, "Iconic devices: Toward an ethnography of physics images," in *Cyborgs and Citadels: Anthropological Interventions in Emerging Sciences and Technologies*, ed. Gary Downey and Joseph Dumit (School of American Research Press, 1997), 103–115.

7. The preceding part of this paragraph is taken largely from Traweek, "Big science as colonialist discourse: Regional differences in Japanese high-energy physics," in *Big Science: The Growth of Large-Scale Research*, ed. Peter Galison and Bruce Hevly (Stanford University Press, 1992), 100–128, on 104.

8. A senior physicist at the lab began to discuss this issue with me then. I mentioned that this apprenticeship model could also be described as Japanese. We talked at length about physics education and *iemoto*. Eventually, he asked me for references to Japanese and American anthropologists who had written on this subject and explained that representatives from several of the national research laboratories and institutes were currently meeting to formulate a presentation on this topic to the agency now called the Ministry of Education, Culture, Sports, Science, Technology (MEXT) of the Japanese government. (The group included anthropologists.) On *ie* and *iemoto*, see, for example, Jane Bachnik, Inside and Outside the Japanese Household (Ie): A Contextual Approach to Japanese Social Organization (PhD dissertation, Harvard University, 1978); the special issue on "*Ie* Society" in *Journal of Japanese Studies* (winter 1985); Francis L. K. Hsu, *Iemoto: The Heart of Japan* (Schenkman, 1975); Isao Kamakura, "The *Iemoto* system in Japanese society," *Japan Foundation Newsletter*, October-November 1981: 1–7; Takie Sugiyama Lebra, *Above the Clouds: Status Culture of the Modern Japanese Nobility* (University of California Press, 1993).

9. For a discussion of Sokendai (the Graduate University for Advanced Studies), see the last section of this essay.

10. On the European history of the concept of individuals and selves, see Amelie Rorty, "Introduction" and "A literary postscript: Characters, persons, selves, individuals," in Amelie Rorty, ed., *The Identities of Persons* (University of California Press, 1976), 1–15, 301–323.

11. Donald Keene, *Bunraku: The Art of the Japanese Puppet Theater* (Kodansha International, 1973); Barbara Adachi, *The Voices and Hands of Bunraku* (Kodansha International, 1978).

12. I also wrote about *hara-gei* in another context: "Bodies of evidence: Law and order, sexy machines, and the erotics of fieldwork among physicists," in *Choreographing History*, ed. Susan Foster (Indiana University Press, 1995), 211–228.

13. On Japanese notions of self, see Nancy Rosenberger, ed., *Japanese Sense of Self* (Cambridge University Press, 1994); Jane Bachnik and Charles Quinn Jr., eds., *Situated Meaning: Inside and Outside in Japanese Self, Society, and Language* (Princeton University Press, 1994). See also Louie Kam,

and Morris Low, eds., *Asian Masculinities: The Meaning and Practice of Manhood in China and Japan* (Routledge, 2003).

14. Dorinne Kondo, "Dissolution and reconstitution of self: Implications for anthropological epistemology," *Cultural Anthropology* 1 (1986), February: 74–88.

15. I discussed these career stages at length in chapter 3 of *Beamtimes and Lifetimes*.

16. That is, after 4 years as an undergraduate, 4–8 years as a graduate student, and 3–6 years as a postdoc, about 75 percent of these physicists leave high-energy physics. Most migrate to other scientific fields; they work in academia, government, and industry. Twenty years ago my informants estimated that about two-thirds left the field.

17. The body of research on this topic is far too extensive for me to cite here. To begin, I suggest the following: Nancy E. Betz, "What stops women and minorities from choosing and completing majors in science and engineering?" (ca. 1990; available from Federation of Behavioral, Psychological and Cognitive Sciences, 1200 Seventeenth St. NW, Washington, DC 20036); Sheila Tobias, *They're Not Dumb, They're Different: Stalking the Second Tier* (Research Corp., 1990); Sheila Tobias, *Revitalizing Undergraduate Science: Why Some Things Work and Most Don't* (Research Corp., 1992).

18. Ibid.

19. Fran Worden Henry, *Toughing It Out at Harvard Business School* (McGraw-Hill, 1984); Tracy Kidder, *The Soul of a New Machine* (Modern Library, 1997 [1981]); Charles LeBaron, *Gentle Vengeance: An Account of the First Year at Harvard Medical School* (Marek, 1981); Scott Turow, *One L: An Insider's Account of Life in The First Year at Harvard Law* (Penguin, 1978); Pepper White, *The Idea Factory: Learning to Think at MIT* (Dutton, 1991); Tom Wolfe, *The Right Stuff* (Bantam Doubleday Dell, 2001 [1979]). Of course, there are many studies of education in the professions, especially in sociology of medicine. I am referring here to the absence of studies about the sentimental education, the education of the sentiments, which is a significant and crucial aspect of that process.

20. Benson R. Snyder, *The Hidden Curriculum* (Knopf, 1971); Charles L. Bosk and Stephen Hilgartner, *Forgive and Remember: Managing Medical Failure* (University of Chicago Press, 1979).

21. See, for example, *Learning in Everyday Places in Japan*, ed. John Singleton (Cambridge University Press, 1994).

22. Two books that support this second hypothesis are Timothy Fry and Timothy Horner, eds., *The Rule of Saint Benedict* (Liturgical Press, 1981) and Janwillem van der Wetering, *The Empty Mirror: Experiences in a Japanese Zen Monastery* (Routledge and Kegan Paul, 1973). See also Janwillem van der Wetering, *A Glimpse of Nothingness: Experiences in an American Zen Community* (St. Martin's Press, 1999 [1976]).

23. David Noble, *A World without Women: The Christian Clerical Culture of Western Science* (Knopf, 1992).

24. Yasunari Kawabata, *The Master of Go*, tr. Edward G. Seidensticker (Vintage, 1996 [1972]).

25. I also have written about generational differences among Japanese physicists in "*Kokusaika, Gaiatsu,* and *Bachigai*: Japanese physicists' strategy for moving into the international political economy of science," in *Naked Science: Anthropological Inquiry into Boundaries, Power, and Knowledge,* ed. Laura Nader (Routledge, 1996), 174–197.

26. This model was followed through most of the world. For a very interesting comparison of the emergence of the Japanese and Australian physics communities during this period, see R. W. Home and Masao Watanabe, "Physics in Australia and Japan to 1914: A comparison," *Annals of Science* 44 (1987): 215–235; idem, "Forming new physics communities: Australia and Japan, 1914–1950," *Annals of Science* 47 (1990): 317–345.

27. See Hideki Yukawa's autobiography, *Tabibito (The Traveler),* tr. L. Brown and R. Yoshida (World Scientific, 1982); Yoshinori Kaneseki, "The elementary particle theory group," in *Science and Society in Modern Japan: Selected Historical Sources,* ed. Shigeru Nakayama, David Swain, and Eri Yagi (University of Tokyo Press, 1974), 221–252. See also *Particle Physics in Japan, 1930–1950,* ed. Laurie Brown, M. Konuma, and Z. Maki (Research Institute for Fundamental Physics of Kyoto University, 1980); Silvan Schweber, "The historical context in the development of the Standard Model," in *The Rise of the Standard Model,* ed. L. Brown, M. Dresden, L. Hoddeson, and M. Riordan (Cambridge University Press, 1997).

28. See Kenji Ito's essay in this volume.

29. On the business world, see Matthew Hamabata, *Crested Kimono: Power and Love in the Japanese Business Family* (Cornell University Press, 1990). On the several layers of the aristocracy (which, of course, intersects with the business, political, and academic worlds), see Lebra, *Above the Clouds.*

30. See Donald Roden, *Schooldays in Imperial Japan: A Study in the Culture of a Student Elite* (University of California Press, 1980).

31. See two books on the village of Suye Mura based on fieldwork conducted during the 1930s when many of the physicists I study who grew up in villages would have been children: Robert J. Smith and Ella Lury Wiswell, *The Women of Suye Mura* (University of Chicago Press, 1982); John Embree, *Suye Mura: A Japanese Village* (University of Chicago Press, 1939). On urban neighboorhoods, see Theodore C. Bestor, *Neighborhood Tokyo* (Stanford University Press, 1989); Jennifer Robertson, *Native and Newcomer: Making and Remaking a Japanese City* (University of California Press, 1991). See also Edward Seidensticker, *Low City, High City: Tokyo from Edo to the Earthquake* (Knopf, 1983).

32. Kaneseki, "Elementary particle group"; Brown et al., *Particle Physics in Japan.*

33. On regional differences in the Japanese physics community, see Traweek, *Beamtimes and Lifetimes.*

34. See Traweek, "*Kokusaika, Gaiatsu,* and *Bachigai.*" On *jinronnihon* and *nihonbunkaron* in Japanese history, see H. D. Harootunian, *Things Seen and Unseen: Discourse and Ideology in Tokugawa Nativism* (University of Chicago Press, 1988). See also Norma Field, "Somehow: The postmodern as atmosphere," in *Postmodernism and Japan,* ed. M. Miyoshi and H. D. Harootunian (Duke University Press,

1989), 169–188; Masao Miyoshi, *Off Center: Power and Culture Relations between Japan and the United States* (Harvard University Press, 1991).

35. On the symbolic meaning of various parts of human bodies in Japan, see Emiko Onuki-Tierny, *The Monkey as Mirror: Symbolic Transformations in Japanese History and Ritual* (Princeton University Press, 1987). See also Traweek, "Bodies of Evidence." On *amae*, see Takeo Doi, *The Anatomy of Dependence*, tr. John Bester (Kodansha International, 1973); idem, *The Anatomy of Self: The Individual versus Society*, tr. Mark A. Harbison (Kodansha International, 1986). On how the capacity to invoke *amae* in others is seen as a mark of maturity and leadership, see Thomas P. Rohlen, "The promise of adulthood in Japanese spiritualism," *Daedalus* (spring 1976): 125–143.

36. I wrote briefly about this issue in "*Kokusaika, Gaiatsu*, and *Bachigai*."

37. Traweek, "Iconic devices"; idem, "*Kokusaika, Gaiatsu*, and *Bachigai*"; idem, "Bachigai (out of place) in Ibaraki: Tsukuba Science City, Japan," in *Late Editions*, volume 2, *Scientific Imaginaries*, ed. George Marcus (forthcoming); idem, "When Eliza Doolittle studies 'enry 'iggins," in *Technoscience and Cyberculture*, ed. Stanley Aronowitz, Barbara Marinsons, Michael Menser, and Jennifer Rich (Routledge, 1996), 37–55; idem, "Bodies of evidence"; idem, "Border crossings: Narrative strategies in science studies and among high-energy physicists at Tsukuba Science City, Japan," in *Science as Practice and Culture*, ed. Andrew Pickering (University of Chicago Press, 1992), 429–465; idem, "Big science as colonialist discourse."

38. See Traweek, *Beamtimes and Lifetimes* and "Iconic devices."

39. See Sherry Turkle, *The Second Self: Computers and the Human Spirit* (Simon and Schuster, 1984).

40. On the iconography of laboratory sites, as well as the iconic signatures of research groups, see "Iconic devices." See also *Beamtimes and Lifetimes*.

41. "Standard model of fundamental particles and interactions" (Contemporary Physics Education Project, 1990). See also the following charts: "Standard model of elementary particles" (Fermilab, 1987); "Quark structure of proton and neutron" (Fermilab, 1987); "Unification of the forces" (Fermilab, 1986).

42. Traweek, "Iconic devices," 105.

43. Traweek, "When Eliza Doolittle studies 'enry 'iggins." See also Traweek, "Big science as colonialist discourse"; idem, "Essay on gender, science, and technology," in *Bulletin of the Institute for Women's Studies*, Ochanomizu Women's University, Japan (in Japanese and English), n.s. 5 (1991) [published in 1993].

44. For data on the number of women in U.S. science, including specific fields, during the twentieth century, see Margaret Rossiter, *Women Scientists in America: Struggles and Strategies to 1940* (Johns Hopkins University Press, 1982); idem, *Women Scientists in America: Before Affirmative Action, 1940–1972* (Johns Hopkins University Press, 1995); *Women in Academe: Progress and Prospects*, ed. Miriam Chamberlain (Russell Sage Foundation, 1988); *Women in Scientific and Engineering Professions*, ed. Violet Haas and Carolyn Perruci (University of Michigan Press, 1984). See also statistical data since the mid 1960s compiled and published by the Division of Science Resources

Studies of the Directorate for Social, Behavioral, and Economic Sciences at the National Science Foundation. Examples include: Susan T. Hill, *Science and Engineering Degrees 1966–2002: Detailed Statistical Tables* (2002); Susan T. Hill, *Undergraduate Origins of Recent (1991–95) Science and Engineering Doctorate Recipients: Special Report* (1996); Kelly H. Kang, *Characteristics of Doctoral Scientists and Engineers in the United States: 1997* (1999). See also the statistical data collected by the National Science Board, including *Science & Engineering Indicators 1991* (U.S. Government Printing Office, 1991); *Science & Engineering Indicators 2002*, volume 2: appendix tables (U.S. Government Printing Office, n.d.). Much of this statistical data can be found at www.nsf.gov. See also dozens of articles on the topic of women scientists in *Science* since the early 1990s; for example, see the following special issues: "Women in Science," March 13, 1992; "Women in Science '93," April 16, 1993. For Japan, see the extensive statistics compiled by the Japanese government throughout the twentieth century on gender in data on students and faculty at all educational institutions. Some of their statistical data from 1948 to the present can be found at the website of the Ministry of Education, Culture, Sports, Science, Technology (www.mext.go.jp). See also their *Nihon tokei nenkan* (*Japan Statistical Yearbook*) compiled and published annually by Sorifu Tokeikyoku (Statistics Division). In addition the Japanese government's Gender Equality Bureau of the Prime Minister's Office has a website with links to many government reports (www.gender.go.jp). Many data can be found in the publications and websites of universities, as well as national and international professional organizations. For example, see Dongqi Li and Beverly Karplus Hartline, "Learning from the world: A report on the IUPAP international conference on women in physics," in *AIP Conference Proceedings*, volume 628, ed. Dongqi Li and Beverly Karplus Hartline.

45. See the articles on the topic of women scientists in *Science* since the early 1990s. Most of the positive examples presented concern women in biology. For example, see the "Women in Science" and "Women in Science '93" issues of *Science* (cited above).

46. This section is taken from Traweek, "Essay on Gender, Science, and Technology."

47. Informal current estimates range from $3 billion to $8 billion, depending on whether the device is built in one or two stages.

48. See definitions and histories of "big science" and "big science" policies in Galison and Hevley, *Big Science* and in Arie Rip, "Fashions, lock-ins, and the heterogeneity of knowledge production," in *The Future of Knowledge Production in the Academy*, ed. Merle Jacob and Thomas Hellström (Open University Press, 2000), 28–39.

49. See Armin Hermann, Lanfranco Belloni, John Krige, Ulrike Mersits, and Dominique Pestre, *History of CERN, Volume 1: Launching the European Organization for Nuclear Research* (North-Holland, 1987); Armin Hermann, John Krige, Ulrike Mersits, Dominique Pestre, and Laura Weiss, *History of CERN, Volume 2: Building and Running the Laboratory* (North-Holland, 1990); John Krige, *History of CERN, Volume 3* (North-Holland, 1996). See also *History of European Scientific and Technological Cooperation*, ed. John Krige and Luca Guzzetti (Official Publications of the European Community, 1997).

50. See country-by-country data compiled by the Organization for Economic Cooperation and Development, available at www.oecd.org.

51. Ministry of Education, Culture, Sports, Science, Technology (www.mext.go.jp).

52. National Research Institutes, operated by Monkasho, available at www.mext.go.jp.

53. For a discussion of this process, see Traweek, "Faultlines: How modern became retro: An historical political economy of knowledge," in *Doing Science + Culture*, ed. Roddey Reid and Sharon Traweek (Routledge, 2000), 19–48.

54. See Hayden White, *Metahistory* (Johns Hopkins University Press, 1976); idem, *The Content of the Form: Narrative Discourse and Historical Representation* (Johns Hopkins University Press, 1987).

# Conclusion
## Kuhn, Foucault, and the Power of Pedagogy

**Andrew Warwick and David Kaiser**

The essays in this volume present examples drawn from the physical sciences and engineering in three continents in the nineteenth and twentieth centuries. Each essay highlights different ways in which scientists' and engineers' training shaped the way they approached problems, adjudicated answers, and established self-understandings of their proper roles in their disciplines and in the world. In all these examples, pupils had to *learn* how to speak and act as scientists and engineers. There was nothing automatic about this process, nor did it unfold the same way across time and space; distinct institutional and intellectual traditions took shape within varying cultural and political settings. Many of the accounts in this volume draw upon a productive tension (whether noted explicitly or not), at once incorporating and critically responding to the works of Thomas Kuhn and Michel Foucault. In this concluding essay we focus directly upon these works as a way of illuminating some of the more radical implications of a pedagogical epistemology of science. We hope this discussion will not only clarify the comparative similarity, difference, and synergy of Kuhn's and Foucault's respective contributions but also provide a useful resource upon which further historical studies of scientific pedagogy might fruitfully draw.

It is of course a risky business to invoke the names of two such well-known and controversial characters. One is likely to be pigeonholed by the casual reader as a mere Kuhnian or Foucauldian with little more to offer than a revamped or illustrated account of well-worn theoretical positions. There are nonetheless several good reasons for taking the risk. First, although there is an enormous literature on the history of education, virtually none of it is concerned with the relationship between training and the production of scientific knowledge. Most historians of education are concerned mainly with the history of institutions, educational reform, or the teaching of notable individuals. The relevance of Kuhn and Foucault is that, while neither is generally regarded as an historian or philosopher of education, their work can be read as pointing to an epistemology of science based on training. It is moreover an advantage of dealing with their work together that, insofar as their approaches to training differ, they are extremely complementary.

A second reason is that Kuhn's 1962 book *The Structure of Scientific Revolutions* is one of the few texts that one can still reasonably expect to be familiar to the majority of those philosophers, historians, sociologists, and anthropologists of science who constitute the field broadly known as science studies. Foucault's work not only plays a similarly inclusive role for literary studies, cultural history, and the history of the human sciences; it also provides some common ground between scholars working in these disciplines and those working in science studies. Third, despite the alleged importance of Kuhn's and Foucault's ideas to the development of science studies, the writings of both men contain major claims concerning the importance of training that have received very little critical discussion or historical evaluation.[1] In particular, we have in mind Kuhn's remarks on the relationship between "exemplars" and "normal science" and Foucault's suggestive comments on the development of discipline, examination, and institutionalized pedagogy in the decades around 1800.

We begin with Kuhn. It should first be noted, in a cautionary vein, that most of Kuhn's examples came from his own first area of training: theoretical physics (in itself a delightful reminder of the power of pedagogy). He certainly expected his observations to pertain to science more broadly, and indeed many of his insights do retain salience beyond this narrow field. Yet we must exercise care when extending Kuhn's ideas to treat examples from across the physical sciences, let alone biomedical or social sciences. At the time he was writing, both Kuhn and the philosophical schools he aimed to criticize (such as the logical positivists) could take theoretical physics as the obvious stand-in for all of science. Today no such synecdoche will do.

One of Kuhn's most valuable resources for writing a pedagogical account of modern science and technology is the fundamental relationship he forged between scientific *training* and professional scientific *practice*.[2] He argued that scientific training is not just, or even primarily, a matter of teaching students the laws, principles, and concepts of their discipline and inculcating a spirit of critical and unbiased inquiry, but is rather a dogmatic and authoritarian process by which they are drilled through exercises to master the craft of the professional practitioner.[3] These claims, Kuhn asserted, were based in part on two observations regarding the way physics students were actually taught. First, he noted that, while it was often assumed that applying the laws and principles of physics to solve practical problems was a straightforwardly rule-governed activity, he could see neither where such rules were explained in standard teaching texts nor even where such frequently applied concepts as space, time, and force were formally defined.[4] Second, he drew attention to the fact that many physics students who believed they had grasped the meaning of the physical laws defined in textbooks subsequently found themselves unable to solve the associated end-of-chapter examples and problems.[5] The clue to resolving these puzzling observations lay in the way students usually overcame their difficulties. They did so not by reflecting further upon the inner meaning of the laws or concepts themselves, but by studying canonical problem solutions—or "exemplars"—and attempting to generate similar solutions to a range of similar problems.[6] Kuhn concluded that the laws of physics were not applied according to a set of explicitly stated rules but through knowledge acquired in the very act of struggling to reproduce those "concrete problem solutions" recognized within the physics community as correct.[7]

This approach to physics pedagogy alters our understanding of the origin and the nature of the physicist's knowledge in several ways. In the first place it subverts the usual hierarchy between the fundamental laws and concepts of the discipline and the routine exercises tackled by students. According to Kuhn, the purpose of working through examples and problems is not merely to indicate whether a student has already absorbed the essential meaning of the relevant laws and concepts, but to *generate* that meaning through extended practice. On this showing the physicist's knowledge does not reside in mental retention of definitions and rules, but in the embodied ability to "perform" the solution to a problem when required to do so. Thus an important concept such as "force" becomes meaningful to students not by sudden insight into its metaphysical essence or role in the formal statement of Newton's laws of motion, but through the process of applying it to solve problems in conjunction with other concepts, physical models, analogies, mathematical methods, appropriate boundary conditions, approximations and so on. Kuhn's model of learning does not even require students or members of the wider physics community to reach explicit agreement on the precise definition and metaphysical meaning of the fundamental laws and concepts they employ, or even on which specific set of characteristics justify the exemplary status of an exemplar. All that is necessary is that they acknowledge the canonical status of the relevant exemplars and know how to work through them.[8] To Kuhn, therefore, the process of learning physics has more in common with the mastery of a craft skill than with simply coming to understand and to believe a collection of formal propositions concerning the nature of the physical world. It was these shared technical skills and practices that Kuhn initially associated with a "paradigm" and later defined in a narrower technical sense as a "disciplinary matrix."[9]

A second way in which Kuhn's approach to scientific pedagogy alters our view of modern science and technology concerns the relationship between training and research. One of the skills that students acquire while wrestling with numerous exemplars is how to see a new problem as a variant form of one they have already successfully solved. This ability is important in uniting different branches of a discipline through common techniques of problem solution, but also enables the best students to tackle increasingly difficult and novel problems. Kuhn argued that it was these same skills that enabled the scientist or engineer engaged in research to recognize what kinds of problem should be amenable to solution and (usually) to reduce them to canonical form.[10]

Kuhn's identification of a strong link between problem solving and research challenged a widely held view that the physical scientist's main preoccupation was the invention and testing of major new theories. He pointed out that the vast majority of physicists were not engaged in activities of this kind, but in expanding the explanatory power of those theories enshrined in the current paradigm or disciplinary matrix.[11] Kuhn described the work of modeling new problem solutions on canonical examples taken from established branches of physics—often likened to "puzzle solving"—as "normal science."[12]

Kuhn's notion of normal science is widely acknowledged as his most enduring contribution to the philosophy of science, but the mechanisms by which normal-science activity is generated require considerable clarification before the notion can be used for historical purposes.[13] Despite the powerful link Kuhn forged between training and knowing, it remains, as Barry Barnes has pointed out, the "thinnest and most weakly substantiated part of his general discussion of science."[14] Kuhn's view of scientific knowledge committed him to the idea that concerted and progressive work within a scientific discipline took place within a global disciplinary matrix, yet he offered only an implicit and very idealistic account of how the matrix became global.[15] In essence he assumed that the distribution of canonical texts such as Newton's *Principia* (or derivative textbooks) was sufficient to generate a community of practitioners engaged in a common project. He had very little to say regarding the role or form of the training regimes within which such texts and textbooks were employed.[16] This almost total reliance on the canonical status of major treatises and textbooks generates a troubling ambiguity in Kuhn's account of how students learn by solving problems. On the one hand, he adduces the authoritative tone of such books when arguing for the dogmatic nature of scientific education; on the other hand, he claims that the formal explanations found in textbooks cannot teach students the very problem-solving skills he identifies as crucial to competent practice. Kuhn points suggestively to a relationship between training and knowing only to impoverish the relationship by locating the scientist's learning experience in inaccessible acts of private study. A much richer account of the pedagogical resources used to train scientists and engineers is required if we are to provide a plausible account of the way training is generative of normal-science activity.

It is also extremely important to recognize that scientific training, like science itself, has a history. According to Kuhn, scientific knowledge progresses with historical time, but the way in which it does so is cyclical and essentially timeless.[17] In order to grasp this point we must look a little more closely at Kuhn's model of scientific change. Kuhn famously argued that any period of normal science will eventually culminate in a crisis because of the gradual accumulation of anomalous problems that cannot be solved using the tools of the dominant disciplinary matrix. Such crises are resolved by the invention of new canonical problem solutions which, although "incommensurable" with the old ones, are adopted by the physics community as the basis of a new period of normal science. Kuhn referred to the process by which one period of normal science is succeeded by another as constituting a "scientific revolution."[18] The notion of scientific revolutions turned out to be the most widely debated and criticized aspect of Kuhn's philosophy of science. It challenged the belief that science developed in a straightforwardly cumulative manner; even more seriously, it was understood by many philosophers to imply that the switch from one disciplinary matrix to another was a mystical process that defied rational analysis.[19] Kuhn argued that scientists working with pre- and post-revolutionary theories lived in different worlds and that the switch from one to the other was analogous to a conversion experience or a Gestalt switch.

Setting aside for the moment the problem of incommensurability, consider the difficulties raised by the cyclical and timeless nature of Kuhn's model of scientific change. It is implicit in this model that, say, the Copernican revolution of the sixteenth and seventeenth centuries constituted a transformation that was very similar in structure to that wrought by Einstein's general theory of relativity more than two and a half centuries later. Yet there are many very significant differences between these cases. The most significant to our present concerns are the speed with which the revolutions occurred and the way in which they were subsequently taught. Copernicanism took around a century to become the orthodox view of the heavens, and, even then, was not taught in technical detail to undergraduates. The general theory of relativity, by contrast, had been adopted by an international community of astronomers, mathematicians, and physicists little more than five years after its publication, and was being taught in technical detail to undergraduates at Cambridge and other universities within a similarly short period.[20] Bearing in mind that general relativity deploys a far wider range of difficult mathematical methods than does Copernican astronomy, Kuhn's model leaves us with an unexplained acceleration in the assessment and transmission of novel scientific work of more than an order of magnitude. What happened between the mid sixteenth century and the early twentieth century to enable more people to assess and to master more technical material far more quickly? The answer to this question lies in the pedagogical revolution that began in the late eighteenth century, and in order to cast additional light on the nature of that revolution we can usefully turn to Foucault's work on the history of institutionalized training.[21] Foucault argued that one of the central characteristics of the Enlightenment and its aftermath was the emergence of a new relationship between the human body and the exercising of "power." The relationship emerged coextensively with new disciplinary regimes in institutions such as schools, factories, barracks, prisons, asylums, and hospitals, enclosures in which the citizen's body was discovered and remade as an "object and target of power."[22] Central to Foucault's account of these "projects of docility" was the scale of change that was accomplished by the application of seemingly "petty forms of coercion." In order for schools to produce pupils, or hospitals patients, it was not a question of treating the body "wholesale," but of "working it 'retail,' individually; of exercising upon it a subtle coercion, of obtaining holds upon it at the level of the mechanism itself—movements, gestures, attitudes, rapidity; an infinitesimal power over the active body." Thus for Foucault such an apparently mundane activity as teaching good handwriting actually presupposed "a gymnastics—a whole routine whose rigorous code invests the body in its entirety, from the points of the feet to the tip of the index finger." Skills of this kind were generated, he argued, through an "uninterrupted constant coercion, supervising the processes of the activity rather than its result" and "exercised according to a codification that partition[ed] as closely as possible time, space [and] movement."[23]

As these remarks imply, there were for Foucault three general and closely integrated mechanisms by which this new disciplinary power obtained its grip on the body. The first was by the partitioning of space. The new institutions were designed and organized to control the way the inmates were segregated or congregated, how they moved, what they could see and hear, and to whom they could speak. Second, the temporal rhythm of the day was organized according to a detailed and strictly implemented timetable. The hours at which one slept, ate, worked, studied, took exercise or recreation, and so on were all stipulated and carefully policed. Third, the inmates were subject to constant surveillance by those charged with supervising or training them. Under what Foucault described as the classificatory, supervisory, or normalizing gaze of an inspector or teacher, the activities of each inmate were continually directed, corrected, and assessed.[24] Foucault's most detailed and best-known study of this process in operation is his description of the Panopticon, Jeremy Bentham's imaginary penitentiary (designed circa 1790). This building consisted of an annular outer block of individual cells, each facing a single tower standing at the center of an inner courtyard. The architecture was contrived so that the lone occupant of each cell—be he "a madman, a patient, a condemned man, a worker or a schoolboy"—could see only the tower (from which, he assumed, he was under surveillance), while the inspector could survey the activity of every occupant (silhouetted against a window in the back of his cell).[25] The effect of this arrangement was to "induce in the inmate a state of conscious and permanent visibility that assures the automatic functioning of power." Foucault concluded that this "marvellous machine," this "laboratory" of behavior, training, and correction, expressed power as a "concerted distribution of bodies, surfaces, lights, [and] gazes" which caught the inmates "in a power situation of which they [were] themselves the bearers."[26]

The last remark highlights one of the most interesting if underdeveloped aspects of Foucault's account of the new disciplinary regimes installed in educational institutions in the late eighteenth century. Their effect was not simply to coerce and subjugate the inmate—to make him "docile"—but to seek capacities and aptitudes in his body that could be enhanced and utilized. Indeed, for Foucault it was the operation of those very mechanisms through which docility and utility were co-produced in the individual that constituted "discipline." In this sense he regarded discipline as a "positive economy" capable of accelerating and enhancing the productive capacity of the student. Through experimentation with different architectures, timetables, and exercises, the disciplinary regime was capable of "extracting from time, ever more available moments and, from each moment, ever more useful forces." In the late eighteenth century, Foucault argued, it was this "disciplinary time that was gradually imposed on pedagogical practice."[27] To illustrate this point he contrasted the carefully contrived techniques of the new pedagogy with the long established method of learning by guild apprenticeship. In the latter case the apprentice learned mainly by participating and assisting in the everyday business of his master, while in the former the tasks of learning were separated from the temporal practice of any craft or profession and taught solely in order to develop specific technical competences as quickly and efficiently as possible.[28]

Foucault also emphasized that the operation of these regimens of institutionalized training relied not just on the manipulation of the student body but also on the field of knowledge to be taught. The students were divided into groups according to their age, ability, and level of attainment, and were set exercises that were judged to be of appropriate difficulty according to these divisions. The training process was also progressive, each student having to master the lowest level before moving on to the ones above.[29] It is implicit in this system that the field of knowledge to be learned could itself be broken down and reformulated as a series of graded subjects, skills, and exercises. Those in charge of student training had, as it were, to devise a new pedagogical taxonomy of the field, and, in doing so, had to make far-reaching decisions when selecting which subjects, methods, and exercises would be taught. What is particularly important here, though Foucault had little to say on the matter with respect to training, is that pedagogical reorganization on the scale he envisaged could, over time, have profound effects on the field itself. Where an individual master working within the guild system might adopt a piecemeal approach to teaching those techniques necessary to work on the specific problems he considered important, the essence of disciplined training was to find an optimal balance between the aptitudes of the students, a daily routine of organized study, and a pedagogically construed field of knowledge. This field defined the discipline in question for the students exposed to its rigors and prepared them to work with no master or project in particular yet any in general.

The device (or "ritual") that linked and controlled the hierarchies of study and syllabus in Foucault's scheme was the examination. Foucault argued that examinations represented a special form of the normalizing and supervisory gaze of the teacher that made it possible "to qualify, to classify and to punish."[30] They defined the kinds of skills to be learned, the performance required to move from one level to the next, and provided an objective and comparative measure of the relative abilities and achievements of individuals. Regular written examinations in particular generated a cumulative archive of reports and marks that were permanently available for future consultation and use. This for Foucault provided each student with a case history, which, like that of a criminal or hospital patient, could be used to diagnose his capacities, strengths and weaknesses, and to praise or blame his efforts. These case histories thus enabled the instructor efficiently to correct any insufficiencies detected in an individual's performance.[31] A student might be required to repeat lessons or exercises, to re-rehearse specific techniques, to work harder or even to give up a course of study altogether in favor of one to which he was better suited. In all cases, the examination, like the rest of the disciplinary apparatus, was aimed at utilizing the "ultimate capacity of an individual" in a specified field of pedagogical endeavor.[32]

Before discussing ways in which Foucault's account of training supplements Kuhn's remarks on normal science, it will be helpful briefly to emphasize one sense in which the philosophical views of both men are similarly radical and mutually supportive.[33] The most striking point of common ground is the claim that truth is not, in Foucault's words, "the reward of free spirits, the child of protracted solitude . . . the privilege of those who have succeeded in liberating themselves," but is rather "a thing of this world . . . produced only by virtue of multiple forms of constraint."[34] Foucault insisted that power in the form of social or cultural organization ought not to be described in negative terms—as excluding, repressing, censoring, abstracting, masking, or concealing—but as an active agency that "produces domains of objects and rituals of truth."[35] Kuhn can be read as expressing similar sentiments in his early remarks on scientific education and creativity.[36] As we have seen, he denied that scientific education simply liberated the student from human biases and cultural influences so that he or she could confront the natural world in a free and objective manner. Kuhn argued that far from passively nurturing such ill-defined activities as "free" or "creative" thought, the training to which physicists are subjected is necessarily regimented and authoritarian. It was through the painstaking mastery of the many skills required to reproduce exemplars in the proper manner that physics students not only became competent practitioners of their discipline but capable of collective and progressive research.[37]

The point bears repeating. Where other analysts such as Pierre Bourdieu cast pedagogy and discipline only in negative terms—Bourdieu called education "symbolic violence" in his early writings—both Kuhn and Foucault provide a framework in which training is constitutive of professional practice, but not in an automatic or unthinking way.[38] Kuhn and Foucault, in other words, provide hints of how to avoid a sterile educational determinism, which would treat the products of pedagogy as mere automata, destined only to mechanically repeat what they had been taught. Scientists' and engineers' research practices are always pedagogically conditioned, though not determined. Consider, as an analogy, jazz musicians. A major component of being a competent jazz musician is the ability to improvise solos, departing from any given written arrangement and, in a sense, making it up as one goes along. Yet in order to be able to improvise, jazz musicians must spend hours upon hours *practicing* certain rudimentary elements upon which they draw when performing their improvisations—certain scales, chords, and so on—often in highly formalized pedagogical settings. Few jazz musicians in the West practice quarter-tone scales, even though such scales are common in other parts of the world; not surprisingly, most jazz improvisation by Western musicians rarely draws on these particular skills or tonal elements. The musicians are certainly improvising, but their improvisations never take place in a vacuum. They draw on certain elements that, by virtue of their hours of practice and rehearsal, have come to seem natural or ready to hand. Improvisation always takes place within a context of practiced skills and techniques. The notion of pedagogically conditioned improvisation likewise helps make sense of what scientists and engineers do when encountering new research problems. Their prior training does not completely determine what they will do in the new setting, but it conditions how they will approach the problem. Certain tools and techniques—those they have practiced using in the past—will seem most ready to hand or even second nature.

The radicalness of these claims lies in the fact that they run counter to more traditional forms of philosophical enquiry into the nature of scientific knowledge.

Philosophers have typically sought to understand the scientist's way of knowing in terms of hidden and universal rules that relate either to a special scientific method or to necessary and sufficient relationships between theoretical and ontological entities.[39] These traditions have had almost nothing to say about scientific training as they tend to depict scientific theories as static and self-contained collections of propositions, and to analyze them according to criteria that are independent of the practical skills taught in formal science courses. Kuhn's and Foucault's approach, by contrast, seeks to explain agreement within the scientific community in terms of shared skills, commitments, and value judgments. They implicitly argue that the scientist's way of knowing is bound inextricably to these shared competencies and cannot be adequately described or explained by appeal to transcendent meta-rules or methods. As Kuhn succinctly stated, what he had in mind was a "manner of knowing which is misconstrued if reconstructed in terms of rules that are first abstracted from exemplars and thereafter function in their stead."[40] This approach points to an analysis of scientific knowledge that begins not with an investigation of ready-made and disembodied theories, but with the student in the classroom being trained to tackle elementary proofs and problems. The analyst's focus is shifted away from completed *theories* as the ultimate products of modern science, toward those conditions that make it possible to know, to develop, and to apply such collections of ideas.[41] This line of enquiry opens the way to a pedagogically oriented cultural history of modern science and technology by emphasizing the constructive power of training to produce knowing subjects. To draw again upon Foucault's evocative turn of phrase, it is not a matter of "knowing what external power imposes itself on science" but of "what effects of power circulate among scientific statements [and] what constitutes, as it were, their internal regime of power."[42]

Turning now to the complementary aspects of Kuhn's and Foucault's work, there are two closely related senses in which Foucault's writings on training can usefully modify our understanding of normal science. To appreciate the first we should note that where Kuhn was concerned almost entirely with the *product* rather than with the *process* of training, Foucault's interests are almost exactly the other way around. As even our thumbnail sketch of his view of education makes clear, Foucault was concerned first and foremost with the relationship between the minutiae of specific training regimes and their ability to locate, develop, and exploit the capabilities of individuals. The first sense in which Foucault's work complements Kuhn's, therefore, is by drawing our attention to the complexity and productive power of training. Where Kuhn's account of pedagogy was confined to occasional and vague references to the contents of canonical treatises and textbooks, Foucault's evoked a much richer and interactive nexus of institutionalized gazes, bodies, gestures, architectures, routines, incitements, examinations, and punishments. The latter approach constitutes a powerful resource in writing a pedagogical history of modern science and technology since it posits training as a general mechanism for the active production of knowing individuals that recognizes no natural distinction between the mind and the body, nor, by implication, between theory and practice. In this sense Foucault points to a level of analysis at which it should be possible to historicize the processes by which specialized technical competencies became the common preserve of widely extended communities of practitioners—the phenomenon Kuhn referred to as normal science.

This brings us to the second sense in which Foucault's approach to pedagogy, although very different from Kuhn's, can nevertheless complement the latter's occasional remarks on training as a prerequisite to normal scientific activity. Where Kuhn saw training mainly as an unchanging medium of interaction between science students and accepted canonical texts, Foucault saw it as a powerful means of historical change in its own right. New forms of surveillance and training were not only the products of a specific time and place but constitutive of the very historical possibility of new human sciences such as penology, psychiatry and forms of clinical medicine.[43] Thus the institutionalized disciplining of the body that occurred in Europe in the decades around 1800 marked a major and unprecedented watershed in the history of the sciences. Unlike Kuhn's cycles of normal and revolutionary science, which are largely unrelated to the methods of training employed by different cultures, Foucault's approach suggests a profound relationship between the history of training and the level and scale of agreement achievable in a technical discipline. We have, in other words, a promising strategy for investigating those dramatic changes in the scale and temporal propagation of technical competence between the Copernican and Einsteinian "revolutions" noted above.

Despite the utility of this reading of Kuhn's and Foucault's work, there nevertheless remain specific themes in their respective philosophies that require additional scrutiny. In Foucault's case the most profound problem is that having confined his own studies to the history of the human sciences he seemed positively to deny the possibility of analyzing the physical ones in a similar fashion. (In this sense his work complements Kuhn's, which saw all of science as effectively patterned on physics. Thus where Kuhn saw physics everywhere, Foucault opted not even to discuss it.) Foucault was, to be fair, little concerned with the history of what he labeled the "noble" sciences, but by arguing, in effect, that only the "dubious" ones were accessible to cultural analysis, he reinforced the very theme in the Western metaphysical tradition that he otherwise sought to challenge.[44] As we have seen, Foucault denied that "truth" and "reason" were the attributes of "free spirits" who had liberated themselves from society, claiming instead that power, discipline, and social engagement were necessary conditions of knowledge production. It seems odd, therefore, that he chose an imaginary building (the Panopticon) to illustrate an ideal disciplinary apparatus, when he might have chosen a real institution such as the École Polytechnique in Paris. The École was one of the very first centers of higher education in which advanced mathematical analysis was taught to a large number of students, and its architecture and regimes of discipline, surveillance, teaching, and examination clearly illustrate Foucault's remarks concerning the relationship between social order and knowledge.[45] It is a further twist of irony that the Panopticon was conceived by Jeremy Bentham's brother after a visit to a Parisian military school, one of the very institutions upon which, as Ken Alder has argued, the École was subsequently modeled.[46] The important point to note here is that Foucault was very traditional in equating greater technical content with greater disengagement from social or cultural settings. One might even go so far as to say that he assumed the human sciences to be explicable as products of their time and place because they were not really sciences at all.

A second and more specific problem with Foucault's work concerns the relationship he posits between training and the production of knowledge. While he rightly draws attention to the complexity and productivity of pedagogical regimes, his analysis remains too narrowly concerned with subjectification and surveillance. Despite his claim that training can find and exploit new capacities in those subjected to its rigors, it is invariably the teacher who is the "knower" while the student remains the "known." Yet the purpose of technical training is not just to manipulate the student's behavior for the purposes of the master, but to reproduce the master's skills in the student. For this process to work effectively the student must *want* to acquire the master's knowledge and be a willing and active participant in his or her own education. In order to display the productive power of pedagogy we must therefore inquire into what motivates students in various settings to submit themselves to its rigors, explore the relationship between student and teacher more closely and with greater historical specificity than Foucault did, and describe those complex regimes of self-discipline which constitute the student's own contribution to the pedagogical process.[47]

In Kuhn's case the problem lies both in the rigidity of his cyclical model of theoretical change and in the meaning he ascribes to the term "incommensurability." Over and above the model's insensitivity to the historical specificity of training, further difficulties reside in the sharp distinction Kuhn draws between periods of normal science and crisis, and between scientific developments of a normal and revolutionary kind. As Joseph Rouse has pointed out, episodes of normal science and crisis often coexist for long periods and are therefore better understood to describe different kinds of scientific activity than to designate necessarily sequential historical events.[48] Where one individual may consider a problem to constitute a crisis requiring a novel form of solution, another may regard the same problem as soluble within the current disciplinary matrix. Moreover, the extent to which new problem solutions are regarded as revolutionary can vary considerably from one individual to another. As Rouse rightly concludes, the "border between revolution and a significant development within an evolving tradition of normal science need not be a sharp one" and depends to some extent on whether the analyst chooses to emphasize continuous or discontinuous aspects of scientific change.[49]

When viewed not from the impossible Olympian heights of global scientific practice but rather from specific, local pedagogical institutions, Kuhn's "incommensurability" boils down to mismatches between mentors' and students' choices of what skills to practice and which techniques to naturalize. Shedding the "mentalism" behind much of Kuhn's notion of "paradigms"—especially when cast as conceptual worldviews—considerably changes the thrust of "incommesurability" for scholars in science studies. Instead of focusing on the incompatibility of the respective terms that appear in successive theories, for example, one should focus on the various meanings ascribed to published work by readers from different pedagogical traditions. The problem then becomes one of understanding how different meanings are generated around the same text and under what conditions the authorial meaning can be preserved and successfully propagated. What is *incommensurable* on this showing is not the essential meaning of scientific theories themselves, but the particular skills, techniques, and assumptions that go into generating a working interpretation of them at different sites. As the examples in this volume reveal, these skills and assumptions are to a large extent the products of local regimes of training. The problem of incommensurability therefore is best construed by historians, sociologists, and anthropologists not so much as one of time in the form of sequential theories but as one of space in the form of pedagogical geography.[50]

This critical and comparative conflation of some of the most interesting and radical themes in Kuhn and Foucault's philosophy—what might be dubbed the 'Foukuhnian' view—is powerful because it makes training a crucial site for exploring the nature of scientific knowledge. This stands in stark contrast to traditional analyses which seek to abstract science from its practice and practitioners in order to depict it as a method, a structure, or a correspondence between theoretical and measured values. An abstracted view of science is not in our view consistent either with the way science is actually mastered or with the skills practitioners use when engaged in research. The approach outlined above not only places science and scientific education in a broader cultural context but attempts to analyze the cultural values of science itself by investigating how they are inculcated and preserved from one generation to the next. Central to this approach are

Kuhn's insights that paper-based, technical knowledge consists largely of skills acquired through carefully crafted routines of training, and that research normally takes place more by applying than by transcending these skills. However, in order to historicize training's place in the sciences we need to transcend Kuhn's idealistic, global, static, and cyclic model of scientific change. We have suggested that this can be accomplished in two stages: first by noting the compatibility of Kuhn's emphasis on skill acquisition with Foucault's insight that power in the form of social relations does not inhibit or conceal knowledge but is necessary to its production; and, second, by building on Foucault's claim that the minutiae of everyday training practices have the power to generate new capabilities in human beings, thereby bringing about significant historical change.

The essays in this volume, by interrogating the practice-practitioner dyad in modern science and engineering, illustrate many features from this critical combination of Kuhn and Foucault. The themes around which the essays have been organized—transferring skills on the one hand, and generating sensibilities on the other—highlight a broad range of pedagogical activities constitutive of modern science and engineering. Building on these examples, we suggest that future studies might usefully explore some of the following issues in science and pedagogy. How do the norms (perhaps peculiar to specific disciplines) by which new scientific knowledge is codified for educational purposes relate to specific sites and times? What prompts these norms to change? In what ways is the research undertaken by young scientists related to their training? How easy is it for scientists from different schools to find meaning in each others' research and to agree on what constitute legitimate problems and solutions? What sensibilities regarding group loyalty, bodily conduct, patterns of work and collaboration, professional aspiration, and professional hierarchy are inculcated by specific training regimes, and what is their pedagogical and wider social significance? Perhaps the single most important theme running through these questions is what might be learned from comparative studies. By comparing the skills and competencies generated through different training regimes we can illuminate those technical skills and sensibilities which, although normally tacit, lie at the very heart of different forms of scientific knowledge. These skills and sensibilities must be taught, learned, and applied; pedagogy is the link connecting these steps through time and space.

## Notes

1. Important exceptions include Hubert Dreyfus and Paul Rabinow, *Michel Foucault: Beyond Structuralism and Hermeneutics*, second edition (University of Chicago Press, 1983); Joseph Rouse, *Knowledge and Power: Toward a Political Philosophy of Science* (Cornell University Press, 1987). Jan Goldstein discusses the relevance of Foucault's work to the history of the professions and disci-

plines, but has almost nothing to say on the history of training: Goldstein, "Foucault among the sociologists: The 'disciplines' and the history of professions," *History and Theory* 23 (1984): 170–192.

2. This interpretation of Kuhn's work is not one with which he would necessarily have agreed. On radical versus conservative readings of Kuhn's work, see Rouse, *Knowledge and Power,* chapter 2.

3. Kuhn's most trenchant expositions of this view are in Thomas Kuhn, *The Essential Tension* (University of Chicago Press, 1977), chapter 9; idem, *The Structure of Scientific Revolutions,* second edition (University of Chicago Press, 1970), chapter 3; idem, "The function of dogma in scientific research," in *Scientific Change,* ed. A. C. Crombie (Heinemann, 1963), 347–369. See also Kuhn, *Essential Tension,* chapter 12.

4. Kuhn, *Essential Tension,* xviii–xix, 318–319.

5. Kuhn, *Structure of Scientific Revolutions,* 189; Kuhn, *Essential Tension,* 305.

6. Kuhn came to see the notion of an "exemplar" as the "most novel" aspect of his philosophy. See Kuhn, *Structure of Scientific Revolutions,* 187–189; Kuhn, *Essential Tension,* 298, 305.

7. Kuhn, *Essential Tension,* xix, 229, and 298.

8. Ibid., xviii–xix, 299. See also Rouse, *Knowledge and Power,* 30–31.

9. Kuhn's reasons for replacing "paradigm" with "disciplinary matrix" are discussed in Kuhn, *Structure of Scientific Revolutions,* 181–183; and in Kuhn, *Essential Tension,* 297–298. The earlier term united a community of believers whereas the latter one designated a collection of shared skills. See also Rouse, *Knowledge and Power,* 36–40.

10. Kuhn, *Essential Tension,* 306–307.

11. Kuhn, *Structure of Scientific Revolutions,* 24, 35–36; Imre Lakatos and Alan Musgrave, eds., *Criticism and the Growth of Knowledge* (Cambridge University Press, 1970), 4–10.

12. Kuhn, *Structure of Scientific Revolutions,* chapters 3 and 4.

13. Even philosophers broadly hostile to Kuhn's approach acknowledged the importance of his notion of normal science. See Lakatos and Musgrave, *Criticism and the Growth of Knowledge,* 26, 52, 212.

14. Barry Barnes, *T. S. Kuhn and Social Science* (Macmillan, 1982), 16.

15. On Kuhn's move from the broad term "paradigm" to the more specific focus on "disciplinary matrices" and "exemplars" see Kuhn, *Structure of Scientific Revolutions,* 182–187; Kuhn, *Essential Tension,* 297–298.

16. Kuhn was not alone in recognizing a link between pedagogy and scientific knowledge but failing to develop or historicize his comments. See for example Ludwik Fleck, *Genesis and Development of a Scientific Fact,* ed. Thaddeus Trenn and Robert K. Merton, tr. Fred Bradley and Thaddeus Trenn (University of Chicago Press, 1979 [1935]).

17. Kuhn, *Structure of Scientific Revolutions,* chapter 13 and 205–207.

18. Kuhn's classic account of paradigm change through crisis and revolution is Kuhn, *Structure of Scientific Revolutions*, chapters 6–10. On the problem of incommensurability see ibid., 103, 112, 198–204.

19. See for example Lakatos and Musgrave, *Criticism and the Growth of Knowledge*, 93.

20. See Andrew Warwick, *Masters of Theory: Cambridge and the Rise of Mathematical Physics* (University of Chicago Press, 2003), chapter 9.

21. As with our discussion of Kuhn, this interpretation of Foucault's work is tailored to our own historical purposes. For an excellent introduction and exegesis of Foucault's work see Dreyfus and Rabinow, *Michel Foucault*. On the application of Foucault's work to laboratory studies see Rouse, *Knowledge and Power*, chapter 7.

22. Michel Foucault, *Discipline and Punish: The Birth of the Prison*, tr. Alan Sheridan (Pantheon, 1977), 136.

23. Foucault, *Discipline and Punish*, 136–137, 139, 152.

24. Ibid., 147, 154, 173, 183. See also Michel Foucault, *Power/Knowledge: Selected Interviews and Other Writings, 1972–1977*, ed. Colin Gordon (Pantheon, 1980).

25. Foucault, *Discipline and Punish*, 200.

26. Ibid., 201, 203–204.

27. Ibid., 137, 154, 159

28. Ibid., 156–157.

29. Ibid., 149–152.

30. Ibid., 184.

31. Ibid., 184–194.

32. Ibid., 160.

33. There are many senses in which Kuhn and Foucault were engaged in very different projects. For a comparative discussion of their work see Dreyfus and Rabinow, *Michel Foucault*, 69–78, 197–202.

34. Foucault, *Power/Knowledge*.

35. Foucault, *Discipline and Punish*, 194.

36. See for example Kuhn, "The function of dogma in scientific research"; Kuhn, *Essential Tension*, chapter 9. The latter essay was written in 1959.

37. On Kuhn's model the vast majority of scientific research was the result of normal scientific activity while the revolutionary changes in practice due to men like Einstein were oddities that required special explanation. See Lakatos and Musgrave, *Criticism and the Growth of Knowledge*, 6.

38. Cf. Pierre Bourdieu and Jean-Claude Passeron, *Reproduction in Education, Society, and Culture*, tr. Richard Nice (Sage, 1977).

39. For critiques of Kuhn's view from more traditional philosophical perspectives see, e.g., Lakatos and Musgrave, *Criticism and the Growth of Knowledge*.

40. Kuhn, *Structure of Scientific Revolutions*, 192.

41. See also David Kaiser, *Drawing Theories Apart: The Dispersion of Feynman Diagrams in Postwar Physics* (University of Chicago Press, 2005), especially chapter 10.

42. Foucault, *Power/Knowledge*, 112.

43. Ibid., 117.

44. Foucault contrasted the "dubious" human sciences with "mathematics, cosmology and physics." The latter he described as "noble sciences, rigorous sciences, sciences of the necessary, all close to philosophy: one can observe in their history the almost uninterrupted emergence of truth and pure reason." Michel Foucault, *The Order of Things: An Archaeology of the Human Sciences* (Pantheon, 1970), ix; idem, *Power/Knowledge*, 109–110.

45. See, for example, Henry Barnard's remarkable description of student life at the École in the middle of the nineteenth century: Henry Barnard, *Military Schools and Courses of Instruction in the Science and Art of War in France, Prussia, Austria, Russia, Sweden, Switzerland, Sardinia, England, and the United States* (Lippincott, 1862).

46. Foucault, *Power/Knowledge*, 147; Ken Alder, "French engineers become professionals; or, How meritocracy made knowledge objective," in *The Sciences in Enlightened Europe*, ed. William Clark, Jan Golinski, and Simon Schaffer (University of Chicago Press, 1999), 94–125, on 98.

47. Cf. Kathryn Olesko, *Physics as a Calling: Discipline and Practice in the Königsberg Seminar for Physics* (Cornell University Press, 1991); Warwick, *Masters of Theory*; Kaiser, "The postwar suburbanization of American physics," *American Quarterly* 56 (2004), no. 4: 851–888.

48. Rouse, *Knowledge and Power*, 34. Paul Feyerabend criticized Kuhn's sequential model of monolithic normal and revolutionary periods, arguing instead that revolutions were prompted by the clash of parallel forms of normal science; see Lakatos and Musgrave, *Criticism and the Growth of Science*, 211–214.

49. Rouse, *Knowledge and Power*, 35.

50. See esp. Warwick, *Masters of Theory*, chapter 8; Kaiser, *Drawing Theories Apart*, chapter 10. Mario Biagioli similarly points out that incommensurability has a history, and that claims not to understand alternative points of view should sometimes be understood as active strategies in the production of new socio-professional identities: Biagioli, *Galileo, Courtier: The Practice of Science in the Culture of Absolutism* (University of Chicago Press, 1993), chapter 4.

## Contributors

**Bernadette Bensaude-Vincent** is a professor of history and philosophy of science at the Université de Paris X. Her fields of research are the history of chemistry and the publics of science. Her recent publications include *L'émergence d'une science des manuels: Les livres de chimie en France (1789–1852)*, co-authored with Antonio García-Belmar and José R. Bertomeu-Sánchez (Archives Contemporaines, 2003). In 1997 she received the Dexter Award from the American Chemical Society for Outstanding Achievement in the History of Chemistry.

**José Ramón Bertomeu-Sánchez** teaches history of science at the University of Valencia. His research is focused on nineteenth-century toxicology, chemical classifications, and textbooks. He is a co-author (with B. Bensaude-Vincent and A. García-Belmar) of *L'émergence d'une science des manuels: Les livres de chimie en France (1789–1852)* (Archives Contemporaines, 2003).

**Antonio García-Belmar** teaches history of science at the University of Alicante. He has published (with José Ramón Bertomeu-Sánchez) *Opening Black Boxes: Scientific Instruments at the University of Valencia* (Valencia University Press, 2002) and (with José Ramón Bertomeu-Sánchez and Bernadette Bensaude-Vincent) *L'émergence d'une science des manuels: Les livres de chimie en France (1789–1852)* (Archives Contemporaines, 2003).

**Graeme Gooday** is a senior lecturer in history and philosophy of science at the University of Leeds. He specializes in the history of electrical technology. His first book, *The Morals of Measurement: Accuracy, Irony and Trust in Late Victorian Electrical Practice*, was published by the Cambridge University Press in 2004.

**Michael D. Gordin** is an assistant professor of the history of science at Princeton University. He recently published the cultural biography *A Well-Ordered Thing: Dmitrii Mendeleev and the Shadow of the Periodic Table* (Basic Books, 2004). He is working on a study of comparative nationalism in German and Russian chemistry in the nineteenth century.

**Hugh Gusterson** is an associate professor of anthropology and science studies at the Massachusetts Institute of Technology. He is the author of *Nuclear Rites* (University of California Press, 1996) and *People of the Bomb* (University of Minnesota Press, 2004) and a co-editor of *Why America's Top Pundits Are Wrong* (University of California Press, 2004).

**Karl Hall** is an assistant professor in the history department at Central European University in Budapest. He is completing a book on the history of Soviet theoretical physics before World War II. He counts Soviet scientific relations with the Warsaw Pact countries among his research interests.

**Kenji Ito** is a research associate at the University of Tokyo. In 2002 he completed a dissertation on the introduction of quantum mechanics into Japan at Harvard University. He is writing on Japan's first female physicist, Yuasa Toshiko. In addition, he works on robotics and videogames in Japan.

**David Kaiser** is an associate professor in the Massachusetts Institute of Technology's Program in Science, Technology, and Society and a lecturer in MIT's department of physics. He has written *Drawing Theories Apart: The Dispersion of Feynman Diagrams in Postwar Physics* (University of Chicago Press, 2005). He is writing a book about physicists' changing roles in the United States during the Cold War.

**Cyrus C. M. Mody** is the Gordon Cain Fellow at the Chemical Heritage Foundation in Philadelphia. His PhD dissertation (Department of Science and Technology Studies, Cornell University, 2004) was on the development of probe microscopy.

**Kathryn Olesko** is an associate professor in the department of history and a member of the core faculty of the BMW Center for German and European Studies at Georgetown University. She began her academic career looking at issues in science pedagogy, but now her research focuses broadly on the social history of science and technology in Europe, especially Germany. She is the author of *Physics as a Calling: Discipline and Practice in the Königsberg Seminar for Physics* (Cornell University Press, 1991).

**Buhm Soon Park** is an associate historian at the National Institutes of Health. He has authored several papers on the history of quantum chemistry. He is working on the history of research policies and programs at the NIH after World War II.

**Sharon Traweek** is an associate professor in the department of history at the University of California, Los Angeles. She is the author of *Beamtimes and Lifetimes: The World of High Energy Physicists* (Harvard University Press, 1988). She has two books nearing completion: *Building Big Science in Japan* and *Who Knows? Crafting Cultural Studies of Physics, History, and Anthropology*.

**Andrew Warwick** teaches history of science at Imperial College, London. His research interests include the history of the physical sciences and mathematics since 1750, the history of medical imaging, and the sociology of scientific knowledge. In 2003 he published *Masters of Theory: Cambridge and the Rise of Mathematical Physics* (University of Chicago Press).

**Inside Technology**

edited by Wiebe E. Bijker, W. Bernard Carlson, and Trevor Pinch

Janet Abbate, *Inventing the Internet*

Charles Bazerman, *The Languages of Edison's Light*

Marc Berg, *Rationalizing Medical Work: Decision-Support Techniques and Medical Practices*

Wiebe E. Bijker, *Of Bicycles, Bakelites, and Bulbs: Toward a Theory of Sociotechnical Change*

Wiebe E. Bijker and John Law, editors, *Shaping Technology/Building Society: Studies in Sociotechnical Change*

Stuart S. Blume, *Insight and Industry: On the Dynamics of Technological Change in Medicine*

Pablo J. Boczkowski, *Emerging Media: Innovation in Online Newspapers*

Geoffrey C. Bowker, *Science on the Run: Information Management and Industrial Geophysics at Schlumberger, 1920–1940*

Geoffrey C. Bowker and Susan Leigh Star, *Sorting Things Out: Classification and Its Consequences*

Louis L. Bucciarelli, *Designing Engineers*

H. M. Collins, *Artificial Experts: Social Knowledge and Intelligent Machines*

Paul N. Edwards, *The Closed World: Computers and the Politics of Discourse in Cold War America*

Herbert Gottweis, *Governing Molecules: The Discursive Politics of Genetic Engineering in Europe and the United States*

Gabrielle Hecht, *The Radiance of France: Nuclear Power and National Identity after World War II*

Kathryn Henderson, *On Line and On Paper: Visual Representations, Visual Culture, and Computer Graphics in Design Engineering*

David Kaiser, editor, *Pedagogy and the Practice of Science: Historical and Contemporary Perspectives*

Peter Keating and Alberto Cambrosio, *Biomedical Platforms: Reproducing the Normal and the Pathological in Late-Twentieth-Century Medicine*

Eda Kranakis, *Constructing a Bridge: An Exploration of Engineering Culture, Design, and Research in Nineteenth-Century France and America*

Pamela E. Mack, *Viewing the Earth: The Social Construction of the Landsat Satellite System*

Donald MacKenzie, *Inventing Accuracy: A Historical Sociology of Nuclear Missile Guidance*

Donald MacKenzie, *Knowing Machines: Essays on Technical Change*

Donald MacKenzie, *Mechanizing Proof: Computing, Risk, and Trust*

Maggie Mort, *Building the Trident Network: A Study of the Enrolment of People, Knowledge, and Machines*

Nelly Oudshoorn and Trevor Pinch, editors, *How Users Matter: The Co-Construction of Users and Technologies*

Paul Rosen, *Framing Production: Technology, Culture, and Change in the British Bicycle Industry*

Susanne K. Schmidt and Raymund Werle, *Coordinating Technology: Studies in the International Standardization of Telecommunications*

Dominique Vinck, editor, *Everyday Engineering: An Ethnography of Design and Innovation*

# Index